江苏大学英文教材基金资助出版

ADVANCED MATHEMATICS II

高等数学　下

主编
杜瑞瑾　　江苏大学
（Du Ruijin）

董高高　　江苏大学
（Dong Gaogao）

杨　洁　　江苏大学
（Yang Jie）

参编
王明刚　南京师范大学
陈　琳　西北工业大学

江苏大学出版社
JIANGSU UNIVERSITY PRESS
镇江

图书在版编目(CIP)数据

高等数学. 下＝Advanced Mathematics Ⅱ：英文 / 杜瑞瑾,董高高,杨洁主编. —镇江：江苏大学出版社，2019.4
 ISBN 978-7-5684-0959-9

Ⅰ.①高… Ⅱ.①杜… ②董… ③杨… Ⅲ.①高等数学－高等学校－教材－英文 Ⅳ.①O13

中国版本图书馆 CIP 数据核字(2019)第 056349 号

高等数学(下)
Advanced Mathematics（Ⅱ）

主　　编 / 杜瑞瑾　董高高　杨　洁
责任编辑 / 吴昌兴
出版发行 / 江苏大学出版社
地　　址 / 江苏省镇江市梦溪园巷 30 号(邮编：212003)
电　　话 / 0511-84446464(传真)
网　　址 / http：// press. ujs. edu. cn
排　　版 / 镇江文苑制版印刷有限责任公司
印　　刷 / 虎彩印艺股份有限公司
开　　本 / 787 mm×1 092 mm　1/16
印　　张 / 17.5
字　　数 / 582 千字
版　　次 / 2019 年 4 月第 1 版　2019 年 4 月第 1 次印刷
书　　号 / ISBN 978-7-5684-0959-9
定　　价 / 56.00 元

如有印装质量问题请与本社营销部联系(电话：0511-84440882)

CONTENTS

Chapter 8   Infinitive series / 1

8.1   The concepts and characters of infinite series / 2
   8.1.1   The concepts of constant term progression / 2
   8.1.2   The convergence and divergence of infinite series / 2
   8.1.3   The characters of convergent series / 5
Exercises 8-1 / 8
8.2   Positive series and its convergence test / 9
   8.2.1   The basic characters of positive series / 9
   8.2.2   The comparison test of positive series / 11
   8.2.3   d'Alembert method of positive series / 14
   8.2.4   The root test of positive series / 17
Exercises 8-2 / 19
8.3   Arbitrary term progression / 20
   8.3.1   Alternating series and Leibniz Principle / 20
   8.3.2   The absolute value method of arbitrary term progression / 22
Exercises 8-3 / 24
8.4   Power series / 25
   8.4.1   Function series / 25
   8.4.2   Power series and its convergent interval / 26
   8.4.3   The properties of power series / 31
Exercises 8-4 / 34
8.5   Unfold functions into power series / 35
   8.5.1   Taylor series / 35
   8.5.2   The method to unfold functions into power series / 37
Exercises 8-5 / 42
Summary / 42
Quiz / 44
Exercises / 44

Chapter 9 Differential equations / 46

9.1 Basic concepts of differential equation / 46
Exercises 9-1 / 49
9.2 First-order differential equation / 50
  9.2.1 Separable differential equation / 50
  9.2.2 Homogeneous differential equation / 53
  9.2.3 First-order linear differential equation / 55
  9.2.4 Bernoulli equation / 59
Exercises 9-2 / 61
9.3 Reducible high-order differential equation / 61
  9.3.1 The form $y^{(n)} = f(x)$ / 61
  9.3.2 The form $y'' = f(x, y')$ / 62
  9.3.3 The form $y'' = f(y, y')$ / 63
Exercises 9-3 / 64
9.4 High-order linear differential equation / 64
  9.4.1 The property and structure of solution of second-order linear differential equation / 65
  *9.4.2 The property and structure of solution of high-order linear differential equation / 69
Exercises 9-4 / 70
9.5 Second-order linear differential equation with constant coefficients / 71
  9.5.1 Second-order homogeneous linear differential equation with constant coefficients / 71
  9.5.2 Second-order inhomogeneous linear differential equation with constant coefficients / 74
  *9.5.3 Vibration equation / 79
Exercises 9-5 / 82
Summary / 83
Quiz / 85
Exercises / 85

Chapter 10 Vectors and analytic geometry of space / 87

10.1 Space right angle coordinate system / 87
  10.1.1 The space right angle coordinate system / 87
  10.1.2 Right angle coordinate of spatial point / 88
  10.1.3 Distance between two points in space / 89
Exercises 10-1 / 91

10.2 Vector algebra / 91
    10.2.1 Concepts of vector / 91
    10.2.2 Linear operations of vector / 92
    10.2.3 Coordinates of vector / 95
    10.2.4 Dot product of two vectors / 99
    10.2.5 Vector product of two vectors / 101
Exercises 10-2 / 102
10.3 Plane or space straight line / 103
    10.3.1 Plane and its equation / 103
    10.3.2 Included angle between two planes / 106
    10.3.3 Distance from point to plane / 107
    10.3.4 Space line and its equation / 107
    10.3.5 Included angle between two straight lines / 109
    10.3.6 Angle between straight line and plane / 110
Exercises 10-3 / 111
10.4 Curved surface and space curve / 112
    10.4.1 Equation of spatial curved surface / 112
    10.4.2 Equation of spatial curve / 115
    10.4.3 Quadric surface / 118
Exercises 10-4 / 120
Summary / 121
Quiz / 124
Exercises / 124

Chapter 11 Differentiation of multivariable function and its application / 126

11.1 The concept of multivariable function / 127
    11.1.1 Plane point set and n-dimensional space / 127
    11.1.2 Multivariable function / 129
    11.1.3 The limit of multivariable function / 133
    11.1.4 Continuity of multivariable function / 135
Exercises 11-1 / 137
11.2 Differential method of multivariate function / 138
    11.2.1 Partial derivative / 138
    11.2.2 Perfect differential and its applications / 149
    11.2.3 Differential method of multivariable compound function / 156
    11.2.4 Derivative of implicit function / 165
Exercises 11-2 / 171

11.3　Direction derivative and gradient　/　174
　　11.3.1　Direction derivative　/　174
　　11.3.2　Gradient　/　176
Exercises 11-3　/　179
11.4　Geometric applications of the differential of multivariable function　/　180
　　11.4.1　Tangent line and normal plane of space curve　/　180
　　11.4.2　Tangent plane and normal line of curve surface　/　184
Exercises 11-4　/　186
11.5　The extreme value of the multivariable function and the maximum and minimum　/　187
　　11.5.1　The extreme value of the multivariable function　/　187
　　11.5.2　The maximum and minimum of multivariable function　/　189
　　11.5.3　Conditional extremum　Lagrange multiplier　/　190
Exercises 11-5　/　193
11.6　Taylor's formula of binary function　/　193
　　11.6.1　Taylor's formula of binary function　/　193
　　11.6.2　The proof of sufficient condition of extreme of binary function　/　197
Exercises 11-6　/　198
Summary　/　198
Quiz　/　202
Exercises　/　203

Chapter 12　Multiple integral　/　205

12.1　The concept and properties of double integrals　/　205
　　12.1.1　Examples　/　205
　　12.1.2　The definition of double integrals　/　207
　　12.1.3　The property of double integrals　/　209
Exercises 12-1　/　210
12.2　Calculation of double integrals　/　211
　　12.2.1　Calculate double integrals in rectangular coordinate system　/　211
　　12.2.2　Calculate double integral in the polar coordinate system　/　216
　　12.2.3　Application in the economic management　/　220
　　12.2.4　Variable substitution　/　222
Exercises 12-2　/　224
12.3　Triple integral and its calculation　/　227
　　12.3.1　The definition and property of triple integral　/　227
　　12.3.2　Evaluate the triple integral by space right angle coordinate　/　228
　　12.3.3　Calculate the triple integral by cylindrical coordinate　/　230
　　12.3.4　Calculate the triple integral by spherical coordinate　/　232
Exercises 12-3　/　234

12.4 Application of multiple integral / 236
   12.4.1 Applications in geometry / 236
   12.4.2 Applications in physics / 239
Exercises 12-4 / 245
*12.5 Integral with parameters / 246
*Exercises 12-5 / 250
Summary / 250
Quiz / 253
Exercises / 254

Answers / 257

Chapter 8 Infinitive series

From the elementary mathematics, We should know that finite real numbers $u_1, u_2, \cdots, u_n$, can be added together and the sum is also a real number. Can infinitely many real numbers be added together? What is the sum? To be specific, let's look at an example first.

Ancient Chinese philosopher Chuang Tzu said that by cutting one wooden stick in half day by day, the stick will never run out. It describes the change of the remaining length after each cutting. What is the total length cut down at last? Suppose the length cutted down after i days is denoted by s_i, we have

the first day $\qquad s_1 = \dfrac{1}{2}$;

the second day $\qquad s_2 = \dfrac{1}{2} + \dfrac{1}{2^2} = \dfrac{3}{4}$;

....

the tenth day $\qquad s_{10} = \dfrac{1}{2} + \dfrac{1}{2^2} + \cdots + \dfrac{1}{2^{10}} = \dfrac{1\,023}{1\,024}$;

....

the n th day $\qquad s_n = \dfrac{1}{2} + \dfrac{1}{2^2} + \cdots + \dfrac{1}{2^n} = 1 - \dfrac{1}{2^n}$; $\qquad$ (1)

If cutting without end, the sum of the length is

$$s = \dfrac{1}{2} + \dfrac{1}{2^2} + \dfrac{1}{2^3} + \cdots + \dfrac{1}{2^n} + \cdots. \qquad (2)$$

In fact,

$$s = \lim_{n \to \infty} s_n = \lim_{n \to \infty} \left(1 - \dfrac{1}{2^n}\right) = 1. \qquad (3)$$

The formula (2) denoting the sum of infinite real numbers is called the infinite series(or series). The formula (1) which is the sum of first n terms of the infinite series(2) is called partial sum. From the formula (3), we see that the sum of the infinite real numbers is equal to the limit of the partial sum as n approaches to infinity and the limit 1 is called the sum of the infinite series. Note that the sum is not calculated by adding total numbers together but by limit.

The sum of infinite terms occurs widely in practical problems, and it is actually the infinite series. Infinite series is a useful tool to express functions, study characters of functions, evaluate functions and solve differential equations. It also plays an important role in the theoretical researches of mathematic and practical application.

In this chapter, we will first discuss infinite series and introduce some basic contents about infinite series. Then we will also talk about power series and discuss how to write a function into a power series.

8.1 The concepts and characters of infinite series

8.1.1 The concepts of constant term progression

An infinite sequence
$$u_1, u_2, u_3, \cdots, u_n, \cdots,$$
the sum of total terms is
$$u_1 + u_2 + u_3 + \cdots + u_n + \cdots,$$
which is called an infinite series (or just series) and is denoted by the symbol $\sum_{n=1}^{\infty} u_n$ or $\sum u_n$ for short, that is,
$$\sum_{n=1}^{\infty} u_n = u_1 + u_2 + u_3 + \cdots + u_n + \cdots, \tag{1}$$
where u_n is called the n th term (or general term). Now, we give some examples.

① $\sum_{n=1}^{\infty} \frac{1}{n} = 1 + \frac{1}{2} + \frac{1}{3} + \cdots + \frac{1}{n} + \cdots$ is called a harmonic series with $u_n = \frac{1}{n}$.

② $\sum_{n=1}^{\infty} aq^{n-1} = a + aq + aq^2 + \cdots + aq^{n-1} + \cdots$ (both a and q are constants) is called a geometric series with $u_n = aq^{n-1}$, where q is the common ratio.

③ $\sum_{n=1}^{\infty} \frac{1}{n^p} = 1 + \frac{1}{2^p} + \frac{1}{3^p} + \cdots + \frac{1}{n^p} + \cdots (p > 0)$ is called a p- series with $u_n = \frac{1}{n^p}$.

④ $\sum_{n=1}^{\infty} (-1)^{n-1} = 1 - 1 + 1 - 1 + \cdots + (-1)^{n-1} + \cdots$ is a series with $u_n = (-1)^{n-1}$.

⑤ $\sum_{n=1}^{\infty} 2n = 2 + 4 + 6 + \cdots + 2n + \cdots$ is a series with $u_n = 2n$.

8.1.2 The convergence and divergence of infinite series

For series(1), how to add the infinite terms? From the example in the introduction of this chapter, we add up the first n terms and observe its characteristic, and then, take the limit to get the sum of the infinite terms. In general, we denote the sum of the first n terms as

$$s_n = u_1 + u_2 + u_3 + \cdots + u_n = \sum_{k=1}^{n} u_k, \qquad (2)$$

which is called partial sum of the first n terms. If $n=1,2,3,\cdots$, we will get a new sequence $s_1, s_2, s_3, \cdots, s_n, \cdots$, which is called the sequence of partial sums denoted by $\{s_n\}$.

Since the partial sums s_n will get closer and closer to the sum of the series as n approaches to infinity, we define the sum of the series by the limit of the partial sums as $n \to \infty$.

Definition If the sequence $\{s_n\}$ is convergent and $\lim\limits_{n\to\infty} s_n = s$ exists as a real number, the series $\sum\limits_{n=1}^{\infty} u_n$ is called convergent, the number s is called the sum of series $\sum\limits_{n=1}^{\infty} u_n$ and we write

$$\sum_{n=1}^{\infty} u_n = u_1 + u_2 + u_3 + \cdots + u_n + \cdots = s.$$

It is also called that the series $\sum\limits_{n=1}^{\infty} u_n$ converges to s. If the sequence $\{s_n\}$ is divergent, the series $\sum\limits_{n=1}^{\infty} u_n$ is called divergent and the divergent series has no sum. In particular, if the sequence $\{s_n\}$ approaches to infinity, we call that the sum is infinity. However, the series is actually divergent.

Therefore, if the series (1) is convergent, the partial sum s_n will be the approximate value of the sums, that is, $s \approx s_n$. We define the difference $r_n = s - s_n = u_{n+1} + u_{n+2} + \cdots$ to be the remainder term of the series (1). The absolute value of remainder term

$$|r_n| = |u_{n+1} + u_{n+2} + \cdots|$$

is the error produced by substituting s_n for s.

Example 1 Is the geometric series

$$\sum_{n=0}^{\infty} aq^n = a + aq + aq^2 + \cdots + aq^n + \cdots \qquad (3)$$

convergent or divergent, where q is the common ratio and $a \neq 0$? (If $a = 0$, each term is 0. It is convergent and the sum is 0.)

Solution If $q \neq 1$, the partial sum is

$$s_n = a + aq + \cdots + aq^{n-1} = \frac{a - aq^n}{1-q} = \frac{a}{1-q} - \frac{aq^n}{1-q}.$$

If $|q| < 1$, we have $\lim\limits_{n\to\infty} s_n = \frac{a}{1-q}$, the series is convergent and its sum is $\frac{a}{1-q}$.

If $|q| > 1$, we have $\lim\limits_{n\to\infty} s_n = \infty$, the series is divergent.

If $q = 1$, we have $s_n = na \to \infty$, the series is divergent.

If $q = -1$, the series is $a - a + a - a + \cdots$, we have s_n is a if n is odd and 0 if n is even, that is, the limit of s_n does not exist, the series is divergent.

Above all, the geometric series (3) is convergent and its sum is $\frac{a}{1-q}$ if $|q|<1$ and divergent if $|q|\geq 1$, that is,

$$\sum_{n=0}^{\infty} aq^n = \begin{cases} \frac{a}{1-q}, & |q|<1, \\ \text{divergent}, & |q|\geq 1. \end{cases}$$

For example, series $1-\frac{1}{2}+\frac{1}{2^2}-\frac{1}{2^3}+\cdots+(-1)^{n-1}\frac{1}{2^{n-1}}+\cdots$ is a geometric series and the common ratio is $q=-\frac{1}{2}$, so the series is convergent and its sum is

$$s=\frac{a}{1-q}=\frac{1}{1-\left(-\frac{1}{2}\right)}=\frac{2}{3}.$$

Example 2 Show that the harmonic series

$$1+\frac{1}{2}+\frac{1}{3}+\cdots+\frac{1}{n}+\cdots \tag{4}$$

is divergent.

Solution The partial sum is

$$s_n=1+\frac{1}{2}+\frac{1}{3}+\cdots+\frac{1}{n}.$$

By the inequation $x>\ln(1+x)$ where $x>0$ ($f(x)=x-\ln(1+x)$ is continuous if $x\geq 0$. If $x>0$, $f'(x)>0$ and $f(0)=0$. Thus if $x>0$, $f(x)>0$), we get that

$$s_n = 1+\frac{1}{2}+\frac{1}{3}+\cdots+\frac{1}{n}$$

$$> \ln(1+1)+\ln\left(1+\frac{1}{2}\right)+\cdots+\ln\left(1+\frac{1}{n}\right)$$

$$= \ln 2 + \ln\frac{3}{2}+\cdots+\ln\frac{n+1}{n}$$

$$= \ln 2 + (\ln 3 - \ln 2) + \cdots + [\ln(n+1) - \ln n]$$

$$= \ln(n+1).$$

Since $\lim_{n\to\infty}\ln(n+1)=+\infty$, we have $\lim_{n\to\infty} s_n=+\infty$. Therefore the harmonic progression is divergent.

Note The character that harmonic progression is divergent plays an important role in the convergence tests of series and the discussion of conditions that series converges.

Example 3 Show that $\frac{1}{1\cdot 2}+\frac{1}{2\cdot 3}+\frac{1}{3\cdot 4}+\cdots+\frac{1}{n(n+1)}+\cdots=1$.

Solution Because

$$s_n = \frac{1}{1\cdot 2}+\frac{1}{2\cdot 3}+\frac{1}{3\cdot 4}+\cdots+\frac{1}{n(n+1)}$$

$$= \left(1-\frac{1}{2}\right)+\left(\frac{1}{2}-\frac{1}{3}\right)+\left(\frac{1}{3}-\frac{1}{4}\right)+\cdots+\left(\frac{1}{n}-\frac{1}{n+1}\right)$$

$$= 1 - \frac{1}{n+1},$$

we have
$$\lim_{n\to\infty} s_n = \lim_{n\to\infty}\left(1 - \frac{1}{n+1}\right) = 1.$$

Thus the series is convergent and its sum is 1.

Then give the necessary condition where the progression is convergent.

Theorem If $\sum_{n=1}^{\infty} u_n$ is convergent, then $\lim_{n\to\infty} u_n = 0$.

Proof Because $\sum_{n=1}^{\infty} u_n$ is convergent, we have $\lim_{n\to\infty} s_n = s$ and $\lim_{n\to\infty} s_{n-1} = s$. It is obvious that $u_n = s_n - s_{n-1}$, so
$$\lim_{n\to\infty} u_n = \lim_{n\to\infty}(s_n - s_{n-1}) = \lim_{n\to\infty} s_n - \lim_{n\to\infty} s_{n-1} = s - s = 0.$$

Note 1 $\lim_{n\to\infty} u_n = 0$ is just the necessary condition of the convergence of $\sum_{n=1}^{\infty} u_n$ but not the sufficient condition, which means that if $\lim_{n\to\infty} u_n = 0$ the series $\sum_{n=1}^{\infty} u_n$ might be convergent or divergent. For example, for the harmonic series $\sum_{n=1}^{\infty} \frac{1}{n}$, we have $\lim_{n\to\infty} u_n = \lim_{n\to\infty} \frac{1}{n} = 0$ but the harmonic series is actually divergent.

Note 2 From Theorem 1, we have that if $\lim_{n\to\infty} u_n \neq 0$, then $\sum_{n=1}^{\infty} u_n$ must be divergent. For example,
$$\sum_{n=1}^{\infty} (-1)^{n-1} = 1 - 1 + 1 - 1 + \cdots,$$
its general term $(-1)^{n-1}$ does not tend to 0, so the series is divergent.

To determine whether a series is convergent or not, we usually first evaluate the limit of the general term. If u_n doesn't tend to 0, the series is divergent definitely.

8.1.3 The characters of convergent series

Because the convergence or divergence of a series is corresponding to the convergence or divergence of the sequence $\{s_n\}$, so we can get the properties of convergent series from the properties of limit of sequence.

Property 1 If $\sum_{n=1}^{\infty} u_n$ is convergent and k is a constant, then the series $\sum_{n=1}^{\infty} ku_n$ is also convergent and $\sum_{n=1}^{\infty} ku_n = k \sum_{n=1}^{\infty} u_n$.

In a word, after multiplying each term of a convergent series by a constant, the new series is still convergent.

Proof For given series $\sum_{n=1}^{\infty} u_n$, $s_n = u_1 + u_2 + \cdots + u_n$, then for the series $\sum_{n=1}^{\infty} k u_n$, we have the partial sums
$$\sigma_n = k u_1 + k u_2 + \cdots + k u_n = k(u_1 + u_2 + \cdots + u_n) = k s_n.$$

Since $\lim_{n \to \infty} s_n = s$, we have $\lim_{n \to \infty} \sigma_n = \lim_{n \to \infty} k s_n = k s$. Thus the series $\sum_{n=1}^{\infty} k u_n$ is also convergent and its sum is ks.

For example, if $|q| < 1$, the geometric series $\sum_{n=1}^{\infty} q^n$ is convergent and its sum is $s = \frac{q}{1-q}$. Thus if $|q| < 1$, $\sum_{n=1}^{\infty} a q^n$ is also convergent and $\sum_{n=1}^{\infty} a q^n = a \sum_{n=1}^{\infty} q^n = \frac{aq}{1-q}$.

Property 2 If series $\sum_{n=1}^{\infty} u_n$ and $\sum_{n=1}^{\infty} v_n$ are both convergent and their sum are s and σ respectively, then $\sum_{n=1}^{\infty} (u_n \pm v_n)$ is also convergent and its sum is $s \pm \sigma$, that is
$$\sum_{n=1}^{\infty} (u_n \pm v_n) = \sum_{n=1}^{\infty} u_n \pm \sum_{n=1}^{\infty} v_n.$$

Proof Assume that the partial sums of series $\sum_{n=1}^{\infty} u_n$ and $\sum_{n=1}^{\infty} v_n$ are respectively
$$s_n = u_1 + u_2 + \cdots + u_n,$$
$$\sigma_n = v_1 + v_2 + \cdots + v_n,$$
then the partial sums of $\sum_{n=1}^{\infty} (u_n \pm v_n)$ is
$$\begin{aligned} \lambda_n &= (u_1 \pm v_1) + (u_2 \pm v_2) + \cdots + (u_n \pm v_n) \\ &= (u_1 + u_2 + \cdots + u_n) \pm (v_1 + v_2 + \cdots + v_n) \\ &= s_n \pm \sigma_n. \end{aligned}$$

Since $\lim_{n \to \infty} s_n = s$, $\lim_{n \to \infty} \sigma_n = \sigma$, then $\lim_{n \to \infty} \lambda_n = \lim_{n \to \infty} (s_n \pm \sigma_n) = s \pm \sigma$. Thus $\sum_{n=1}^{\infty} (u_n \pm v_n)$ is also convergent and its sum is $s \pm \sigma$.

Example 4 Show that series $\sum_{n=1}^{\infty} \left(\frac{1}{2^{n-1}} + \frac{5}{3^n} \right)$ is convergent and find the sum.

Solution The series $\sum_{n=1}^{\infty} \frac{1}{2^{n-1}}$ is a geometric series with $a=1$ and $q=\frac{1}{2}$, so $\sum_{n=1}^{\infty} \frac{1}{2^{n-1}}$ is convergent. The series $\sum_{n=1}^{\infty} \frac{1}{3^n}$ is a geometric series with $a=\frac{1}{3}$ and $q=\frac{1}{3}$, so $\sum_{n=1}^{\infty} \frac{1}{3^n}$ is also convergent. Then we have $\sum_{n=1}^{\infty} \left(\frac{1}{2^{n-1}} + \frac{5}{3^n} \right)$ is convergent and
$$\sum_{n=1}^{\infty} \left(\frac{1}{2^{n-1}} + \frac{5}{3^n} \right) = \sum_{n=1}^{\infty} \frac{1}{2^{n-1}} + \sum_{n=1}^{\infty} \frac{5}{3^n}$$

$$= \sum_{n=1}^{\infty} \frac{1}{2^{n-1}} + \frac{5}{3} \sum_{n=1}^{\infty} \frac{1}{3^{n-1}}$$
$$= \frac{1}{1-\frac{1}{2}} + \frac{5}{3} \cdot \frac{1}{1-\frac{1}{3}} = 2 + \frac{5}{2} = \frac{9}{2}.$$

Property 3 Remove, add or change finite number of terms, the new series is still convergent and its sum may differ from the original one.

Proof First, we will show that the new series generated by removing finite number of terms is still convergent.

We should prove that the series which removes one term is still convergent. Actually, the process of removing finite number of terms is same to removing one term for finite times. Assumed that we remove the term u_k from the convergent series $\sum_{n=1}^{\infty} u_n$, then the new series is

$$\sum_{n=1}^{\infty} u_n' = u_1 + u_2 + \cdots + u_{k-1} + u_{k+1} + \cdots. \tag{5}$$

Let s_n be the partial sums of the original series and σ_n be the partial sums of the new series (5). If $n > k$, it is obvious that

$$s_n = \sigma_{n-1} + u_k,$$

where u_k is a constant irrelevant to n. Thus we have that the sequence $\{\sigma_n\}$ is convergent if the sequence $\{s_n\}$ is convergent as $n \to \infty$. Therefore, the new series $\sum_{n=1}^{\infty} u_n'$ is convergent.

In the similar way, we can prove that adding or changing finite number of terms does not influence the convergence of series.

Property 4 If a series is convergent, the new series obtained by adding brackets to some terms is also convergent and has the same sum of the original series.

Proof Assumed that $u_1 + u_2 + u_3 + u_4 + \cdots$ is convergent, that is, the limit of the partial sums $s_1, s_2, s_3, \cdots$ exists. The new series obtained by adding brackets to the original one is

$$(u_1 + \cdots + u_{n_1}) + (u_{n_1+1} + \cdots + u_{n_2}) + \cdots + (u_{n_{k-1}+1} + \cdots + u_{n_k}) + \cdots = \sum_{k=1}^{\infty} v_k,$$

where $v_k = u_{n_{k-1}+1} + u_{n_{k-1}+2} + \cdots + u_{n_k}$, which means that every bracket is one term of the new series. Obviously, there are the following relations between the partial sums of the new series and the original series:

$$A_1 = u_1 + \cdots + u_{n_1} = s_{n_1},$$
$$A_2 = (u_1 + \cdots + u_{n_1}) + (u_{n_1+1} + \cdots + u_{n_2}) = s_{n_2},$$
$$\cdots\cdots\cdots\cdots$$

$$A_k = (u_1 + \cdots + u_{n_1}) + (u_{n_1+1} + \cdots + u_{n_2}) + \cdots + (u_{n_{k-1}+1} + \cdots + u_{n_k}) = s_{n_k},$$
....

It is obvious that the sequence $\{A_k\}$ is a subsequence of the sequence $\{s_n\}$. From the relationship between convergent sequence and its subsequences, since the sequence $\{s_n\}$ is convergence, we have the sequence $\{A_k\}$ is convergent and
$$\lim_{k \to \infty} A_k = \lim_{n \to \infty} s_n.$$

So the new series obtained by adding brackets converges to the same sum of the original series.

Note Convergent series cannot remove brackets optionally. In fact, the new series generated by removing brackets may not be convergent. For example, the series
$$(1-1) + (1-1) + \cdots + (1-1) + \cdots = 0 + 0 + \cdots + 0 + \cdots = 0$$
is convergent, but the series which is removed brackets
$$1 - 1 + 1 - 1 + \cdots + (-1)^{n-1} + \cdots$$
is divergent.

Exercises 8-1

1. Find the general terms of the following series and rewrite the series as the form $\sum$:

   (1) $\dfrac{1!}{2} + \dfrac{2!}{5} + \dfrac{3!}{10} + \dfrac{4!}{17} + \cdots$;

   (2) $-\dfrac{1}{2} + \dfrac{2}{2^2} - \dfrac{3}{2^3} + \dfrac{4}{2^4} - \cdots$;

   (3) $\dfrac{a^2}{3} - \dfrac{a^3}{5} + \dfrac{a^4}{7} - \dfrac{a^5}{9} + \cdots$;

   (4) $\dfrac{\sqrt{a}}{2} + \dfrac{a}{2 \cdot 4} + \dfrac{a\sqrt{a}}{2 \cdot 4 \cdot 6} + \dfrac{a^2}{2 \cdot 4 \cdot 6 \cdot 8} + \cdots$.

2. Given that the partial sum of $\sum\limits_{n=1}^{\infty} u_n$ is $s_n = \dfrac{2n}{n+1}$.

   (1) Find the general term.

   (2) Find the first five terms.

   (3) Find the sum s.

3. For the series $\sum\limits_{n=1}^{\infty} \dfrac{1}{(2n-1)(2n+1)}$, u_n can be rewritten as $u_n = \dfrac{1}{2}\left(\dfrac{1}{2n-1} - \dfrac{1}{2n+1}\right)$.

   (1) Find the partial sum s_n.

   (2) Determine whether the series is convergent or divergent. If it is convergent, find its sum.

4. For the convergent series $\sum\limits_{n=1}^{\infty} u_n$, discuss the convergence or divergence of the following series:

   (1) if $\sum\limits_{n=1}^{\infty} v_n = \sum\limits_{n=1}^{\infty} (u_n + 0.01)$, $\sum\limits_{n=1}^{\infty} v_n$ is _____;

   (2) if $\sum\limits_{n=1}^{\infty} v_n = \sum\limits_{n=1}^{\infty} u_{n+100}$, $\sum\limits_{n=1}^{\infty} v_n$ is _____;

   (3) if $\sum\limits_{n=1}^{\infty} v_n = \sum\limits_{n=1}^{\infty} \dfrac{1}{u_n}$ ($u_n \neq 0$), $\sum\limits_{n=1}^{\infty} v_n$ is _____.

5. By the definition of series' convergence or divergence, determine whether the following series is convergent or divergent.

(1) For series $\sum_{n=1}^{\infty} \frac{n}{3^n}$, show that $\frac{2}{3}s_n = s_n - \frac{1}{3}s_n = \frac{1}{2}\left(1 - \frac{1}{3^n}\right) - \frac{n}{3^{n+1}}$ and find the sum.

(2) For series $\sum_{n=1}^{\infty} \arctan \frac{1}{2n^2}$, show that by the mathematical induction $s_n = \arctan \frac{n}{n+1}$ and find the sum.

(3) For series $\sum_{n=1}^{\infty}(\sqrt{n+1} - \sqrt{n})$, find the partial sum s_n and show that the series is divergent.

6. By the convergence and divergence of geometric series and harmonic series as well as the properties of series, determine the convergence or divergence of the following series:

(1) $\frac{\ln 2}{2} + \frac{\ln^2 2}{2^2} + \frac{\ln^3 2}{2^3} + \cdots + \frac{\ln^n 2}{2^n} + \cdots$;   (2) $\frac{1}{1\,001} + \frac{1}{2\,001} + \frac{1}{3\,001} + \cdots + \frac{1}{1\,000n+1} + \cdots$;

(3) $\sum_{n=1}^{\infty}\left(\frac{1}{2^n} + \frac{1}{10n}\right)$;   (4) $\sum_{n=1}^{\infty}\left(\frac{1}{2^n} + \frac{1}{3^n}\right)$;

(5) $\sum_{n=1}^{\infty} 10^{10} \frac{1}{a^n}$ $(a > 0)$;   (6) $\sum_{n=1}^{\infty} \frac{1}{\sqrt[n]{5}}$.

8.2 Positive series and its convergence test

One of the core problems is how to judge the convergence or divergence of a series because we can just do operations, find sum and approximate value of convergent series. The method to determine whether a positive series is convergent or divergent is the basic method for other series.

8.2.1 The basic characters of positive series

If each term of the series $\sum_{k=1}^{\infty} u_k$ is nonnegative real number, that is, $u_n \geqslant 0 (n=1,2,\cdots)$, then the series is called positive series. Positive series is not only the most basic, but also very important. Later we will see that many series' problems of convergence or divergence come down to the problems of positive series.

Assume that $\sum_{k=1}^{\infty} u_k$ is a positive series and s_n is the partial sum, then $s_{n+1} - s_n = u_{n+1} \geqslant 0$ $(n=1,2,\cdots)$, and the partial sums $\{s_n\}$ is strictly increasing. If the sequence $\{s_n\}$ has upper bound, then $\{s_n\}$ is convergent since a monotonic bounded sequence has a limit, and the positive series is convergent. On the other hand, if the positive series is convergent, the partial sums $\{s_n\}$ is also convergent. Therefore we have the following theorem of the convergent positive series.

Theorem 1 The positive series $\sum_{k=1}^{\infty} u_k$ is convergent if and only if its partial sums $\{s_n\}$ has upper bound.

Then discuss another important series—p-series. p-series is similar to the harmonic series and the geometric series. It is regarded as the standard to judge convergence or divergence of other progressions.

The p-series is defined by

$$\sum_{n=1}^{\infty}\frac{1}{n^p} = 1 + \frac{1}{2^p} + \frac{1}{3^p} + \cdots + \frac{1}{n^p} + \cdots, \tag{1}$$

where $p>0$ and p-series is positive series. It has been proved that the p-series is divergent if $p=1$ while it is convergent if $p=2$ (see Example 2 in Chapter 8.1). In general, there is the following conclusion about the convergence or divergence of p-series.

Example 1 For p-series(1), show that the series is divergent if $p \leqslant 1$ and convergent if $p>1$.

Proof If $p \leqslant 1$, the partial sum of p-series is

$$s_n = 1 + \frac{1}{2^p} + \frac{1}{3^p} + \cdots + \frac{1}{n^p} \geqslant 1 + \frac{1}{2} + \frac{1}{3} + \cdots + \frac{1}{n} = \sigma_n.$$

Known from the proof of Example 2 (Harmonic progression is divergent) in chapter 8.1 that $s_n \geqslant \sigma_n > \ln(1+n)$, therefore the partial sums $\{s_n\}$ has no upper bound and the series $\sum_{n=1}^{\infty}\frac{1}{n^p}$ is divergent.

If $p>1$, we firstly substitute continuous x for n in general term $\frac{1}{n^p}$, and it is the function $\frac{1}{x^p}$. As shown in Figure 8-1, sketch the graph of the function $y=\frac{1}{x^p}$ on the interval $1 \leqslant x < +\infty$. Therefore the partial sum of p-series

$$1 + \frac{1}{2^p} + \frac{1}{3^p} + \cdots + \frac{1}{n^p}$$

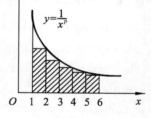

Figure 8-1

is equal to the sum of the shaded area of rectangle which is smaller than that of trapezoid with curved side on the interval $[1, n]$. Thus the partial sum of p-series is

$$s_n = 1 + \left(\frac{1}{2^p} + \frac{1}{3^p} + \cdots + \frac{1}{n^p}\right)$$

$$< 1 + \int_1^n \frac{1}{x^p}dx = 1 + \left[\frac{1}{1-p}x^{1-p}\right]_1^n$$

$$= 1 + \frac{1}{p-1}\left(1 - \frac{1}{n^{p-1}}\right) < 1 + \frac{1}{p-1} = M.$$

Since s_n has upper bound, p-series is convergent.

In summary,

$$\sum_{n=1}^{\infty}\frac{1}{n^p} = \begin{cases} \text{convergent}, & p > 1, \\ \text{divergent}, & p \leqslant 1. \end{cases}$$

Known from Example 1 that series $\sum_{n=1}^{\infty}\frac{1}{n^{\frac{3}{2}}}$ is convergent while series $\sum_{n=1}^{\infty}\frac{1}{n^{\frac{1}{2}}}$ is divergent. Basic Theorem 1 is seldom used directly to judge the convergence or divergence of positive series, but it has high theoretical value that can infer many convergence tests of positive series. Then introduce some common convergence tests of positive series.

8.2.2 The comparison test of positive series

Theorem 2 (Comparison test) For positive series $\sum_{n=1}^{\infty} u_n$,

(1) if there is a convergent positive series $\sum_{n=1}^{\infty} v_n$ and $u_n \leqslant v_n (n=1,2,3,\cdots)$, then $\sum_{n=1}^{\infty} u_n$ is convergent;

(2) if there is a divergent positive series $\sum_{n=1}^{\infty} v_n$ and $u_n \geqslant v_n (n=1,2,3,\cdots)$, then $\sum_{n=1}^{\infty} u_n$ is divergent.

Proof (1) Since $\sum_{n=1}^{\infty} v_n$ is convergent, the partial sum $v_1+v_2+v_3+\cdots+v_n$ has upper bound M. Because

$$0 \leqslant s_n = u_1+u_2+u_3+\cdots+u_n \leqslant v_1+v_2+v_3+\cdots+v_n \leqslant M \ (n=1,2,3,\cdots),$$

s_n also has upper bound. Thus positive series $\sum_{n=1}^{\infty} u_n$ is also convergent.

(2) Since $\sum_{n=1}^{\infty} v_n$ is divergent, the partial sum $v_1+v_2+v_3+\cdots+v_n$ has no upper bound. Because

$$s_n = u_1+u_2+u_3+\cdots+u_n \geqslant v_1+v_2+v_3+\cdots+v_n \ (n=1,2,3,\cdots),$$

s_n also has no upper bound. Thus positive series $\sum_{n=1}^{\infty} u_n$ is divergent.

By the comparison test, three important series introduced before (geometric series, p-series and harmonic series) are often regarded as comparison objects.

Example 2 Given that $a>1$, determine whether $\sum_{n=1}^{\infty} \dfrac{1}{a^n+1}$ is convergent or divergent.

Solution Assume that $u_n = \dfrac{1}{a^n+1} < \left(\dfrac{1}{a}\right)^n = v_n (n=1,2,3,\cdots)$. Since $a>1$, $\dfrac{1}{a}<1$, $\sum_{n=1}^{\infty} v_n$ is a convergent geometric series with the common ratio $q<1$. By the comparison test, $\sum_{n=1}^{\infty} \dfrac{1}{a^n+1}$ is convergent.

Example 3 Determine whether $\sum_{n=1}^{\infty} \dfrac{1}{\sqrt{n(n+2)}}$ is convergent or divergent.

Solution Since $n(n+2) \leqslant (n+1)^2$, we have
$$u_n = \frac{1}{\sqrt{n(n+2)}} \geqslant \frac{1}{n+1}, \quad n=1,2,3,\cdots,$$
and the series
$$\sum_{n=1}^{\infty} \frac{1}{n+1} = \frac{1}{2} + \frac{1}{3} + \cdots + \frac{1}{n+1} + \cdots$$
is divergent. Thus $\sum_{n=1}^{\infty} \frac{1}{\sqrt{n(n+2)}}$ is divergent.

Notice that multiplying each term by a nonzero constant, adding or removing finite number of terms will not influence the convergence or divergence of series. Therefore, there are the following inferences through comparison test.

Corollary 1 Assume that $\sum_{n=1}^{\infty} u_n$ and $\sum_{n=1}^{\infty} v_n$ are both positive series.

(1) If the series $\sum_{n=1}^{\infty} v_n$ is convergent and there exists natural number N such that $u_n \leqslant k v_n (k>0)$ if $n \geqslant N$, then the series $\sum_{n=1}^{\infty} u_n$ is convergent.

(2) If the series $\sum_{n=1}^{\infty} v_n$ is divergent and there exists natural number N such that $u_n \geqslant k v_n (k>0)$ if $n \geqslant N$, then the series $\sum_{n=1}^{\infty} u_n$ is divergent.

Example 4 Determine whether $\sum_{n=1}^{\infty} \frac{\ln n}{n}$ is convergent or divergent.

Solution Obviously after the third term, there will be $\frac{\ln n}{n} > \frac{1}{n}$. Since $\sum_{n=1}^{\infty} \frac{1}{n}$ is divergent, $\sum_{n=1}^{\infty} \frac{\ln n}{n}$ is divergent.

Because using comparison test needs to notice the inequality relation of each term after some term, it is used inconveniently under the condition that the expression of u_n is complex. Then we give more practical methods on the basis of the comparison test.

Corollary 2 (Comparison test in limit form) Assume that series $\sum_{n=1}^{\infty} u_n$ and $\sum_{n=1}^{\infty} v_n$ are both positive series.

(1) If $\lim_{n \to \infty} \frac{u_n}{v_n} = l \ (0 \leqslant l < +\infty)$ and series $\sum_{n=1}^{\infty} v_n$ is convergent, then series $\sum_{n=1}^{\infty} u_n$ is convergent.

(2) If $\lim\limits_{n\to\infty}\dfrac{u_n}{v_n}=l\ (0<l<+\infty)$ or $\lim\limits_{n\to\infty}\dfrac{u_n}{v_n}=+\infty$, and series $\sum\limits_{n=1}^{\infty}v_n$ is divergent, then series $\sum\limits_{n=1}^{\infty}u_n$ is divergent.

Proof (1) Known from the definition of limit, there exists natural number N for $\varepsilon=1$. If $n>N$, there is the inequation $\left|\dfrac{u_n}{v_n}-l\right|<1$. Thus if $n>N$, we have $u_n<(l+1)v_n$. By Corollary 1, we can obtain the convergence.

(2) From the condition $\lim\limits_{n\to\infty}\dfrac{v_n}{u_n}=l(0\leqslant l<+\infty)$. If series $\sum\limits_{n=1}^{\infty}u_n$ is convergent, series $\sum\limits_{n=1}^{\infty}v_n$ must be convergent. But given that series $\sum\limits_{n=1}^{\infty}v_n$ is divergent, so series $\sum\limits_{n=1}^{\infty}u_n$ cannot possibly be convergent. Thus series $\sum\limits_{n=1}^{\infty}u_n$ is divergent.

Corollary 2 illustrate that if $\lim\limits_{n\to\infty}\dfrac{u_n}{v_n}=l(0<l<+\infty)$, the two positive series $\sum\limits_{n=1}^{\infty}u_n$ and $\sum\limits_{n=1}^{\infty}v_n$ are convergent or divergent at the same time.

Example 5 Judge the convergence or divergence of $\sum\limits_{n=1}^{\infty}\sin\dfrac{\pi}{2^n}$ and $\sum\limits_{n=1}^{\infty}\ln\dfrac{n+2}{n+1}$.

Solution As $n\to\infty$, $\sin\dfrac{\pi}{2^n}\sim\dfrac{\pi}{2^n}$, $\ln\dfrac{n+2}{n+1}=\ln\left(1+\dfrac{1}{n+1}\right)\sim\dfrac{1}{n+1}$, it is easy to know $\sum\limits_{n=1}^{\infty}\dfrac{\pi}{2^n}$ converges and $\sum\limits_{n=1}^{\infty}\dfrac{1}{n+1}$ diverges.

Thus $\sum\limits_{n=1}^{\infty}\sin\dfrac{\pi}{2^n}$ is convergent and $\sum\limits_{n=1}^{\infty}\ln\left(\dfrac{n+2}{n+1}\right)$ is divergent.

In Corollary 2, if we let $v_n=\dfrac{1}{n^p}\ (n=1,2,\cdots)$, the p-series. We can obtain the following conclusions according to the convergence and divergence of p-series.

Corollary 3 Assume that $\sum\limits_{n=1}^{\infty}u_n$ is positive series.

(1) If $p\leqslant 1$, $\lim\limits_{n\to\infty}n^p\cdot u_n=A>0$ (or $\lim\limits_{n\to\infty}nu_n=+\infty$), then $\sum\limits_{n=1}^{\infty}u_n$ is divergent.

(2) If $p>1$, $\lim\limits_{n\to\infty}n^p\cdot u_n=A(0\leqslant A<+\infty)$, then $\sum\limits_{n=1}^{\infty}u_n$ is convergent.

Example 6 Determine whether $\sum\limits_{n=1}^{\infty}\dfrac{n+1}{\sqrt{n^3+2n+1}}$ is convergent or divergent.

Solution Let $p=\dfrac{1}{2}$, since

$$\lim_{n\to\infty} n^{\frac{1}{2}} \cdot \dfrac{n+1}{\sqrt{n^3+2n+1}} = 1 > 0,$$

so $\sum\limits_{n=1}^{\infty} \dfrac{n+1}{\sqrt{n^3+2n+1}}$ is divergent.

Example 7 Determine whether $\sum\limits_{n=1}^{\infty} \sin\dfrac{\pi}{n^2}$ is convergent or divergent.

Solution Let $p=2>1$, since

$$\lim_{n\to\infty} n^2 \sin\dfrac{\pi}{n^2} = \pi < +\infty,$$

so $\sum\limits_{n=1}^{\infty} \sin\dfrac{\pi}{n^2}$ is convergent.

8.2.3 d'Alembert method of positive series

Judging the convergence or divergence of positive series $\sum\limits_{n=1}^{\infty} u_n$ through comparison test by another positive series $\sum\limits_{n=1}^{\infty} v_n$. It is often difficult to choose $\sum\limits_{n=1}^{\infty} v_n$, so d'Alembert used comparison test to obtain the d'Alembert method. It utilizes the character of positive series to judge its convergence or divergence.

Theorem 3 (d'Alembert method) Assume that each term of positive series $\sum\limits_{n=1}^{\infty} u_n$ is nonzero, that is $u_n > 0 (n=1,2,3,\cdots)$.

(1) If $\lim\limits_{n\to\infty} \dfrac{u_{n+1}}{u_n} = \rho < 1$, then $\sum\limits_{n=1}^{\infty} u_n$ is convergent.

(2) If $\lim\limits_{n\to\infty} \dfrac{u_{n+1}}{u_n} = \rho > 1$ or $\lim\limits_{n\to\infty} \dfrac{u_{n+1}}{u_n} = +\infty$, then $\sum\limits_{n=1}^{\infty} u_n$ is divergent and $\lim\limits_{n\to\infty} u_n = +\infty$.

(3) If $\lim\limits_{n\to\infty} \dfrac{u_{n+1}}{u_n} = \rho = 1$, then $\sum\limits_{n=1}^{\infty} u_n$ might be convergent or divergent. The d'Alembert method is invalid and we need to use other methods.

Proof (1) According to the definition of limit, there exists natural number N for $\varepsilon = \dfrac{1-\rho}{2} > 0$. Since

$$\lim_{n\to\infty} \dfrac{u_{n+1}}{u_n} = \rho < 1 \quad \text{and} \quad \left|\dfrac{u_{n+1}}{u_n} - \rho\right| < \dfrac{1-\rho}{2},$$

if $n > N$, we have $0 \leqslant \dfrac{u_{n+1}}{u_n} < \dfrac{1+\rho}{2} = r < 1$. Thus

$$u_{N+2} < r u_{N+1}, \quad u_{N+3} < r u_{N+2} < r^2 u_{N+1}, \quad u_{N+4} < r u_{N+3} < r^3 u_{N+1}, \cdots.$$

Because the positive geometric series

$$u_{N+1}+u_{N+1}r+u_{N+1}r^2+u_{N+1}r^3+\cdots \quad (r<1)$$

is convergent, the series

$$u_{N+1}+u_{N+2}+u_{N+3}+u_{N+4}+\cdots$$

is also convergent by the comparison test. Series $\sum_{n=1}^{\infty} u_n$ will not change the convergence after adding finite number of terms(for example N terms)

$$u_1+u_2+\cdots+u_N,$$

so $\sum_{n=1}^{\infty} u_n$ is convergent.

(2) Assume that $\rho>1$, since as $n\to\infty$, $\frac{u_{n+1}}{u_n}\to\rho>1$, there exists some natural number N, if $n>N$, $\left|\frac{u_{n+1}}{u_n}-\rho\right|<\frac{\rho-1}{2}$ and therefore

$$\frac{u_{n+1}}{u_n}>\frac{\rho+1}{2}=r>1.$$

Thus we have

$$u_{n+1}>u_n r>u_{n-1}r^2>\cdots>u_{N+1}r^{n-N},$$

if $n>N$. Since $r>1$, we obtains that $\lim_{n\to\infty}u_n=+\infty$. Known from the necessary condition of convergent series, the series $\sum_{n=1}^{\infty} u_n$ is divergent.

(3) If $\rho=1$ and $\lim_{n\to\infty}\frac{u_{n+1}}{u_n}=1$, the series $\sum_{n=1}^{\infty} u_n$ may be convergent or divergent. For instance, the p-series, for whatever the value of p, there is always

$$\lim_{n\to\infty}\frac{u_{n+1}}{u_n}=\lim_{n\to\infty}\frac{\frac{1}{(1+n)^p}}{\frac{1}{n^p}}=1.$$

But we know that the p-series is convergent if $p>1$ and divergent if $p\leqslant 1$. Thus we cannot determine the convergence or divergence if $\rho=1$ by the d'Alembert method.

Example 8 Determine the convergence or divergence of the following series:

(1) $\sum_{n=1}^{\infty} \frac{a^n}{n!}(a>0)$; (2) $\sum_{n=1}^{\infty} \frac{n!}{5^n}$; (3) $\sum_{n=1}^{\infty} \frac{n!}{n^n}$.

Solution (1) Because

$$\lim_{n\to\infty}\frac{u_{n+1}}{u_n}=\lim_{n\to\infty}\left[\frac{a^{n+1}}{(n+1)!}\times\frac{n!}{a^n}\right]=\lim_{n\to\infty}\frac{a}{n+1}=0<1,$$

the series $\sum_{n=1}^{\infty} \frac{a^n}{n!}$ is convergent.

(2) Because

$$\lim_{n\to\infty}\frac{u_{n+1}}{u_n}=\lim_{n\to\infty}\left[\frac{(n+1)!}{5^{n+1}}\times\frac{5^n}{n!}\right]=\lim_{n\to\infty}\frac{n+1}{5}=+\infty,$$

the series $\sum_{n=1}^{\infty} \frac{n!}{5^n}$ is divergent.

(3) Because
$$\lim_{n\to\infty}\frac{u_{n+1}}{u_n}=\lim_{n\to\infty}\left[\frac{(n+1)!}{(n+1)^{n+1}}\times\frac{n^n}{n!}\right]=\lim_{n\to\infty}\left(\frac{n}{n+1}\right)^n=\lim_{n\to\infty}\frac{1}{\left(1+\frac{1}{n}\right)^n}=\frac{1}{e}<1,$$

the series $\sum_{n=1}^{\infty}\frac{n!}{n^n}$ is convergent.

In addition, we can also obtain that $\lim_{n\to\infty}\frac{a^n}{n!}=0(a>0)$ according to the Theorem 1 and comparison test.

Example 9 Determine whether the positive series $\sum_{n=1}^{\infty}\frac{(\lambda-e)^2\lambda^n\cdot n!}{n^n}(\lambda>0)$ is convergent or divergent.

Solution Let $u_n=\frac{(\lambda-e)^2\lambda^n\cdot n!}{n^n}$, if $\lambda\neq e$, then $u_n>0(n=1,2,3,\cdots)$ and

$$\frac{u_{n+1}}{u_n}=u_{n+1}\cdot\frac{1}{u_n}=\frac{(\lambda-e)^2\lambda^{n+1}\cdot(n+1)!}{(n+1)^{n+1}}\cdot\frac{n^n}{(\lambda-e)^2\lambda^n\cdot n!}$$
$$=\lambda\left(\frac{n}{n+1}\right)^n=\frac{\lambda}{\left(1+\frac{1}{n}\right)^n},$$

thus
$$\lim_{n\to\infty}\frac{u_{n+1}}{u_n}=\lim_{n\to\infty}\frac{\lambda}{\left(1+\frac{1}{n}\right)^n}=\frac{\lambda}{e}.$$

Therefore, ① if $0<\lambda<e$, the series is convergent;
② if $\lambda=e$, $u_n=0(0,1,2,\cdots,)$, the series is convergent;
③ if $\lambda>e$, the series is divergent.

Example 10 Show that the series
$$1+\frac{1}{1}+\frac{1}{1\cdot 2}+\frac{1}{1\cdot 2\cdot 3}+\cdots+\frac{1}{1\cdot 2\cdot 3\cdot\cdots\cdot(n-1)}+\cdots$$

is convergent and estimate the error $|r_n|$ generated by substituting partial sum s_n for sum s.

Solution Since
$$\frac{u_{n+1}}{u_n}=\frac{1}{1\cdot 2\cdot 3\cdot\cdots\cdot n}\bigg/\frac{1}{1\cdot 2\cdot 3\cdot\cdots\cdot(n-1)}=\frac{1}{n},$$
$$\lim_{n\to\infty}\frac{u_{n+1}}{u_n}=\lim_{n\to\infty}\frac{1}{n}=0<1,$$

by d'Alembert method, the series is convergent.

The error $|r_n|$ generated by substituting partial sum s_n for sum s_n is

$$|r_n|=\frac{1}{n!}+\frac{1}{(n+1)!}+\frac{1}{(n+2)!}+\cdots$$
$$=\frac{1}{n!}\left[1+\frac{1}{n+1}+\frac{1}{(n+1)(n+2)}\right]+\cdots$$
$$<\frac{1}{n!}\left(1+\frac{1}{n}+\frac{1}{n^2}+\cdots\right)=\frac{1}{n!}\cdot\frac{1}{1-\frac{1}{n}}=\frac{1}{(n-1)\cdot(n-1)!}.$$

Thus we see that the error $|r_n|$ will be smaller than any positive number if n is large enough.

Example 11 Determine whether $\sum\limits_{n=1}^{\infty} \dfrac{1}{(2n-1) \cdot 2n}$ is convergent or divergent.

Solution Since
$$\lim_{n \to \infty} \frac{u_{n+1}}{u_n} = \lim_{n \to \infty} \frac{(2n-1) \cdot 2n}{(2n+1)(2n+2)} = 1,$$
the limit is 1, the d'Alembert method is invalid (it is always such conditions if u_n is the rational expression of u), we have to use other methods.

Because $2n > 2n - 1 \geq n$, we have
$$u_n = \frac{1}{(2n-1) \cdot 2n} < \frac{1}{n^2} = v_n,$$

Then $\sum\limits_{n=1}^{\infty} v_n$ is the convergent p-series with $p = 2$. According to the comparison test, the given series is convergent.

At last we give an example to show that d'Alembert method is often combined with the comparison test to use.

Example 12 Determine whether $\sum\limits_{n=1}^{\infty} \dfrac{n^3 [\sqrt{2} + (-1)^n]^n}{3^n}$ is convergent or divergent.

Solution Notice that it is a positive series. Considering that there is $(-1)^n$ in u_n, we first use the comparison test,
$$\frac{n^3 [\sqrt{2} + (-1)^n]^n}{3^n} \leq \frac{n^3 [\sqrt{2} + 1]^n}{3^n} = v_n.$$

Then use d'Alembert method to the series $\sum\limits_{n=1}^{\infty} v_n$, we have
$$\lim_{n \to \infty} \frac{v_{n+1}}{v_n} = \frac{(n+1)^3 [\sqrt{2}+1]^{n+1}}{3^{n+1}} \cdot \frac{3^n}{n^3 [\sqrt{2}+1]^n}$$
$$= \lim_{n \to \infty} \left(\frac{n+1}{n} \right)^3 \cdot \frac{\sqrt{2}+1}{3}$$
$$= \frac{\sqrt{2}+1}{3} < 1.$$

By the comparison test, $\sum\limits_{n=1}^{\infty} \dfrac{n^3 [\sqrt{2}+1]^n}{3^n}$ is convergent and $\sum\limits_{n=1}^{\infty} \dfrac{n^3 [\sqrt{2}+(-1)^n]^n}{3^n}$ is convergent.

8.2.4 The root test of positive series

From the comparison test of positive series, we can also obtain the root test of positive series.

Theorem 4 (Root test or Cauchy's convergence test) For positive series $\sum\limits_{n=1}^{\infty} u_n$,

(1) if $\lim\limits_{n \to \infty} \sqrt[n]{u_n} = \rho < 1$, then $\sum\limits_{n=1}^{\infty} u_n$ is convergent;

(2) if $\lim\limits_{n\to\infty} \sqrt[n]{u_n} = \rho > 1$, then $\sum\limits_{n=1}^{\infty} u_n$ is divergent and $\lim\limits_{n\to\infty} u_n = +\infty$;

(3) if $\lim\limits_{n\to\infty} \sqrt[n]{u_n} = 1$, the series $\sum\limits_{n=1}^{\infty} u_n$ might be convergent or divergent. The root test is invalid.

Proof (1) From the definition of sequence limit we can obtain that there exists $N > 0$ for $\varepsilon = \dfrac{1-\rho}{2} > 0$. Because $\lim\limits_{n\to\infty} \sqrt[n]{u_n} = \rho < 1$ and $|\sqrt[n]{u_n} - \rho| < \dfrac{1-\rho}{2}$ if $n > N$, we have that if $n > N$,

$$0 \leqslant \sqrt[n]{u_n} < \frac{1-\rho}{2} + \rho = \frac{1+\rho}{2} = r < 1,$$

then $0 \leqslant u_n < r^n$.

Because if $0 \leqslant r < 1$, $r^{N+1} + r^{N+2} + r^{N+3} + \cdots$ is convergent. By the comparison test, we obtain that $\sum\limits_{k=1}^{\infty} u_{N+k}$ is convergent. Thus $\sum\limits_{n=1}^{\infty} u_n$ is convergent.

(2) If $\rho > 1$, there exists $N > 0$ for $\varepsilon = \dfrac{\rho-1}{2} > 0$. If $n > N$, then $|\sqrt[n]{u_n} - \rho| < \dfrac{\rho-1}{2}$ and

$$\sqrt[n]{u_n} > -\frac{\rho-1}{2} + \rho = \frac{\rho+1}{2} = r > 1,$$

which means $u_n > r^n$ $(r > 1)$. Therefore $\lim\limits_{n\to\infty} u_n = +\infty \neq 0$ and then $\sum\limits_{n=1}^{\infty} u_n$ is divergent.

(3) For any ρ, there will be $\sqrt[n]{u_n} \to 1 (n \to \infty)$ for p-series $\sum\limits_{n=1}^{\infty} \dfrac{1}{n^p}$. If $p = 1$, $\sum\limits_{n=1}^{\infty} \dfrac{1}{n}$ is divergent while if $p = 2$, $\sum\limits_{n=1}^{\infty} \dfrac{1}{n^2}$ is convergent. It means that series may be convergent or divergent if $\rho = 1$.

Example 13 Determine the convergence or divergence of the following series:

(1) $\sum\limits_{n=1}^{\infty} \left(\dfrac{n}{2n+1}\right)^n$;   (2) $\sum\limits_{n=1}^{\infty} \dfrac{3 + (-1)^n}{3^n}$.

Solution (1) Let $u_n = \left(\dfrac{n}{2n+1}\right)^n$. Since

$$\sqrt[n]{u_n} = \sqrt[n]{\left(\frac{n}{2n+1}\right)^n} = \frac{n}{2n+1} \to \frac{1}{2} < 1 \ (n \to \infty),$$

the series $\sum\limits_{n=1}^{\infty} \left(\dfrac{n}{2n+1}\right)^n$ is convergent.

(2) Since

$$\lim_{n\to\infty} \sqrt[n]{u_n} = \lim_{n\to\infty} \frac{1}{3} \sqrt[n]{3 + (-1)^n} = \frac{1}{3} < 1,$$

we have that the series $\sum\limits_{n=1}^{\infty} \dfrac{3 + (-1)^n}{3^n}$ is convergent by the root test.

Example 14 Assume that a_n, b and a are all positive numbers and $a_n \to a (n \to \infty)$. Determine the convergence or divergence of $\sum_{n=1}^{\infty} \left(\dfrac{b}{a_n}\right)^n$.

Solution Since
$$\lim_{n\to\infty} \sqrt[n]{u_n} = \lim_{n\to\infty} \sqrt[n]{\left(\dfrac{b}{a_n}\right)^n} = \lim_{n\to\infty} \dfrac{b}{a_n} = \dfrac{b}{a},$$
we will discuss the different conditions about the parameters a, b.

If $b < a$, that is $\dfrac{b}{a} < 1$, the series is convergent.

If $b > a$, that is $\dfrac{b}{a} > 1$, the series is divergent.

If $b = a$, that is $\dfrac{b}{a} = 1$, this method is invalid. Since $u_n = \left(\dfrac{b}{a_n}\right)^n \to 1 \neq 0 (n \to \infty)$, the series is divergent.

Exercises 8-2

1. Use the comparison test to judge the convergence or divergence of following series:

   (1) $\sum_{n=1}^{\infty} \dfrac{1}{1+\sqrt{n}}$;

   (2) $\sum_{n=1}^{\infty} \dfrac{1+n}{1+n^2}$;

   (3) $\sum_{n=1}^{\infty} \dfrac{\ln n}{n}$;

   (4) $\sum_{n=1}^{\infty} 2^n \sin \dfrac{\pi}{3^n}$;

   (5) $\sum_{n=1}^{\infty} \dfrac{1}{2n^2 - n + 1}$.

2. Use d'Alembert method to judge the convergence or divergence of following series:

   (1) $\sum_{n=1}^{\infty} \dfrac{n!}{n^n}$;

   (2) $\sum_{n=1}^{\infty} n \tan \dfrac{\pi}{2^{n+1}}$;

   (3) $\sum_{n=1}^{\infty} \dfrac{(a+1)(2a+1)\cdots(na+1)}{(b+1)(2b+1)\cdots(nb+1)}$ $(a > 0, b > 0)$.

3. Use the root test to judge the convergence or divergence of following series:

   (1) $\sum_{n=1}^{\infty} \left(1 - \dfrac{1}{n}\right)^{n^2}$;

   (2) $\sum_{n=1}^{\infty} \dfrac{1}{2^n} \left(1 + \dfrac{1}{n}\right)^{n^2}$;

   (3) $\sum_{n=1}^{\infty} \left(\dfrac{n}{3n-1}\right)^{2n-1}$;

   (4) $\sum_{n=1}^{\infty} \dfrac{n^{n-1}}{(2n^2 + n + 1)^{\frac{n+1}{2}}}$.

4. Judge the convergence or divergence of the following series:

   (1) $\sum_{n=1}^{\infty} \left(1 - \cos \dfrac{\pi}{n}\right)$;

   (2) $\sum_{n=1}^{\infty} \dfrac{1}{n \sqrt[n]{n}}$;

   (3) $\sum_{n=1}^{\infty} \dfrac{[(n+1)!]^2}{2!4!\cdots(2n)!}$;

   (4) $\sum_{n=1}^{\infty} \int_0^{\frac{1}{n}} \dfrac{\sqrt[3]{x}}{4+x^2} dx$;

   (5) $\sum_{n=1}^{\infty} (\sqrt{n+1} - \sqrt{n}) \sin \dfrac{4\pi}{n}$;

   (6) $\sum_{n=1}^{\infty} \dfrac{a^n}{n^s}$ $(a > 0, s > 0)$.

8.3 Arbitrary term progression

We define those progressions whose terms have no same symbol as arbitrary term progression or general term progression. The convergence and divergence of arbitrary term progressions could be judged by rules of positive series.

8.3.1 Alternating series and Leibniz Principle

If the terms of a series are alternatively positive and negative, the series is called an alternating series. Assume that $u_n > 0 (n=1,2,3,\cdots)$, then the series

$$\sum_{n=1}^{\infty}(-1)^{n-1}u_n = u_1 - u_2 + u_3 - u_4 + \cdots + (-1)^{n-1}u_n + \cdots, \tag{1}$$

is alternating series. The other alternating series is

$$-u_1 + u_2 - u_3 + u_4 - \cdots + (-1)^n u_n + \cdots,$$

which will become the form (1) after multiplying each term by -1. So we just need to discuss the convergence or divergence of the series (1).

For the alternating series (1), the partial sum s_n changes unstably (Figure 8-2):

$$s_1 = u_1, s_2 = s_1 - u_2, s_3 = s_2 + u_3, \cdots.$$

The range of changes constitutes the sequence

$$u_1, u_2, u_3, u_4, \cdots.$$

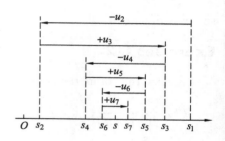

Figure 8-2

If the range of changes diminishes constantly and approaches 0, s_n must approach one number s from the Figure 8-2. The following principle can be understood easily by geometric induction.

Theorem 1 (Leibniz Principle) If alternating series $\sum_{n=1}^{\infty}(-1)^{n-1}u_n$ satisfies the following conditions,

(1) $u_n \geqslant u_{n+1}$ ($n=1, 2, 3, \cdots$);

(2) $\lim\limits_{n \to \infty} u_n = 0$,

then

(i) the series is convergent;

(ii) the sum $s \leqslant u_1$;

(iii) $|r_n| \leqslant u_{n+1}$, where r_n is the remainder term.

Condition (1) and (2) are called Leibniz conditions.

Proof We will prove the limit of the partial sums $\{s_{2n}\}$ exists first.

According to $0 < s_{2n} = (u_1 - u_2) + (u_3 - u_4) + \cdots + (u_{2n-1} - u_{2n})$ and condition (1), we see that $\{s_{2n}\}$ is strictly increasing. Then because

$$s_{2n} = u_1 - (u_2 - u_3) - \cdots - (u_{2n-2} - u_{2n-1}) - u_{2n} < u_1,$$

$\{s_{2n}\}$ has the upper bound u_1. So $\{s_{2n}\}$ exists limit which is written as $\lim\limits_{n\to\infty} s_{2n} = s$ and $s \leqslant u_1$.

Then prove the limit of the partial sums $\{s_{2n+1}\}$ exists.

According to $s_{2n+1} = s_{2n} + u_{2n+1}$ and condition (2) ($\lim\limits_{n\to\infty} u_{2n+1} = 0$), we obtain that

$$\lim_{n\to\infty} s_{2n+1} = \lim_{n\to\infty} s_{2n} = s.$$

Since the partial sums of odd terms or even terms both tend to the same limit s, that is, $\lim\limits_{n\to\infty} s_n = s$. So we obtain (i) $\sum\limits_{n=1}^{\infty} (-1)^{n-1} u_n$ is convergent and (ii) $s \leqslant u_1$.

At last we will prove that (iii) $|r_n| \leqslant u_{n+1}$. In fact,

$$r_n = \pm (u_{n+1} - u_{n+2} + \cdots),$$

and its absolute value

$$|r_n| = u_{n+1} - u_{n+2} + \cdots.$$

The right of the formula above is also an alternating series which satisfies Leibniz conditions. Known from the proof above that the series $|r_n|$ is convergent and its sum is smaller than the first term, that is, $|r_n| \leqslant u_{n+1}$.

Convergent series always have a sum but it is usually difficult to find the sum. Therefore in practical application, people often use the partial sums s_n to approximate the sum s and leave out the error produced by substituting s_n for s, which is defined as truncation error. Thus the truncation error actually is the absolute value of the remainder r_n. Therefore for the alternating series that satisfies Leibniz conditions, it is easy to estimate the truncation error since $|r_n| \leqslant u_{n+1}$, and we just need to calculate u_{n+1}.

Example 1 Show that the series $\sum\limits_{n=1}^{\infty} (-1)^{n-1} \dfrac{1}{n^p} (p>0)$ is convergent and estimate its remainder term.

Proof The series is an alternating series. Because

$$u_n = \frac{1}{n^p} \geqslant \frac{1}{(n+1)^p} = u_{n+1},$$

$$u_n = \frac{1}{n^p} \to 0 \ (n\to\infty),$$

so $\sum\limits_{n=1}^{\infty} (-1)^{n-1} \dfrac{1}{n^p} \ (p>0)$ is convergent and

$$|r_n| \leqslant u_{n+1} = \frac{1}{(n+1)^p}.$$

by the Leibniz Principle.

In particular, if $p=1$, we can obtain the convergent series

$$\sum_{n=1}^{\infty} \frac{(-1)^{n-1}}{n} = 1 - \frac{1}{2} + \frac{1}{3} - \frac{1}{4} + \cdots + \frac{(-1)^{n-1}}{n} + \cdots.$$

And we will find the sum $\ln 2$ later.

Example 2 Determine the convergence or divergence of

$$\sum_{n=1}^{\infty} (-1)^{n-1} \frac{\ln n}{n}.$$

Solution It is an alternating series and we will use the Leibniz Principle. Let $u_n = \dfrac{\ln n}{n} = f(n)$, we can prove that $f(x) = \dfrac{\ln x}{x}$ is strictly decreasing on the interval $[3, +\infty)$ ($f'(x) < 0$). Therefore if $n \geq 3$, then $u_n \geq u_{n+1}$ and
$$\lim_{n \to \infty} u_n = \lim_{n \to \infty} \dfrac{\ln n}{n} = 0.$$

Thus by the Leibniz Principle, $\sum\limits_{n=1}^{\infty} (-1)^{n-1} \dfrac{\ln n}{n}$ is convergent.

Example 3 Determine the convergence or divergence of the alternating series
$$\sum_{n=1}^{\infty} (-1)^{n-1} (\sqrt{n+1} - \sqrt{n}).$$

Solution Assume that $f(x) = \sqrt{x+1} - \sqrt{x}$, then
$$f'(x) = \dfrac{1}{2\sqrt{x+1}} - \dfrac{1}{2\sqrt{x}} = \dfrac{1}{2} \dfrac{\sqrt{x} - \sqrt{x+1}}{\sqrt{x(x+1)}} < 0.$$

We have that $u_n = \sqrt{n+1} - \sqrt{n} > \sqrt{n+2} - \sqrt{n+1} = u_{n+1}$. And since
$$\lim_{n \to \infty} (\sqrt{n+1} - \sqrt{n}) = \lim_{n \to \infty} \dfrac{1}{\sqrt{n+1} + \sqrt{n}} = 0,$$

by the Leibniz Principle, the given series is convergent.

8.3.2 The absolute value method of arbitrary term progression

The absolute values of each term of any series $\sum\limits_{n=1}^{\infty} u_n$ is a positive series $\sum\limits_{n=1}^{\infty} |u_n|$. The relationships between the two series are as follows.

Theorem 2 (The absolute value method) If series $\sum\limits_{n=1}^{\infty} |u_n|$ is convergent, then $\sum\limits_{n=1}^{\infty} u_n$ is convergent.

Proof Let $v_n = \dfrac{1}{2}(u_n + |u_n|)$ ($n = 1, 2, \cdots$). Obviously, $0 \leq v_n \leq |u_n|$ ($n = 1, 2, \cdots$).

By the comparison test of positive series, $\sum\limits_{n=1}^{\infty} v_n$ is convergent, then $\sum\limits_{n=1}^{\infty} 2v_n$ is also convergent. Because $u_n = 2v_n - |u_n|$, we have
$$\sum_{n=1}^{\infty} u_n = \sum_{n=1}^{\infty} (2v_n - |u_n|).$$

So, $\sum\limits_{n=1}^{\infty} u_n$ is convergent.

The absolute value method translates many problems of judging the convergence and divergence of general term series into that for positive series.

Notice that the converse of absolute value method does not stand up that if $\sum\limits_{n=1}^{\infty} u_n$ is convergent, $\sum_{n=1}^{\infty} |u_n|$ may be divergent.

For example, in the Example 1,
$$\sum_{n=1}^{\infty} (-1)^{n-1} \frac{1}{n} = 1 - \frac{1}{2} + \frac{1}{3} - \frac{1}{4} + \cdots$$
is convergent while
$$\sum_{n=1}^{\infty} \left| (-1)^{n-1} \frac{1}{n} \right| = \sum_{n=1}^{\infty} \frac{1}{n} = 1 + \frac{1}{2} + \frac{1}{3} + \frac{1}{4} + \cdots$$
is divergent.

Thus convergent series could be divided into two categories. One is that $\sum_{n=1}^{\infty} u_n$ and $\sum_{n=1}^{\infty} |u_n|$ are both convergent and $\sum_{n=1}^{\infty} u_n$ is called absolutely convergence, while the other is that $\sum_{n=1}^{\infty} u_n$ is convergent but $\sum_{n=1}^{\infty} |u_n|$ is divergent, and $\sum_{n=1}^{\infty} u_n$ is called conditionally convergence.

For example, $\sum_{n=1}^{\infty} (-1)^{n-1} \frac{1}{n^2}$ is absolutely convergent series and $\sum_{n=1}^{\infty} (-1)^{n-1} \frac{1}{n}$ is conditionally convergent series.

Example 4 Determine the convergence or divergence of $\sum_{n=1}^{\infty} \frac{\sin \frac{n\pi}{4}}{n^{\frac{5}{4}}}$.

Solution This is an arbitrary term series. Since
$$\left| \frac{\sin \frac{n\pi}{4}}{n^{\frac{5}{4}}} \right| \leqslant \frac{1}{n^{\frac{5}{4}}}$$
and the positive series $\sum_{n=1}^{\infty} \frac{1}{n^{\frac{5}{4}}}$ is a convergent p-series with $p = \frac{5}{4} > 1$, so $\sum_{n=1}^{\infty} \left| \frac{\sin \frac{n\pi}{4}}{n^{\frac{5}{4}}} \right|$ is also convergent. Therefore, $\sum_{n=1}^{\infty} \frac{\sin \frac{n\pi}{4}}{n^{\frac{5}{4}}}$ is absolutely convergent.

When using the absolute value method, we need to notice that though we cannot get the divergence of $\sum_{n=1}^{\infty} u_n$ by the divergence of $\sum_{n=1}^{\infty} |u_n|$, we can judge that $\sum_{n=1}^{\infty} u_n$ is divergent if we judge that $\sum_{n=1}^{\infty} |u_n|$ (Positive series) is divergent by d'Alembert method or Cauchy's test. By the two methods, if $\sum_{n=1}^{\infty} |u_n|$ is divergent, then $|u_n| \to +\infty \, (n \to \infty)$ and $u_n \not\to 0$ $(n \to \infty)$, thus $\sum_{n=1}^{\infty} u_n$ is divergent. Similarly, if $\lim_{n \to \infty} \frac{|u_{n+1}|}{|u_n|} = \rho > 1$ or $\lim_{n \to \infty} \sqrt[n]{|u_n|} = \rho > 1$,

$\sum\limits_{n=1}^{\infty} u_n$ must be divergent.

Example 5 Determine the convergence or divergence of
$$\sum_{n=1}^{\infty}(-1)^n \frac{1}{2^n}\left(1+\frac{1}{n}\right)^{n^2}.$$

Solution Since $|u_n|=\frac{1}{2^n}\left(1+\frac{1}{n}\right)^{n^2}$, we have
$$\sqrt[n]{|u_n|}=\frac{1}{2}\left(1+\frac{1}{n}\right)^{n} \to \frac{e}{2}\ (n\to\infty).$$

Since $\frac{e}{2}>1$, then $|u_n|\to +\infty\ (n\to\infty)$. Thus $\sum\limits_{n=1}^{\infty}(-1)^n \frac{1}{2^n}\left(1+\frac{1}{n}\right)^{n^2}$ is divergent.

Example 6 Determine the convergence or divergence of
$$\sum_{n=1}^{\infty}(-1)^n \frac{1}{n}x^n\ (x>0).$$

Solution Since $|u_n|=\frac{x^n}{n}$, we have
$$\lim_{n\to\infty}\left|\frac{u_{n+1}}{u_n}\right|=\lim_{n\to\infty}\frac{x^{n+1}}{n+1}\cdot\frac{n}{x^n}=x.$$

By d'Alembert method, we know that

if $x<1$, $\sum\limits_{n=1}^{\infty}(-1)^n \frac{1}{n}x^n$ is absolutely convergent;

if $x>1$, $\sum\limits_{n=1}^{\infty}(-1)^n \frac{1}{n}x^n$ is divergent;

if $x=1$, $\sum\limits_{n=1}^{\infty}\left|\frac{(-1)^n}{n}\right|$ is divergent while $\sum\limits_{n=1}^{\infty}(-1)^n \frac{1}{n}$ is convergent. Thus $\sum\limits_{n=1}^{\infty}(-1)^n \frac{1}{n}x^n$ is conditionally convergent.

Exercises 8-3

1. Determine whether the following series is convergent or divergent. If it is convergent, is it absolutely convergent or conditionally convergent?

(1) $\sum\limits_{n=1}^{\infty}\sin\left(n\pi+\frac{1}{\ln n}\right)$;

(2) $\sum\limits_{n=1}^{\infty}(-1)^{n-1}\frac{2n+1}{2^n}$;

(3) $\sum\limits_{n=1}^{\infty}(-1)^n(\sqrt{n+1}-\sqrt{n})$;

(4) $\sum\limits_{n=1}^{\infty}(-1)^{n+1}\frac{2n^2}{n!}$;

(5) $\sum\limits_{n=1}^{\infty}(-1)^n \frac{(n+1)!}{n^{n+1}}$;

(6) $\sum\limits_{n=1}^{\infty}(-1)^{n-1}\frac{1}{3^n}\left(1+\frac{1}{n}\right)^{n^2}$.

2. Calculate the limits of the following series through necessary conditions of convergence:

(1) $\lim\limits_{n\to\infty}\frac{n^n}{(n!)^2}$;

(2) $\lim\limits_{n\to\infty}\frac{11\cdot 12\cdot\cdots\cdot(n+10)}{2\cdot 5\cdot\cdots\cdot(3n-1)}$.

8.4 Power series

8.4.1 Function series

Given the sequence of function
$$u_1(x), u_2(x), u_3(x), \cdots, u_n(x), \cdots,$$
whose domain is set E, then
$$\sum_{n=1}^{\infty} u_n(x) = u_1(x) + u_2(x) + u_3(x) + \cdots + u_n(x) + \cdots \tag{1}$$
is called the function series on set E or series in short.

For function series, if x is substituted by a constant, it becomes a constant series that we could judge its convergence or divergence.

Definition If $\sum_{n=1}^{\infty} u_n(x_1)$ is convergent when $x_1 \in E$, then x_1 is called series' convergent point and the set includes all convergent points is called the convergent domain. If $\sum_{n=1}^{\infty} u_n(x_2)$ is divergent when $x_2 \in E$, then x_2 is called series' divergent point. We denote the convergent domain of $\sum_{n=1}^{\infty} u_n(x)$ by D and for $x \in D$, $\sum_{n=1}^{\infty} u_n(x)$ has the sum that is written as $\sum_{n=1}^{\infty} u_n(x) = s(x)$. Thus we defined the function $s(x)$ on the set D, which is called sum function, then $\sum_{n=1}^{\infty} u_n(x)$ is convergent to $s(x)$.

Example 1 Consider
$$\sum_{n=1}^{\infty} x^{n-1} = 1 + x + x^2 + \cdots + x^{n-1} + \cdots,$$
which could be deemed as a geometric series with the common ratio x (x is a variable). According to the conclusions of geometric series, the series is convergent when $|x| < 1$, while the series is divergent when $|x| \geq 1$. Then all of convergent points belong to the set $-1 < x < 1$, so the convergent domain is $(-1, 1)$ and
$$\sum_{n=1}^{\infty} x^{n-1} = \frac{1}{1-x}, \quad x \in (-1, 1).$$
Thus on the interval $(-1, 1)$, the sum function of $\sum_{n=1}^{\infty} x^{n-1}$ is $\frac{1}{1-x}$.

Notice that $\frac{1}{1-x}$ is the sum function of $\sum_{n=1}^{\infty} x^{n-1}$ just within the convergent domain $(-1, 1)$. Thus we must indicate its convergent domain when writing the equation of convergent series and its sum function.

Example 2 Find the convergent domain of

$$\sum_{n=1}^{\infty} \frac{\sin nx}{n^2} = \frac{\sin x}{1^2} + \frac{\sin 2x}{2^2} + \cdots + \frac{\sin nx}{n^2} + \cdots.$$

Solution Because $\left|\frac{\sin nx}{n^2}\right| \leqslant \frac{1}{n^2}$, $x \in (-\infty, +\infty)$ and $\sum_{n=1}^{\infty} \frac{1}{n^2}$ is convergent, the convergent domain is $(-\infty, +\infty)$.

Then we will discuss an important function series, power series.

8.4.2 Power series and its convergent interval

For function series, the easiest and most important series is power series that plays an important role in mathematical theory and practice. We regard the form of series like

$$\sum_{n=0}^{\infty} a_n (x-x_0)^n = a_0 + a_1(x-x_0) + a_2(x-x_0)^2 + \cdots + a_n(x-x_0)^n + \cdots \quad (2)$$

as the power series of $(x-x_0)$. If $x_0 = 0$, the series

$$\sum_{n=0}^{\infty} a_n x^n = a_0 + a_1 x + a_2 x^2 + \cdots + a_n x^n + \cdots \quad (3)$$

is the power series of x.

In the Formula (2), if let $t = x - x_0$, then we have

$$\sum_{n=0}^{\infty} a_n (x-x_0)^n = \sum_{n=0}^{\infty} a_n t^n,$$

so we can write power series (2) into the form of power series (3).

Thus we will discuss the characters of power series (3). Which points does power series (3) converge at? What's the character of convergent domain? Obviously when $x=0$, power series (3) is convergent, which means power series always has convergent point. From Example 1, the convergent domain of $\sum_{n=1}^{\infty} x^n$ is $(-1, 1)$. For further study, we give another example.

Example 3 Find the convergent domain of power series

$$\sum_{n=1}^{\infty} \frac{1}{n \cdot 2^n} x^n.$$

Solution By the d'Alembert method, and we have

$$\left|\frac{u_{n+1}(x)}{u_n(x)}\right| = \left|\frac{x^{n+1}}{(n+1) \cdot 2^{n+1}} \cdot \frac{n \cdot 2^n}{x^n}\right| = \frac{n}{n+1} \frac{|x|}{2} \to \frac{|x|}{2} \quad (n \to \infty).$$

If $|x| < 2$, the series is absolutely convergent while the series is divergent when $|x| > 2$.

If $x = 2$, the series becomes $1 + \frac{1}{2} + \frac{1}{3} + \cdots + \frac{1}{n} + \cdots$, it is divergent.

If $x = -2$, the series becomes $-1 + \frac{1}{2} - \frac{1}{3} + \cdots + (-1)^n \frac{1}{n} + \cdots$, it is convergent.

So the convergent domain of the series is $-2 \leqslant x < 2$.

From Example 1 and 3, we know that the convergent domain is a symmetric interval about $x = 0$ apart from endpoints. Actually for the power series (3), the conclusion also stands up.

Chapter 8 Infinitive series

Theorem 1(Abel's Theorem) If power series (3) converges at the point $x_1 \neq 0$, then power series (3) is absolutely convergent at x that satisfies inequation $|x| < |x_1|$, if power series (3) diverges at x_2, then power series (3) is also divergent at x that satisfies inequation $|x| > |x_2|$.

Proof Because $\sum\limits_{n=1}^{\infty} a_n x_1^n$ is convergent, from the necessary condition we have $\lim\limits_{n \to \infty} a_n x_1^n = 0$ and the sequence $\{a_n x_1^n\}$ has bounds, that is, there exists positive number M such that $|a_n x_1^n| \leqslant M$, $n = 0, 1, 2, \cdots$. Therefore when $|x| < |x_1|$,

$$|a_n x^n| = \left| a_n x_1^n \cdot \left(\frac{x}{x_1}\right)^n \right| < M \left| \frac{x}{x_1} \right|^n, \quad n = 0, 1, 2, \cdots.$$

Because $\left| \dfrac{x}{x_1} \right| < 1$, the geometric series

$$M + M \left| \frac{x}{x_1} \right| + M \left| \frac{x}{x_1} \right|^2 + \cdots + M \left| \frac{x}{x_1} \right|^n + \cdots$$

is convergent, by the comparison tests, $\sum\limits_{n=0}^{\infty} |a_n x^n|$ is convergent when $|x| < |x_1|$ and thus power series (3) is absolutely convergent.

Then use proof by contradiction to show that the second part of the theorem. Assume that there exists x_1 such that $|x_1| > |x_2|$ and the power series (3) is convergent at x_1. According to the proof of the first part, the power series (3) is convergent at x_2 for $|x_2| < |x_1|$, which is contrary to the assumption. Thus the proposition stands up.

Now assume that the power series (3) converges at x_1 and diverges at x_2. According to the Abel Theorem, if $|x_1| < |x_2|$, the power series (3) is convergent when $|x| < |x_1|$, while it is divergent when $|x| > |x_2|$ (Figure 8-3). Then observe the convergent and divergent points of series (3) between $|x_1|$ and $|x_2|$. By the Abel theorem, there exists one point x_0 such that the series (3) is convergent when $|x| < |x_0|$ and divergent when $|x| > |x_0|$. Let $|x_0| = R$, we define $(-R, R)$ to be the convergent interval of the series $\sum\limits_{n=0}^{\infty} a_n x^n$ and R is called the convergent radius.

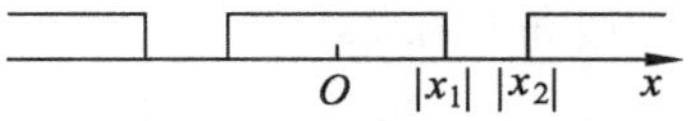

Figure 8-3

If power series $\sum\limits_{n=0}^{\infty} a_n x^n$ is convergent on the interval $(-\infty, +\infty)$, then $(-\infty, +\infty)$ is called the convergent interval and $R = +\infty$ is the convergent radius. If power series $\sum\limits_{n=0}^{\infty} a_n x^n$ is only convergent on $x = 0$, then $R = 0$ is called the convergent radius. Therefore, all power series have the convergent radius.

For power series (2) $\sum\limits_{n=0}^{\infty} a_n (x - x_0)^n$, let $t = x - x_0$, then we have the power series $\sum\limits_{n=0}^{\infty} a_n t^n$. Thus if (2) has convergent point and divergent point apart from x_0, there must exist positive number $R>0$ such that (2) is absolutely convergent when $|x-x_0|<R$ and is divergent when $|x-x_0|>R$. The interval (x_0-R, x_0+R) is the convergent interval and R is called the convergent radius of power series (2).

Note The convergent interval and the convergent domain are different concepts. Convergent interval is an open interval which excludes endpoints. However, if power series is convergent at one endpoint, then its convergent domain will include this endpoint. Thus the convergent domain may have one or two endpoints. If the convergent radius is $+\infty$, the convergent interval $(-\infty, +\infty)$ is also the convergent domain.

Because the power series is determined by its coefficients $a_n(n=0,1,2,\cdots)$, so the convergent radius R is also determined by these coefficients. Then we give the formulas to calculate the convergent radius of power series.

Theorem 2 For the series $\sum_{n=0}^{\infty} a_n x^n$, suppose $a_n \neq 0 (n=0,1,2,\cdots)$. If

$$\lim_{n\to\infty} \left| \frac{a_{n+1}}{a_n} \right| = \rho \left(\text{or } \lim_{n\to\infty} \left| \frac{a_{n+1}}{a_n} \right| = +\infty, \text{ denoted as } \rho = +\infty \right), \tag{4}$$

then

(i) when $0<\rho<+\infty$, the convergent radius is $R=1/\rho$;

(ii) when $\rho=0$, the convergent radius is $R=+\infty$;

(iii) when $\rho=+\infty$, the convergent radius is $R=0$.

Proof Consider the series generated by taking the absolute value of each term of original series,

$$|a_0|+|a_1 x|+|a_2 x^2|+|a_3 x^3|+\cdots+|a_n x^n|+\cdots. \tag{5}$$

Since

$$\lim_{n\to\infty} \left| \frac{u_{n+1}(x)}{u_n(x)} \right| = \lim_{n\to\infty} \left| \frac{a_{n+1} x^{n+1}}{a_n x^n} \right| = \lim_{n\to\infty} \left| \frac{a_{n+1}}{a_n} \right| \cdot |x| = \rho \cdot |x|,$$

by the d'Alembert method:

(i) If $0<\rho<+\infty$, then when $\rho|x|<1$ or $|x|<\frac{1}{\rho}$, $\sum_{n=0}^{\infty} a_n x^n$ is absolutely convergent, while when $\rho|x|>1$ or $|x|>\frac{1}{\rho}$, $\sum_{n=0}^{\infty} a_n x^n$ is divergent. Thus the convergent radius is $R=\frac{1}{\rho}$.

(ii) If $\rho=0$, then $\rho|x|=0<1$, the series is convergent for any x, so $R=+\infty$.

(iii) If $\rho=+\infty$ or $\lim_{n\to\infty} \left| \frac{a_{n+1}}{a_n} \right| = +\infty$, there is

$$\lim_{n\to\infty} \left| \frac{u_{n+1}(x)}{u_n(x)} \right| = \lim_{n\to\infty} \left| \frac{a_{n+1}}{a_n} \right| \cdot |x| = +\infty$$

for $x \neq 0$. Thus the series is divergent for $x \neq 0$ and $R=0$.

Chapter 8 Infinitive series

Example 4 Given that $\sum_{n=0}^{\infty} \dfrac{x^n}{n!}$, find the convergent interval.

Solution Because $\left|\dfrac{a_{n+1}}{a_n}\right| = \dfrac{1}{(n+1)!} \cdot \dfrac{n!}{1} = \dfrac{1}{n+1} \to 0 \ (n \to \infty)$, $\rho = 0$, so the convergent radius is $R = +\infty$ and the convergent interval is $(-\infty, +\infty)$.

Example 5 For $\sum_{n=0}^{\infty} n! x^n$, find its convergent radius, convergent interval and convergent domain.

Solution Because

$$\left|\dfrac{a_{n+1}}{a_n}\right| = \dfrac{(n+1)!}{n!} = n+1 \to +\infty \ (n \to \infty),$$

then the convergent radius is $R = 0$. Power series is divergent for $x \neq 0$ and only convergent at $x = 0$, so the convergent domain is just one point $x = 0$.

Example 6 Given that $\sum_{n=1}^{\infty} \dfrac{3^n + (-2)^n}{n} x^n$, find its convergent radius, convergent interval and convergent domain.

Solution Since

$$\rho = \lim_{n \to \infty} \left|\dfrac{a_{n+1}}{a_n}\right| = \lim_{n \to \infty} \dfrac{3^{n+1} + (-2)^{n+1}}{n+1} \cdot \dfrac{n}{3^n + (-2)^{n+1}} = 3,$$

the convergent radius is $R = \dfrac{1}{\rho} = \dfrac{1}{3}$ and the convergent interval is $\left(-\dfrac{1}{3}, \dfrac{1}{3}\right)$.

Then consider the endpoints of convergent interval. When $x = -\dfrac{1}{3}$, the power series is

$$\sum_{n=1}^{\infty} \dfrac{3^n + (-2)^n}{n} \left(-\dfrac{1}{3}\right)^n = \sum_{n=1}^{\infty} \dfrac{(-1)^n + \left(\dfrac{2}{3}\right)^n}{n}.$$

Since $\sum_{n=1}^{\infty} \dfrac{(-1)^n}{n}$ is also convergent (by the Leibniz Principle) and $\sum_{n=1}^{\infty} \dfrac{1}{n}\left(\dfrac{2}{3}\right)^n$ is also convergent (by the d'Alembert method), the original series is convergent at $x = -\dfrac{1}{3}$.

When $x = \dfrac{1}{3}$, the power series is

$$\sum_{n=1}^{\infty} \dfrac{3^n + (-2)^n}{n} \left(\dfrac{1}{3}\right)^n = \sum_{n=1}^{\infty} \dfrac{1 + \left(-\dfrac{2}{3}\right)^n}{n}.$$

Since $\sum_{n=1}^{\infty} \dfrac{1}{n}$ is divergent and $\sum_{n=1}^{\infty} \dfrac{1}{n}\left(-\dfrac{2}{3}\right)^n$ is convergent (absolutely convergence), the original series is divergent at $x = \dfrac{1}{3}$.

Therefore the convergent domain of the original series is $\left[-\dfrac{1}{3}, \dfrac{1}{3}\right)$.

29

Example 7 For $\sum_{n=1}^{\infty} \frac{(x+1)^n}{n^2}$, find its convergent radius, convergent interval and convergent domain.

Solution This is a power series with $x_0 = -1$. Since
$$\left|\frac{a_{n+1}}{a_n}\right| = \frac{1}{(n+1)^2} \cdot \frac{n^2}{1} \to 1,$$
the convergent radius is $R=1$ and the convergent interval is $|x+1|<1$ which means $-1 < x+1 < 1$ or $x \in (-2, 0)$.

If $x=-2$, the power series becomes
$$\sum_{n=1}^{\infty} \frac{(-1)^n}{n^2} = -1 + \frac{1}{2^2} - \frac{1}{3^2} + \cdots + (-1)^n \frac{1}{n^2} + \cdots,$$
it is also convergent.

If $x=0$, the power series becomes
$$\sum_{n=1}^{\infty} \frac{1}{n^2} = 1 + \frac{1}{2^2} + \frac{1}{3^2} + \cdots + \frac{1}{n^2} + \cdots,$$
it is also convergent.

Thus the convergent domain is $[-2, 0]$.

Note Since a_n is the divisor in the formula above, we cannot use this formula if all a_n behind one term are zero. For example, for the power series $\sum_{n=0}^{\infty} a_{2n} x^{2n}$, $a_1 = a_3 = a_5 = \cdots = 0$, and for $\sum_{n=0}^{\infty} a_{2n-1} x^{2n-1}$, $a_0 = a_2 = a_4 = \cdots = 0$, we cannot directly apply the formula above to calculate the convergent radius, but we can use the d'Alembert method to calculate the convergent radius.

Example 8 For the series $\sum_{n=1}^{\infty} \frac{x^{2n-1}}{3^{n-1}(2n-1)}$, find its convergent radius, convergent interval and convergent domain.

Solution The coefficients of even power terms are all 0, so we can just use the d'Alembert method to calculate the convergent radius,
$$\left|\frac{u_{n+1}(x)}{u_n(x)}\right| = \left|\frac{x^{2n+1}}{3^n(2n+1)} \cdot \frac{3^{n-1}(2n-1)}{x^{2n-1}}\right| = \frac{1}{3} \cdot \frac{2n-1}{2n+1}|x|^2 \to \frac{1}{3}|x|^2 \quad (n \to \infty).$$

If $\frac{1}{3}|x|^2 < 1$ or $|x| < \sqrt{3}$, the power series is absolutely convergent.

If $\frac{1}{3}|x|^2 > 1$ or $|x| > \sqrt{3}$, the power series is divergent. Thus the convergent radius is $R = \sqrt{3}$ and the convergent interval is $|x| < \sqrt{3}$ or $(-\sqrt{3}, \sqrt{3})$.

If $x = -\sqrt{3}$, the power series becomes
$$\sum_{n=1}^{\infty} (-1)^{2n-1} \frac{\sqrt{3}}{2n-1} = \sum_{n=1}^{\infty} \frac{-\sqrt{3}}{2n-1}$$
and it is divergent.

If $x=\sqrt{3}$, the power series becomes $\sum_{n=1}^{\infty} \dfrac{-\sqrt{3}}{2n-1}$ and it is divergent.

Therefore, the convergent domain of this power series is $(-\sqrt{3},\sqrt{3})$.

8.4.3 The properties of power series

1) The algebraic properties of power series

Assume that there are two power series

$$\sum_{n=0}^{\infty} a_n x^n = a_0 + a_1 x + a_2 x^2 + \cdots + a_n x^n + \cdots, \tag{6}$$

$$\sum_{n=0}^{\infty} b_n x^n = b_0 + b_1 x + b_2 x^2 + \cdots + b_n x^n + \cdots, \tag{7}$$

we denote the convergent domain of (6) by I_1 and the convergent radius by R_1, and denote the convergent domain of (7) by I_2 and the convergent radius by R_2.

Because power series (6) and (7) are both convergent on $I_1 \cap I_2$, we can obtain properties of convergent series.

Property 1 If the convergent domains of the series (6) and (7) are respectively I_1 and I_2, then the sum(or difference) of the two series is convergent on $I_1 \cap I_2$ and

$$\sum_{n=0}^{\infty} a_n x^n \pm \sum_{n=0}^{\infty} b_n x^n = \sum_{n=0}^{\infty} (a_n \pm b_n) x^n,\ x \in I = I_1 \cap I_2.$$

Note $I_1 \cap I_2$ is an interval with the center 0 and may include the endpoints.

Because power series on the convergent interval is absolutely convergent, we can apply multiplication for polynomial on the common part of convergent interval:

$$(a_0 + a_1 x + a_2 x^2 + \cdots)(b_0 + b_1 x + b_2 x^2 + \cdots)$$
$$= a_0 b_0 + (a_0 b_1 + a_1 b_0) x + (a_0 b_2 + a_1 b_1 + a_2 b_0) x^2 + \cdots,\ |x| < R = \min\{R_1, R_2\}.$$

Property 2 If the convergent radiuses of power series (6) and (7) are respectively R_1 and R_2, then the product of the two power series is convergent on the interval $(-R, R)$ and

$$\left(\sum_{n=0}^{\infty} a_n x^n\right) \cdot \left(\sum_{n=0}^{\infty} b_n x^n\right) = \sum_{n=0}^{\infty} c_n x^n,\ x \in (-R, R).$$

where $R = \min\{R_1, R_2\}$ and $c_n = \sum_{k=0}^{n} a_k b_{n-k}$.

Two power series can also do division operation,

$$\frac{a_0 + a_1 x + a_2 x^2 + \cdots + a_n x^n + \cdots}{b_0 + b_1 x + b_2 x^2 + \cdots + b_n x^n + \cdots} = c_0 + c_1 x + c_2 x^2 + \cdots + c_n x^n + \cdots, \tag{8}$$

assume that $b_0 \neq 0$ here. In order to determine the coefficients $c_0, c_1, c_2, \cdots, c_n, \cdots$, we can multiply series $\sum_{n=0}^{\infty} b_n x^n$ by $\sum_{n=0}^{\infty} c_n x^n$, and we have

$$a_0 = b_0 c_0,$$
$$a_1 = b_1 c_0 + b_0 c_1,$$
$$a_2 = b_2 c_0 + b_1 c_1 + b_0 c_2,$$
$$\cdots\cdots$$

By these equations, we can calculate $c_0, c_1, c_2, \cdots, c_n, \cdots$ in order.

2) The analytic properties of power series

The analytic properties refers to: continuity, derivation, quadrature. We give the following conclusions without proof.

Property 3 Sum function $s(x)$ of the series $\sum\limits_{n=0}^{\infty} a_n x^n$ is continuous on the convergent domain.

For any x_0 in the convergent domain, by the property 3,
$$\lim_{x \to x_0} \sum_{n=0}^{\infty} a_n x^n = \lim_{x \to x_0} s(x) = s(x_0) = \sum_{n=0}^{\infty} a_n x_0^n,$$
or
$$\lim_{x \to x_0} \sum_{n=0}^{\infty} a_n x^n = \sum_{n=0}^{\infty} (\lim_{x \to x_0} a_n x^n). \tag{9}$$

The Formula (9) indicates that the symbol of limit can exchange orders with $\sum$ (within the limit above, if x_0 is the endpoint of convergence interval, limit should be substituted by the corresponding one-sided limit).

Property 4 Assume that the sum function of power series $\sum\limits_{n=0}^{\infty} a_n x^n$ is $s(x)$ and x is one point on the convergence interval $(-R, R)$, then $s(x)$ can be integrated on $[0, x]$ (or $[x, 0]$) and there is the formula
$$\int_0^x s(x) \mathrm{d}x = \int_0^x \left(\sum_{n=0}^{\infty} a_n x^n\right) \mathrm{d}x = \sum_{n=0}^{\infty} \int_0^x a_n x^n \mathrm{d}x = \sum_{n=0}^{\infty} \frac{a_n}{n+1} x^{n+1}, \tag{10}$$
which means that sign of integration can exchange orders with $\sum$ and the power series obtained by integrating term by term has the same convergent radius as original power series (but the convergent domain may extend), and its sum function is the integration of original progression's sum function.

If consider the power series with $x_0 \neq 0$, the integration often select x_0 as the lower limit of integral,
$$\int_{x_0}^x \left[\sum_{n=0}^{\infty} a_n (x - x_0)^n\right] \mathrm{d}x = \sum_{n=0}^{\infty} \int_{x_0}^x a_n (x - x_0)^n \mathrm{d}x = \sum_{n=0}^{\infty} \frac{a_n}{n+1} (x - x_0)^{n+1}. \tag{11}$$

Property 5 The sum function $s(x)$ of power series $\sum_{n=0}^{\infty} a_n x^n$ is derivable on the convergent interval $(-R, R)$ and there is formula

$$s'(x) = \left(\sum_{n=0}^{\infty} a_n x^n\right)' = \sum_{n=0}^{\infty} (a_n x^n)' = \sum_{n=1}^{\infty} n a_n x^{n-1}, \tag{12}$$

which means that derivative symbols can exchange orders with $\sum$ and the power series obtained by differentiating term by term has the same convergent radius as original power series.

For function series, we call the progression obtained by differentiating term by term the derivative series of original series. Therefore, the derivative series and original series have the same convergent radius (but the convergent domain may be lessen) and its sum function is the derivative of primitive sum function. In addition, if the derivative series is convergent at one endpoint on the convergent interval, then the Formula (12) also stands up.

Example 9 Given that when $|x| < 1$,

$$1 + x + x^2 + \cdots + x^n + \cdots = \frac{1}{1-x}.$$

(1) Integrate both sides of the equation and then obtain that,

$$\int_0^x (1 + x + x^2 + \cdots + x^n + \cdots) dx = x + \frac{x^2}{2} + \cdots + \frac{1}{n+1} x^{n+1} + \cdots$$

$$= \int_0^x \frac{1}{1-x} dx = -\ln(1-x).$$

If $|x| < 1$, $x + \frac{x^2}{2} + \cdots + \frac{1}{n+1} x^{n+1} + \cdots = -\ln(1-x)$.

Note The convergent domain of original series is $(-1, 1)$, but after integrating the convergent domain of the series is $[-1, 1)$, which means the series will be convergent at the left endpoint of interval.

(2) When $|x| < 1$, differentiate both sides of the equation and then we obtain

$$1 + 2x + 3x^2 + \cdots + nx^{n-1} + \cdots = \frac{1}{(1-x)^2},$$

the convergent radius and convergent domain are both the same to (1).

Example 10 Find the sum function of $\sum_{n=0}^{\infty} \frac{x^n}{n+1}$ on the interval $(-1, 1)$.

Solution Assume that the sum function is $s(x)$, then $s(x) = \sum_{n=0}^{\infty} \frac{x^n}{n+1}$ and $s(0) = 1$.

Since $xs(x) = \sum_{n=0}^{\infty} \frac{x^{n+1}}{n+1}$, through the property that power series can integrate term by term on the convergent interval, we obtain that

$$[xs(x)]' = \sum_{n=0}^{\infty} \left(\frac{x^{n+1}}{n+1}\right)' = \sum_{n=0}^{\infty} x^n = \frac{1}{1-x}, \quad x \in (-1, 1).$$

Integrate each side from 0 to x, then

$$xs(x) = \int_0^x \frac{1}{1-x}dx = -\ln(1-x).$$

Therefore, when $x \neq 0$, $s(x) = -\frac{1}{x}\ln(1-x)$, then

$$s(x) = \begin{cases} -\frac{1}{x}\ln(1-x), & 0 < |x| < 1, \\ 1, & x = 0. \end{cases}$$

Known from the continuity of power series' sum function, this sum function $s(x)$ is continuous on $x = 0$. It's easy to get

$$\lim_{x \to 0} s(x) = \lim_{x \to 0} \left[-\frac{1}{x}\ln(1-x) \right] = 1.$$

Example 11 Find $\lim\limits_{n \to \infty} \left[2^{\frac{1}{3}} \cdot 4^{\frac{1}{9}} \cdot 8^{\frac{1}{27}} \cdot \cdots \cdot (2^n)^{\frac{1}{3^n}} \right]$.

Solution $2^{\frac{1}{3}} \cdot 4^{\frac{1}{9}} \cdot 8^{\frac{1}{27}} \cdot \cdots \cdot (2^n)^{\frac{1}{3^n}} = 2^{\frac{1}{3}} \cdot 2^{\frac{2}{3^2}} \cdot 2^{\frac{3}{3^3}} \cdot \cdots \cdot 2^{\frac{n}{3^n}} = 2^{\frac{1}{3} + \frac{2}{3^2} + \frac{3}{3^3} + \cdots + \frac{n}{3^n}}$.

Consider the series $\sum\limits_{n=1}^{\infty} \frac{n}{3^n}$, we first find the sum function of the series $\sum\limits_{n=1}^{\infty} n \cdot x^n$.

When $x = \frac{1}{3}$, the power series is the constant term series $\sum\limits_{n=1}^{\infty} \frac{n}{3^n}$.

We know

$$s(x) = \sum_{n=1}^{\infty} nx^n = x \sum_{n=1}^{\infty} nx^{n-1}.$$

And we know when $x \in (-1, 1)$, from the result of Example 9(2) that

$$\sum_{n=1}^{\infty} nx^{n-1} = \frac{1}{(1-x)^2}.$$

Therefore, $\qquad s(x) = \frac{x}{(1-x)^2}$, $s\left(\frac{1}{3}\right) = \frac{\frac{1}{3}}{\left(1 - \frac{1}{3}\right)^2} = \frac{3}{4}$,

thus $\qquad \lim\limits_{n \to \infty} \left[2^{\frac{1}{3}} \cdot 4^{\frac{1}{9}} \cdot 8^{\frac{1}{27}} \cdot \cdots \cdot (2^n)^{\frac{1}{3^n}} \right] = 2^{\lim\limits_{n \to \infty} \sum\limits_{k=1}^{n} \frac{k}{3^k}} = 2^{\frac{3}{4}} = \sqrt[4]{8}.$

Exercises 8-4

1. Find the convergent domains of the following function series:

   (1) $\sum\limits_{n=1}^{\infty} \left(\frac{\ln x}{2} \right)^n$; (2) $\sum\limits_{n=1}^{\infty} \frac{1}{1+x^n} \ (x \neq -1)$.

2. Find the sum functions and convergent intervals of the following geometric series:

   (1) $1 - \frac{x}{2} + \left(\frac{x}{2}\right)^2 + \cdots + (-1)^n \left(\frac{x}{2}\right)^n + \cdots$;

   (2) $3 + 3\left(\frac{x}{4}\right)^2 + 3\left(\frac{x}{4}\right)^4 + \cdots + 3\left(\frac{x}{4}\right)^{2n} + \cdots$.

3. Find the convergent radiuses and convergent intervals of the power series:

   (1) $\sum\limits_{n=1}^{\infty} (-1)^{n-1} (2x)^n$; (2) $\sum\limits_{n=1}^{\infty} \frac{2^n}{1+n^2} x^n$;

(3) $\sum_{n=1}^{\infty}(-1)^{n-1}\dfrac{(x-1)^n}{5n}$;

(4) $\sum_{n=1}^{\infty}\dfrac{\ln(n+1)}{n+1}x^{n-1}$;

(5) $\sum_{n=1}^{\infty}\dfrac{n}{2^n}x^{2n}$;

(6) $\sum_{n=1}^{\infty}\dfrac{1}{3n+1}\left(\dfrac{1+x}{x}\right)^n$.

4. By the method of differentiating term by term, find the sum function of series $\sum_{n=0}^{\infty}\dfrac{x^{2n+1}}{2n+1}$:

5. By the method of integrating term by term, find the sum function of series $x+2x^2+3x^3+\cdots+nx^n+\cdots$. (Tip: do not consider $s(x)$ directly but $\dfrac{s(x)}{x}$.)

6. Find $\lim\limits_{n\to\infty}\dfrac{1}{n}\sum_{k=1}^{n}\dfrac{1}{3^k}\left(1+\dfrac{1}{k}\right)^{k^2}$.

8.5 Unfold functions into power series

Power series not only have simple forms, but also have great algebraic and analytic properties. We consider some power series on its convergent domain, and its sum is a function of x. Whereas, if given a function, can it be expressed as the sum function of some power series? Can it unfold into power series? It has important significance for researching the properties of function.

8.5.1 Taylor series

Assume that function $f(x)$ can be rewritten into power series on the neighborhood of x_0, $U(x_0)$, and for any $x\in U(x_0)$, there is
$$f(x)=a_0+a_1(x-x_0)+a_2(x-x_0)^2+\cdots+a_n(x-x_0)^n+\cdots. \qquad (1)$$
Then by differentiating term by term, $f(x)$ has derivatives of arbitrary order on $U(x_0)$ and $f(x)$ will satisfy
$$f^{(n)}(x)=n!a_n+(n+1)!a_{n+1}(x-x_0)+\cdots \quad (n=0,1,2,\cdots). \qquad (2)$$
Let $x=x_0$ on either side of Formula (2), we have
$$f(x_0)=a_0, \quad f'(x_0)=1!a_1, \quad f''(x_0)=2!a_2, \quad \cdots.$$
Solve these equations for $a_i (i=0,1,2,\cdots)$, we have
$$a_0=f(x_0), \quad a_1=\dfrac{f'(x_0)}{1!}, \quad a_2=\dfrac{f''(x_0)}{2!}, \quad \cdots, \quad a_n=\dfrac{f^{(n)}(x_0)}{n!}, \quad \cdots. \qquad (3)$$
If a function $f(x)$ has all-order derivatives on x_0, we can write out the following power series
$$\sum_{n=0}^{\infty}\dfrac{f^{(n)}(x_0)}{n!}(x-x_0)^n = f(x_0)+f'(x_0)(x-x_0)+\dfrac{f''(x_0)}{2!}(x-x_0)^2+\cdots+\dfrac{f^{(n)}(x_0)}{n!}(x-x_0)^n+\cdots. \qquad (4)$$

We define the power series (4) to be the Taylor series of $f(x)$ on x_0 and the coefficients $a_n=\dfrac{f^{(n)}(x_0)}{n!} (n=0,1,2,\cdots)$ is called Taylor coefficient.

Now the problem is that whether function $f(x)$ equals to its Taylor series (4) on some neighborhood $U(x_0)$. From the Taylor formula of single variable differential calculus, if $f(x)$ has the derivative to $(n+1)$ th-order on the $U(x_0)$, we have

$$f(x)=f(x_0)+f'(x_0)(x-x_0)+\frac{f''(x_0)}{2!}(x-x_0)^2+\cdots+$$
$$\frac{f^{(n)}(x_0)}{n!}(x-x_0)^n+R_n(x), \tag{5}$$

where $R_n(x)=\frac{f^{(n+1)}(\xi)}{(n+1)!}(x-x_0)^{n+1}$ (ξ is a value between x and x_0) is called Lagrange remainder term. Furthermore, we can prove the following conclusions.

Theorem Assume that function $f(x)$ has all-order derivatives on a neighborhood $U(x_0)$, then $f(x)$ can be written into Taylor series $\sum_{n=0}^{\infty}\frac{f^{(n)}(x)}{n!}(x-x_0)^n$ on $U(x_0)$ if and only if the limit of the remainder term $R_n(x)$ is 0 as $n\to\infty$, that is, $\lim_{n\to\infty}R_n(x)=0$ ($x\in U(x_0)$).

Proof Firstly prove the necessity. Assume that $f(x)$ can be written into Taylor series on $U(x_0)$, that is,

$$f(x)=f(x_0)+f'(x_0)(x-x_0)+\frac{f''(x_0)}{2!}(x-x_0)^2+\cdots+\frac{f^{(n)}(x_0)}{n!}(x-x_0)^n+\cdots \tag{6}$$

for any $x\in U(x_0)$. By the Taylor Formula (5),

$$f(x)=s_{n+1}(x)+R_n(x),$$

where

$$s_{n+1}(x)=f(x_0)+f'(x_0)(x-x_0)+\frac{f''(x_0)}{2!}(x-x_0)^2+\cdots+\frac{f^{(n)}(x_0)}{n!}(x-x_0)^n.$$

By the Formula (6), the constant term series on the right side is convergent to $f(x)$ on $x\in U(x_0)$ and then $\lim_{n\to\infty}s_{n+1}(x)=f(x)$ for any $x\in U(x_0)$. Thus

$$\lim_{n\to\infty}R_n(x)=\lim_{n\to\infty}[f(x)-s_{n+1}(x)]=f(x)-f(x)=0,$$

which shows that the condition is necessary.

Then prove the sufficiency. According to the n th-order Taylor Formula (5) of $f(x)$ for $x\in U(x_0)$, we know from the condition that

$$\lim_{n\to\infty}|f(x)-s_{n+1}(x)|=\lim_{n\to\infty}|R_n(x)|=0.$$

Therefore, the series

$$f(x_0)+f'(x_0)(x-x_0)+\frac{f''(x_0)}{2!}(x-x_0)^2+\cdots+\frac{f^{(n)}(x_0)}{n!}(x-x_0)^n+\cdots$$

is convergent to $f(x)$ and then $f(x)$ can be written into Taylor series $\sum_{n=0}^{\infty}\frac{f^{(n)}(x)}{n!}(x-x_0)^n$ on the neighborhood $U(x_0)$.

Note 1 $f(x)$ can unfold into Taylor series $f(x)=\sum_{n=0}^{\infty}\frac{f^{(n)}(x_0)}{n!}(x-x_0)^n$ only on the convergent domain of $\sum_{n=0}^{\infty} \frac{f^{(n)}(x)}{n!}(x-x_0)^n$. For example, $\frac{1}{1-x} = \sum_{n=0}^{\infty} x^n$ only on the interval $(-1,1)$. When $|x| \geqslant 1$, the power series $\sum_{n=0}^{\infty} x^n$ is divergent and has no sum function, so it is impossible to find a function $f(x)$ such that $f(x) = \sum_{n=0}^{\infty} x^n$.

Note 2 If function $f(x)$ can unfold into power series on some neighborhood $U(x_0)$, that is, $f(x) = \sum_{n=0}^{\infty} a_n(x-x_0)^n$, then this power series is unique. Actually, we can know from the Formula (3) that this power series is Taylor series.

For Taylor series (4), if $x_0 = 0$, then the power series is

$$f(0) + f'(0)x + \frac{f''(0)}{2!}x^2 + \cdots + \frac{f^{(n)}(0)}{n!}x^n + \cdots, \tag{7}$$

which is called the Maclaurin series for function $f(x)$. Later we mainly unfold functions into the Maclaurin series (x's power series).

8.5.2 The method to unfold functions into power series

1) Method to unfold directly (also called Taylor Series Approach)

On the basis of Theorem 1, we can unfold the function directly by the following steps.

Step 1 Find $f(x)$'s all-order derivatives

$$f'(x), f''(x), \cdots, f^{(n)}(x), \cdots.$$

Note If certain-order derivative doesn't exist on $x=0$, we should stop. It means $f(x)$ cannot unfold into x's power series according to this theorem. For example, the third-order derivative of $f(x) = x^{\frac{7}{3}}$ doesn't exist on $x=0$, so it cannot unfold into x's power series.

Step 2 Find values of function and all-order derivatives on $x=0$, $f'(0), f''(0), \cdots, f^{(n)}(0), \cdots$.

Step 3 Write out the power series produced by $f(x)$,

$$f(0) + f'(0)x + \frac{f''(0)}{2!}x^2 + \cdots + \frac{f^{(n)}(0)}{n!}x^n + \cdots,$$

and then find the convergent radius R.

Step 4 On the interval $(-R, R)$, find the limit of $R_n(x)$,

$$\lim_{n \to \infty} R_n(x) = \lim_{n \to \infty} \frac{f^{(n+1)}(\xi)}{(n+1)!} x^{n+1} \quad (\xi \text{ is between } 0 \text{ and } x),$$

which is 0 or not. If it is 0, then the expansion equation is

$$f(x) = f(0) + f'(0)x + \frac{f''(0)}{2!}x^2 + \cdots + \frac{f^{(n)}(0)}{n!}x^n + \cdots \quad (-R < x < R).$$

Note The third step is important because all functions which have all-order derivatives can produce Maclaurin series. However, power series only equals to function $f(x)$ on the convergent domain.

Example 1 Unfold the function $f(x) = e^x$ into the Maclaurin series.

Solution Because $f^{(n)}(x) = e^x$ $(n=1, 2, 3, \cdots)$, so $f^{(n)}(0) = 1$ $(n=0, 1, 2, \cdots)$ and $f(0) = 1$. Thus we can write out the power series produced by $f(x) = e^x$,

$$1 + x + \frac{1}{2!}x^2 + \frac{1}{3!}x^3 + \cdots + \frac{1}{n!}x^n + \cdots,$$

and its convergent radius is $R = +\infty$.

For any number x, there is ξ (ξ is between 0 and x) such that

$$|R_n(x)| = \left|\frac{e^\xi}{(n+1)!}x^{n+1}\right| < e^{|x|} \cdot \frac{|x|^{n+1}}{(n+1)!}.$$

For this constant x, $e^{|x|}$ is bounded. Then $\frac{|x|^{n+1}}{(n+1)!}$ is the general term of the convergent series $\sum_{n=0}^{\infty} \frac{|x|^{n+1}}{(n+1)!}$, so $e^{|x|} \cdot \frac{|x|^{n+1}}{(n+1)!} \to 0$ as $n \to \infty$. Thus $\lim_{n \to \infty} R_n(x) = 0$ and then we have

$$e^x = 1 + x + \frac{1}{2!}x^2 + \frac{1}{3!}x^3 + \cdots + \frac{1}{n!}x^n + \cdots \quad (-\infty < x < +\infty). \tag{8}$$

Example 2 Unfold the function $f(x) = \sin x$ into the Maclaurin series.

Solution Since $\sin^{(n)} x = \sin\left(x + n \cdot \frac{\pi}{2}\right)$ $(n=1, 2, \cdots)$, the values of $f^{(n)}(0)$ are periodically $0, 1, 0, -1, \cdots$ $(n=0, 1, 2, 3, \cdots)$ in order. Thus the power series produced by $f(x) = \sin x$ is

$$x - \frac{x^3}{3!} + \frac{x^5}{5!} - \cdots + (-1)^{n-1} \frac{x^{2n-1}}{(2n-1)!} + \cdots,$$

and its convergent radius is $R = +\infty$.

For any number x, the absolute value of $R_n(x)$ is

$$|R_n(x)| = \left|\frac{\sin\left[\xi + \frac{(n+1)}{2}\pi\right]}{(n+1)!}x^{n+1}\right| \leq \frac{|x|^{n+1}}{(n+1)!} \to 0 \; (n \to \infty),$$

and ξ is between 0 and x. Therefore we can obtain the expansion

$$\sin x = x - \frac{x^3}{3!} + \frac{x^5}{5!} - \cdots + (-1)^{n-1} \frac{x^{2n-1}}{(2n-1)!} + \cdots$$

$$= \sum_{n=1}^{\infty} (-1)^{n-1} \frac{x^{2n-1}}{(2n-1)!}, \; x \in (-\infty, +\infty). \tag{9}$$

The equation above can be rewritten as

$$\sin x = \sum_{n=0}^{\infty} (-1)^n \frac{x^{2n+1}}{(2n+1)!}, \; x \in (-\infty, +\infty).$$

Similarly, we get the expansion of function $f(x) = (1+x)^m$ (m is any real number)

$$(1+x)^m = 1 + mx + \frac{m(m-1)}{2!}x^2 + \cdots + \frac{m(m-1)\cdots(m-n+1)}{n!}x^n + \cdots. \tag{10}$$

This expansion normally stands up on $-1 < x < 1$; then when $m > -1$, it also stands up on $x = 1$; and when $m > 0$, it also stands up on $x = -1$. The Formula (10) is called binomial expansion and the series on the right of Formula (10) is called binomial series.

The binomial expansion for $m=\frac{1}{2}, -\frac{1}{2}$.

$\sqrt{1+x}=(1+x)^{\frac{1}{2}}$
$=1+\frac{1}{2}x-\frac{1}{2\cdot 4}x^2+\frac{1\cdot 3}{2\cdot 4\cdot 6}x^3-\frac{1\cdot 3\cdot 5}{2\cdot 4\cdot 6\cdot 8}x^4+\cdots \quad (-1\leqslant x\leqslant 1),$

$\frac{1}{\sqrt{1+x}}=(1+x)^{-\frac{1}{2}}$

$=1+\left(-\frac{1}{2}\right)x+\frac{\left(-\frac{1}{2}\right)\left(-\frac{1}{2}-1\right)}{2!}x^2+$

$\frac{\left(-\frac{1}{2}\right)\left(-\frac{1}{2}-1\right)\left(-\frac{1}{2}-2\right)}{3!}x^3+\cdots+$

$\frac{\left(-\frac{1}{2}\right)\left(-\frac{1}{2}-1\right)\left(-\frac{1}{2}-2\right)\cdots\left(-\frac{1}{2}-n+1\right)}{n!}x^n+\cdots$

$=1-\frac{1}{2}x+\frac{1\cdot 3}{2^2\cdot 2!}x^2-\frac{1\cdot 3\cdot 5}{2^3\cdot 3!}x^3+\frac{1\cdot 3\cdot 5\cdot 7}{2^4\cdot 4!}x^4+\cdots+$

$(-1)^n\frac{1\cdot 3\cdot 5\cdot\cdots\cdot(2n-1)}{2^n\cdot n!}x^n+\cdots \quad (-1<x\leqslant 1).$

2) Method to unfold indirectly

It is very difficult to rewrite the function in a form of Taylor series since we needs to find all-order derivatives and discuss whether the remainder term tends to 0 or not. The method to unfold indirectly is to utilize the known expansion of some functions or by variable substitution, arithmetic, identical deformation, term by term differentiation and term by term integration to find the expansion of the function.

The important expansions of functions are as follows.

$\frac{1}{1-x}=1+x+x^2+\cdots+x^{n-1}+\cdots(-1<x<1);$

$\frac{1}{1+x}=1-x+x^2-x^3+\cdots+(-1)^{n-1}x^{n-1}+\cdots(-1<x<1);$

$e^x=1+x+\frac{x^2}{2!}+\frac{x^3}{3!}+\cdots+\frac{x^n}{n!}+\cdots(-\infty<x<+\infty);$

$\sin x=x-\frac{x^3}{3!}+\frac{x^5}{5!}-\frac{x^7}{7!}+\cdots+(-1)^{n-1}\cdot\frac{x^{2n-1}}{(2n-1)!}+\cdots(-\infty<x<+\infty);$

$(1+x)^m=1+mx+\frac{m(m-1)}{2!}x^2+\cdots+\frac{m(m-1)\cdots(m-n+1)}{n!}x^n+\cdots(-1<x<1).$

(1) Term by term differentiation, term by term integration.

Example 3 Unfold $f(x)=\cos x$ into the Maclaurin series.

Solution $\cos x=(\sin x)'=\left[x-\frac{x^3}{3!}+\frac{x^5}{5!}+\cdots+(-1)^{n-1}\frac{x^{2n-1}}{(2n-1)!}+\cdots\right]'$

$=1-\frac{x^2}{2!}+\frac{x^4}{4!}-\frac{x^6}{6!}+\cdots+(-1)^n\frac{x^{2n}}{(2n)!}+\cdots$

$$= \sum_{n=0}^{\infty}(-1)^n \frac{x^{2n}}{(2n)!} \quad (-\infty < x < +\infty). \tag{11}$$

Example 4 Unfold $\ln(1+x)$, $\arctan x$ into the Maclaurin series and find the series expressions of $\ln 2$, $\frac{\pi}{4}$.

Solution When $-1 < x < 1$, there are expansions that

(i) $\frac{1}{1+x} = 1 - x + x^2 - \cdots + (-1)^{n-1} x^{n-1} + \cdots$,

(ii) $\frac{1}{1+x^2} = 1 - x^2 + x^4 - \cdots + (-1)^{n-1} x^{2n-2} + \cdots$.

For $|x| < 1$, we can obtain by integrating (i), (ii) term by term from 0 to x,

$$\ln(1+x) = \int_0^x \frac{1}{1+x} dx = x - \frac{x^2}{2} + \frac{x^3}{3} - \cdots + (-1)^{n-1} \frac{x^n}{n} + \cdots,$$

$$\arctan x = \int_0^x \frac{1}{1+x^2} dx = x - \frac{x^3}{3} + \frac{x^5}{5} - \cdots + (-1)^{n-1} \frac{x^{2n-1}}{2n-1} + \cdots.$$

The convergent radius of two power series on the right side are both $R=1$.
When $x=-1$, $x=1$, the first power series becomes

$-1 - \frac{1}{2} - \frac{1}{3} - \cdots - \frac{1}{n} - \cdots$, which is divergent;

$1 - \frac{1}{2} + \frac{1}{3} - \cdots + (-1)^{n-1} \frac{1}{n} + \cdots$, which is convergent.

Thus convergent domain of the first series is $-1 < x \leqslant 1$.
When $x=-1$, $x=1$, the second power series becomes

$-1 + \frac{1}{3} - \frac{1}{5} + \cdots + (-1)^n \frac{1}{2n-1} + \cdots$, which is convergent;

$1 - \frac{1}{3} + \frac{1}{5} - \cdots + (-1)^{n-1} \frac{1}{2n-1} + \cdots$, which is convergent.

Thus convergent domain of the second series is $-1 \leqslant x \leqslant 1$.
Therefore there are formulas that

$$\ln(1+x) = x - \frac{x^2}{2} + \frac{x^3}{3} - \cdots + (-1)^{n-1} \frac{x^n}{n} + \cdots \quad (-1 < x \leqslant 1). \tag{12}$$

$$\arctan x = x - \frac{x^3}{3} + \frac{x^5}{5} - \cdots + (-1)^{n-1} \frac{x^{2n-1}}{2n-1} + \cdots \quad (-1 \leqslant x \leqslant 1). \tag{13}$$

Because when $x=1$, $\ln(1+1) = \ln 2$, $\arctan 1 = \frac{\pi}{4}$, we have

$$\ln 2 = 1 - \frac{1}{2} + \frac{1}{3} - \cdots + (-1)^{n-1} \frac{1}{n} + \cdots,$$

$$\frac{\pi}{4} = 1 - \frac{1}{3} + \frac{1}{5} - \cdots + (-1)^{n-1} \frac{1}{2n-1} + \cdots.$$

(2) Variation substitution.

Example 5 Unfold function $\frac{1}{3-x^2}$ into the Maclaurin series and write out the convergent interval.

Solution Rewrite the function and by variation substitution of $t=\frac{x^2}{3}$, we have

$$\frac{1}{3-x^2} = \frac{1}{3} \cdot \frac{1}{1-\frac{x^2}{3}} = \frac{1}{3} \cdot \frac{1}{1-t} = \frac{1}{3}(1+t+t^2+\cdots+t^{n-1}+\cdots)$$

$$= \frac{1}{3}\left[1+\frac{x^2}{3}+\frac{x^4}{3^2}+\cdots+\frac{x^{2n-2}}{3^{n-1}}+\cdots\right] = \frac{1}{3}+\frac{x^2}{3^2}+\frac{x^4}{3^3}+\cdots+\frac{x^{2n-2}}{3^n}+\cdots.$$

Since $-1<t<1$ equals to $-1<\frac{x^2}{3}<1$, the convergent interval is $-\sqrt{3}<x<\sqrt{3}$.

Example 6 Unfold function e^{-x} into Taylor series on $x_0=-1$ (or unfold into power series of $(x+1)$) and write out the convergent interval.

Solution $e^{-x} = e \cdot e^{[-(x+1)]} = e \cdot \sum_{n=0}^{\infty} \frac{1}{n!}[-(x+1)]^n = \sum_{n=0}^{\infty} \frac{(-1)^n e}{n!}(x+1)^n$

$$= e - e(x+1) + \frac{e}{2!}(x+1)^2 - \frac{e}{3!}(x+1)^3 + \cdots + \frac{(-1)^n e}{n!}(x+1)^n + \cdots$$

where $-\infty<x<+\infty$.

(3) Arithmetic.

Example 7 Unfold the function $\sin\left(x+\frac{\pi}{4}\right)$ into the Maclaurin series.

Solution Since $\sin\left(x+\frac{\pi}{4}\right) = \frac{\sqrt{2}}{2}(\sin x + \cos x)$, we can obtain from expansions (9) and (11) that

$$\sin\left(x+\frac{\pi}{4}\right) = \frac{\sqrt{2}}{2}\left[\sum_{n=0}^{\infty}(-1)^n \frac{x^{2n+1}}{(2n+1)!} + \sum_{n=0}^{\infty}(-1)^n \frac{x^{2n}}{(2n)!}\right]$$

$$= \frac{\sqrt{2}}{2}\left[1+x-\frac{1}{2!}x^2-\frac{1}{3!}x^3+\frac{1}{4!}x^4+\frac{1}{5!}x^5-\cdots\right] \quad (-\infty<x<+\infty).$$

Example 8 Unfold function $f(x)=\frac{x}{x^2-2x-3}$ into Taylor series on $x_0=2$.

Solution We need to unfold the function into power series of $(x-2)$. Because

$$f(x) = \frac{x}{(x+1)(x-3)} = \frac{1}{4} \cdot \frac{1}{x+1} + \frac{3}{4} \cdot \frac{1}{x-3}$$

$$= \frac{1}{4} \cdot \frac{1}{3+(x-2)} - \frac{3}{4} \cdot \frac{1}{1-(x-2)}$$

$$= \frac{1}{12} \cdot \frac{1}{1+\frac{x-2}{3}} - \frac{3}{4} \cdot \frac{1}{1-(x-2)},$$

we have

$$f(x) = \frac{1}{12}\sum_{n=0}^{\infty}(-1)^n\left(\frac{x-2}{3}\right)^n - \frac{3}{4}\sum_{n=0}^{\infty}(x-2)^n$$

$$= \sum_{n=0}^{\infty}\left[\frac{(-1)^n}{12 \cdot 3^n} - \frac{3}{4}\right](x-2)^n$$

$$= \sum_{n=0}^{\infty}\frac{1}{4}\left[\frac{(-1)^n - 3^{n+2}}{3^{n+1}}\right](x-2)^n$$

$$= -\frac{2}{3} - \frac{7}{9}(x-2) - \frac{20}{27}(x-2)^2 - \cdots - \frac{1}{4}\left[\frac{3^{n+2}-(-1)^n}{3^{n+1}}\right](x-2)^n - \cdots.$$

The convergent interval is the common part of $\left|\frac{x-2}{3}\right| < 1$ and $|x-2| < 1$, so it is $1 < x < 3$.

Exercises 8-5

1. Use Taylor series approach to unfold function $f(x) = x^2 e^x$ into the Maclaurin series.

2. Write out the n th-order Maclaurin formula for $f(x) = a^x$.

3. Find the convergent domain and sum function of $\sum_{n=0}^{\infty} \frac{x^{2n}}{(2n)!}$.

4. Unfold the following functions into Maclaurin series and write out the convergent interval:

(1) $y = \ln(a+x)$ $(a>0)$;

(2) $y = \frac{1}{(1-x)^2}$;

(3) $y = \frac{x}{\sqrt{1+x^2}}$;

(4) $y = \ln(x + \sqrt{x^2+1})$.

5. Unfold function $f(x) = \cos x$ into power series of $\left(x + \frac{\pi}{3}\right)$.

6. Unfold function $f(x) = \frac{2+x}{1+2x-x^2-2x^3}$ into the Maclaurin series.

7. Use constant term series to find the sum of $\sum_{n=1}^{\infty} \frac{1}{2^n(2n-1)}$.

8. Use the expansion of e^x to unfold $f(x) = \frac{d}{dx}\left(\frac{e^x-1}{x}\right)$ into x's power series and calculate the sum of $\sum_{n=1}^{\infty} \frac{n}{(n+1)!}$.

Summary

1. Main contents

The concept is divided into two parts: infinite series of constant terms and power series.

(1) Infinite series of constant terms.

The positive series and alternating series etc. are infinite series of constant terms.

The main test of convergence: comparison test, root test, integral test, Leibniz's test, etc..

The limit form of the comparison test is easier to use than the non-limit form. When we use the comparison test to judge the convergence and divergence of positive series, we need to select the positive series with known convergence or divergence to compare with the given series. Commonly used as comparison: $\sum_{n=1}^{\infty} aq^{n-1} (a \neq 0)$, $\sum_{n=1}^{\infty} \frac{1}{n}$, $\sum_{n=1}^{\infty} \frac{1}{n^p}$ (p is a positive constant).

Discuss the convergence of the alternating series $\sum_{n=1}^{\infty} (-1)^n u_n (u_n > 0)$, to compare u_n and u_{n+1}. There are several common methods:

① Ratio test: $\frac{u_{n+1}}{u_n}$ is great than or equal to 1;

② D-value test: $u_n - u_{n+1}$ is less than or equal to 0, etc..

Note Only positive series can use comparison test and its limit form, root test and integral test.

(2) Power series.

Power series is a class of function series with a wide range of applications.

The convergent domain of $\sum_{n=0}^{\infty} a_n (x - x_0)^n$ is a symmetric interval centered on x_0 and the convergent radius is

$$R = \lim_{n \to \infty} \left| \frac{a_n}{a_{n+1}} \right| = \frac{1}{\rho}, \text{ where } \rho = \lim_{n \to \infty} \left| \frac{a_{n+1}}{a_n} \right|.$$

Whether the power series converge at the endpoint of the convergent interval can be judged by the convergent method of constant term series.

The methods of expanding function into power series include direct expansion method and indirect expansion method (using expansion formula and term by item derivation or integral etc.). Memorize some function expansions:

$$e^x, \sin x, \cos x, \ln(1+x), (1+x)^a, \frac{1}{1-x}.$$

2. Basic requirements

(1) Understand the convergence and divergence of infinite series and the concept of partial sum, and know the basic properties of infinite series and the necessary conditions for convergence.

(2) Discuss the geometric series and p-series's convergence.

(3) Know the comparison test of positive series.

(4) Know the alternating series's Leibniz Principle, and can estimate the truncation error of alternating series.

(5) Know the concepts of absolute convergence and conditional convergence of infinte series and the relation between absolute convergence and conditional convergence.

(6) Know the functional series's convergent domain, and the concept of sum function.

(7) Know the method for finding convergent interval of simple power series (convergence of interval endpoint cannot be solved).

(8) Know the basic properties of power series in its convergent interval.

(9) Know the necessary condition and the sufficient condition for the expansion of

functions to Taylor Series.

Quiz

1. Find the sum of the following series:
$$\frac{1}{3}+\frac{3}{3^2}+\frac{5}{3^3}+\cdots+\frac{2n-1}{3^n}+\cdots.$$

2. Determine the convergence or divergence of the following series:
   (1) $\sum_{n=1}^{\infty}(-1)^n(\sqrt{n+1}-\sqrt{n})$;
   (2) $\sum_{n=1}^{\infty}\sin\left(n\pi+\frac{1}{\ln n}\right)$.

3. Determine the convergence or divergence of the following series:
$$1-\frac{1}{2^{\alpha}}+\frac{1}{3}-\frac{1}{4^{\alpha}}+\cdots+\frac{1}{2n-1}-\frac{1}{(2n)^{\alpha}}+\cdots \quad (\alpha>0).$$

(Hint: discuss the different conditions about the value of parameter α including that $\alpha=1$, $\alpha>1$ and $0<\alpha<1$.)

4. Find the convergent radius and convergent domain of the following power series:
   (1) $\sum_{n=1}^{\infty}\frac{\ln(n+1)}{n+1}x^{n-1}$;
   (2) $\sum_{n=1}^{\infty}(-1)^n\frac{(x-2)^{2n+1}}{2n+1}$.

5. Find the sum function of the following power series:
   (1) $\sum_{n=1}^{\infty}\frac{x^n}{n(n+1)}$;
   (2) $\sum_{n=1}^{\infty}\frac{(-1)^{n+1}}{n(2n-1)}x^{2n}$.

6. Unfold the following functions into power series of $(x-x_0)$:
   (1) $f(x)=\frac{1}{x^2+7x+6}$, $x_0=-4$;
   (2) $f(x)=\arctan\frac{1+x}{1-x}$, $x_0=0$.

Exercises

1. Determine the convergence or divergence of the following series:
   (1) $\sum_{n=1}^{\infty}n^{\alpha}\cdot\beta^n$ refers to arbitrary number and β refers to nonnegative real number;
   (2) $\sum_{n=1}^{\infty}\frac{1}{\sqrt{n}}\ln\frac{n+1}{n}$;
   (3) $\sum_{n=2}^{\infty}\frac{(-1)^n}{\sqrt{n}+(-1)^n}$.

2. Determine the convergence or divergence of series $\sum_{n=1}^{\infty}\frac{1}{(2n^2+n)^m}$ (m refers to constant).

3. Determine the convergence or divergence of series $\sum_{n=1}^{\infty}\frac{1!+2!+\cdots+n!}{(n+3)!}$.

4. Determine the convergence or divergence of the following series:

(1) $\sum_{n=1}^{\infty}(-1)^{n-1}\dfrac{\sqrt{n+1}}{n+10}$; (2) $\sum_{n=1}^{\infty}(-1)^{\frac{n(n-1)}{2}}\dfrac{\sqrt{n+1}}{n}$.

5. Determine the convergence or divergence of series. If it is convergent, is it absolutely convergence or conditionally convergence?

(1) $\sum_{n=1}^{\infty}\dfrac{\sin n\alpha}{(\ln 10)^n}$; (2) $\sum_{n=1}^{\infty}\dfrac{(-1)^{n-1}}{n+(-1)^{n-1}}$.

6. Find the following limits:

(1) $\lim\limits_{n\to\infty}\dfrac{1}{n}\sum_{k=1}^{n}\dfrac{1}{3^k}\left(1+\dfrac{1}{k}\right)^{k^2}$; (2) $\lim\limits_{n\to\infty}\left[2^{\frac{1}{3}}\cdot 4^{\frac{1}{9}}\cdot 8^{\frac{1}{27}}\cdot\cdots\cdot(2^n)^{\frac{1}{3^n}}\right]$.

7. Find convergent domains of the following series:

(1) $\sum_{n=1}^{\infty}(-1)^{n-1}\dfrac{x^{2n}}{a^{\sqrt{n}}}(a>0)$; (2) $\sum_{n=1}^{\infty}\left(\sin\dfrac{1}{2n}\right)\left(\dfrac{1+2x}{2-x}\right)^n$;

(3) $\sum_{n=1}^{\infty}\left(1+\dfrac{1}{n}\right)^{n^2}x^n$; (4) $\sum_{n=1}^{\infty}\dfrac{n}{2^n}x^{2n}$.

8. Find the sum functions of the following power series:

(1) $\sum_{n=1}^{\infty}\dfrac{x^{2n+1}}{2n}$; (2) $\sum_{n=1}^{\infty}n(n+1)x^n$;

(3) $\sum_{n=1}^{\infty}(3n-2)x^{2n-1}$.

9. Use power series method to calculate the sum of series:

(1) $\sum_{n=1}^{\infty}\dfrac{n+1}{n!}\left(\dfrac{1}{2}\right)^n$; (2) $\sum_{n=1}^{\infty}\dfrac{n^2}{n!}$.

10. Unfold the following functions into power series of x:

(1) $\dfrac{1}{(2-x)^2}$; (2) $\dfrac{1}{(1+x)(1+x^2)(1+x^4)}$.

Chapter 9 Differential equations

Function is the internal relationship among objects on quantity and it can be used to research the regularity of the objects. For practical problems, it is normally required to find the corresponding relations according to conditions of the problems. However, for many problems, we cannot obtain the required function directly but an equality of independent variables, unknown functions and its derivatives (or differential), which is called the differential equation. Thus differential equation is an important tool to solve these practical problems. This chapter mainly introduces some basic concepts of differential equation and some techniques to solve differential equations.

9.1 Basic concepts of differential equation

Now we will illustrate the basic concepts of differential equation through two specific examples.

Example 1 Given that a curve passing through the point $(2, 3)$ and the slope of the tangent at any point (x, y) is three times its abscissa x. Find the equation of the curve.

Solution Assume that the equation is $y = f(x)$. Since the slope of the tangent is equal to the derivative of the function, $y = f(x)$ satisfies that

$$\frac{\mathrm{d}y}{\mathrm{d}x} = 3x. \tag{1}$$

In addition, $y = f(x)$ also satisfies

$$y\big|_{x=2} = 3. \tag{2}$$

Integrate either side of the Equation (1), we have

$$y = \frac{3}{2}x^2 + C. \tag{3}$$

By (2) and (3), we get $C = -3$. Thus the equation is

$$y = \frac{3}{2}x^2 - 3. \tag{4}$$

Example 2 Assume that the object whose quality is m starts to fall down from the height s_0 (Leave out the air friction), the height from the ground after t seconds is denoted by $s(t)$. Find the expression of the function $s(t)$.

Solution It is a rectilinear motion. Suppose we regard upward direction as positive direction of the s-axis (Figure 9-1), then the force on the rectilinear object is $f=-mg$. According to the Newton's second law $f=ma$, we know that $a=-g$. And the acceleration is $a=\dfrac{d^2 s}{dt^2}$, so there is

$$\frac{d^2 s}{dt^2}=-g. \qquad (5)$$

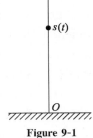

Figure 9-1

In addition, $s(t)$ also satisfies the condition

$$\begin{cases} \dfrac{ds}{dt}\bigg|_{t=0}=0, \\ s\big|_{t=0}=s_0. \end{cases} \qquad (6)$$

Integrate either side of the Equation (5), we have

$$\frac{ds}{dt}=-gt+C_1, \qquad (7)$$

$$s=-\frac{1}{2}gt^2+C_1 t+C_2, \qquad (8)$$

From (6) we get that $C_1=0$, $C_2=s_0$. Thus the expression of function $s(t)$ is

$$s(t)=-\frac{1}{2}gt^2+s_0. \qquad (9)$$

The Equations (1) and (5) both include derivatives of unknown functions, so they are both differential equations. In general, the equation containing an unknown function and one or more its derivatives is called a differential equation. The differential equation which has only one independent variable is called ordinary differential equation. In this chapter we just discuss the ordinary differential equations.

The order of the highest derivative in differential equation is called the order of differential equation. For instance, the Equation (1) is a first-order differential equation and the Equation (5) is a second-order differential equation. Then the equation

$$xy'''+3x^2 y''-x^3 yy'=4x^2+1$$

is a third-order differential equation and the equation

$$x^2 y^{(4)}-x^3 y=\sin 2x$$

is a forth-order differential equation.

In general, the form of n th-order differential equation is

$$F(x,y,y',\cdots,y^{(n)})=0. \qquad (10)$$

Notice that $y^{(n)}$ must exist in the n th-order differential Equation (10) (otherwise, the order of differential equation is not n), but the terms such as $x,y,y',\cdots,y^{(n-1)}$ can exist or not.

If substituting function $y=\varphi(x)$ for y of the differential equation will make the equation satisfied, $y=\varphi(x)$ is called a solution of the differential equation. Exactly, assume that $y=\varphi(x)$ has n th-order derivative on the interval I. On the interval I, if

$$F[x,\varphi(x),\varphi'(x),\cdots,\varphi^{(n)}(x)]\equiv 0,$$

then $y=\varphi(x)$ is called a solution of the differential Equation (10) on the interval I.

For example, the function (3) is the solution of the differential Equation (1) and the function (8) is the solution of differential Equation (5). Please notice that the function (3) has one arbitrary constant C and function (8) includes two arbitrary constants C_1 and C_2. So the number of the solution of some differential equation is not just one but infinite.

If the number of arbitrary constants of the solution is equal to the order of the differential equation, the solution is called general solution. For example, the function (3) is the general solution of the differential Equation (1) and the function (8) is the general solution of the differential Equation (5). Although function (4) is the solution to differential Equation (1) and function (9) is the solution to differential Equation (5), they aren't the general solutions since function (4) and (5) don't include arbitrary constants.

Then for second-order differential equation
$$y''+y=0, \tag{11}$$
we can prove easily that
$$y=C_1\sin x+C_2\cos x$$
is its solution. Since the solution includes two independent arbitrary constants and the Equation (11) is second-order, the solution $y=C_1\sin x+C_2\cos x$ is the general solution. In addition, we can prove that $y=C\sin x$ is also the solution of the Equation (11). Pay attention that though $y=C\sin x$ has arbitrary constant, it isn't the general solution because the number of arbitrary constants is less than the order of the differential equation.

Since the general solution includes arbitrary constants, it cannot reflect the regularity of objective things exactly. If the values of these constants could be determined, the regularity is clear. For the general solution, if arbitrary constant is a certain value, we can obtain an unique solution without any uncertain arbitrary constants. The solution that doesn't contain any arbitrary constants is called the particular solution of differential equation.

For example, the function (4) is a particular solution of the Equation (1) and the function (9) is a particular solution of the Equation (5). The function $y=C\sin x$ is not the particular solution of the Equation (11) since it includes arbitrary constant, and as mentioned before, it is also not the general solution of (11) because it only includes one arbitrary constant.

In order to obtain the particular solution from the general solution, we have to determine the values of the arbitrary constants in the general solution, so the conditions that determine these constants are required.

Assume that the function $y=\varphi(x)$ satisfies the differential equation, and if the differential equation is first-order, the condition that determines arbitrary constant is
$$y\big|_{x=x_0}=y_0.$$

If the differential equation is second-order, the conditions that determine arbitrary constants will be

Chapter 9 Differential equations

$$y|_{x=x_0}=y_0, \quad y'|_{x=x_0}=y_0^1.$$

Generally, if the differential equation is n th-order, the conditions that determine arbitrary constants are

$$y|_{x=x_0}=y_0, \quad y'|_{x=x_0}=y_0^1, \quad \cdots, \quad y^{(n-1)}|_{x=x_0}=y_0^{n-1},$$

where $x_0, y_0, y_0^1, \cdots, y_0^{n-1}$ are given values. These conditions above are called the initial conditions of differential equation.

The graph of a differential equation's solution is a curve, which is called an integral curve. The solution of first-order differential equation which satisfies initial condition $y|_{x=x_0}=y_0$ is the equation of the integral curve passing through the point (x_0, y_0). And the solution of second-order differential equation which satisfies initial conditions $y|_{x=x_0}=y_0$ and $y'|_{x=x_0}=y_0^1$ is the equation of the integral curve passing through the point (x_0, y_0) with the tangent slope y_0^1.

Example 3 Show that the function $y=C_1 e^{2x}+C_2 e^{-2x}$ is the general solution of the second-order differential equation

$$y''-4y=0,$$

and find the particular solution that satisfies the initial conditions

$$y|_{x=0}=0, \quad y'|_{x=0}=1.$$

Solution For function $y=C_1 e^{2x}+C_2 e^{-2x}$, we have

$$y'=2C_1 e^{2x}-2C_2 e^{-2x},$$
$$y''=4C_1 e^{2x}+4C_2 e^{-2x},$$

then

$$y''-4y=4C_1 e^{2x}+4C_2 e^{-2x}-4(C_1 e^{2x}+C_2 e^{-2x})=0.$$

Thus function $y=C_1 e^{2x}+C_2 e^{-2x}$ is the solution of equation $y''-4y=0$. Because it includes two arbitrary constants, it is the general solution. Then by the initial conditions, we have

$$\begin{cases} C_1+C_2=0, \\ 2C_1-2C_2=1, \end{cases}$$

so $C_1=\dfrac{1}{4}$, $C_2=-\dfrac{1}{4}$. Therefore, the particular solution is $y=\dfrac{1}{4}e^{2x}-\dfrac{1}{4}e^{-2x}$.

Exercises 9-1

1. Find the order of following differential equations:

   (1) $x\dfrac{dy}{dx}+y^2 \sin x=3x$;

   (2) $y''+2(y')^2 y+2x^3=1$;

   (3) $xy'''+2y''+x^2 y'+3y=0$;

   (4) $x^2 dy=2ydx$;

   (5) $\dfrac{d^3 x}{dt^3}+t\left(\dfrac{d^2 x}{dt^2}\right)^3+2x=t^3$;

   (6) $\sin\left(\dfrac{d^2 y}{dx^2}\right)+e^y \dfrac{dy}{dx}=x$.

2. Judge whether the function on the left column is the solution of corresponding differential equation on the right columns, general solution or particular solution?

Function	Differential equation	Answer
$y = e^{-3x} + \dfrac{1}{3}$	$y' + 3y = 1$	
$y = 5\cos 3x + \dfrac{x}{9} + \dfrac{1}{8}$	$y'' - 9y = x + \dfrac{1}{2}$	
$y^2(1+x^2) = C$	$xy\,dx + (1+x^2)\,dy = 0$	
$y = x + \displaystyle\int_0^x e^{-t^2}\,dt$	$y'' + 2xy' = x$	
$y = C_1 e^{2x} + C_2 e^{3x}$	$y'' - 5y' + 6y = 0$	

3. Show that the following functions are solutions of the differential equations:

(1) $y = \dfrac{\sin x}{x} - x,\ xy' + y = \cos x$;

(2) $y = 2 + C\sqrt{1-x^2}$ (C is an arbitrary constant), $(1-x^2)y' + xy = 2x$;

(3) $y = e^x,\ y'' - 3y' + 2y = 0$;

(4) $y = x^2 + 1,\ y' = y^2 - (x^2+1)y + 2x$;

(5) $y = -\dfrac{g(x)}{f(x)},\ y' = \dfrac{f'(x)}{g(x)}y^2 - \dfrac{g'(x)}{f(x)}$.

4. Given that the first-order differential equation is $\dfrac{dy}{dx} = 2x$.

(1) Find its general solution.

(2) Find the solution that satisfies the condition $y|_{x=1} = 4$.

(3) Find the solution that is tangent to the straight line $y = 2x + 3$.

(4) Find the solution that satisfies the condition $\displaystyle\int_0^1 y\,dx = 2$.

5. One particle with quality m motions in a straight line. Assume that one pull force which is in direct proportion to time works on it and the particle also gain resistance that is in direct proportion to velocity. Try to find the differential equation that velocity changes along with time.

6. The abscissa of the intersection point of any tangent and abscissa axis is half of the abscissa of the point of tangent in the curve. Try to find the differential equation of the curve.

9.2 First-order differential equation

In this chapter, we will discuss how to solve the first-order differential equation
$$y' = F(x, y). \tag{1}$$

9.2.1 Separable differential equation

For differential Equation (1), if function $F(x,y)$ can be written into the product of function $f_1(y)$ and function $f_2(x)$, and $f_1(y) \neq 0$, we can obtain that

Chapter 9 Differential equations

$$\frac{dy}{f_1(y)} = f_2(x)dx.$$

Let $g(y) = \dfrac{1}{f_1(y)}$ and $f(x) = f_2(x)$, then

$$g(y)dy = f(x)dx.$$

In general, if the first-order differential equation can be written into the form

$$g(y)dy = f(x)dx \qquad (2)$$

the original equation is called a separable differential equation and (2) is called the separated differential equation. The general solution of the original differential equation can be found by solving the separated differential Equation (2).

Then discuss the solution of the differential Equation (2). Assume that $y = \varphi(x)$ is the solution of Equation (2), plug it into (2) and we obtain

$$g[\varphi(x)]\varphi'(x)dx = f(x)dx.$$

Assume that the function $g(y)$ and $f(x)$ are both continuous, then integrate either side of the equation above,

$$\int g(y)dy = \int f(x)dx.$$

where $y = \varphi(x)$.

Assume that the antiderivatives of $g(y)$ and $f(x)$ are $G(y)$ and $F(x)$ respectively, we have

$$G(y) = F(x) + C. \qquad (3)$$

Therefore, the solution $y = \varphi(x)$ of Equation (2) satisfies (3). On the other hand, if $y = \psi(x)$ is the implicit function determined by equality (3), then plug it into (3) and differentiate either side. We have

$$G'[\psi(x)]d\psi(x) = F'(x)dx.$$

Notice that $G(y)$ and $F(x)$ are derivatives of $g(y)$ and $f(x)$ respectively, so

$$g[\psi(x)]d\psi(x) = f(x)dx.$$

That is to say $y = \psi(x)$ satisfies

$$g(y)dy = f(x)dx.$$

Thus $y = \psi(x)$ is the solution to differential Equation (2).

Above all, for separated differential Equation (2), by integrating either side, we get the equality (3), which is called the implicit solution of (2). Then because (3) includes one arbitrary constant, the implicit solution (3) is the general solution of the differential Equation (2). Thus equality (3) is also called the implicit general solution of the differential Equation (2). If we can solve the equality (3) for y (or solve (3) for x), we will get the explicit general solution of differential Equation (2).

Example 1 Find the general solution of differential equation

$$\frac{dy}{dx} = x^2 y.$$

Solution This is a separable equation. After separating variables, we obtain

$$\frac{dy}{y} = x^2 dx.$$

Integrate both sides
$$\int \frac{dy}{y} = \int x^2 dx,$$

then we obtain
$$\ln |y| = \frac{1}{3} x^3 + C_1.$$

Thus
$$y = \pm e^{C_1} e^{\frac{1}{3} x^3}.$$

Since $C_2 = \pm e^{C_1}$ is an arbitrary nonzero constant and $y = 0$ is obviously the solution of the differential equation, the general solution is
$$y = C e^{\frac{1}{3} x^3}.$$

Example 2 Find the solution of differential equation
$$2y dy = 3(x-1)^2 (1+y^2) dx$$
and satisfies the initial condition $y|_{x=0} = 0$.

Solution This is also a separable differential equation. After separating variables, we obtain
$$\frac{2y}{1+y^2} dy = 3(x-1)^2 dx.$$

Integrate both sides
$$\int \frac{2y}{1+y^2} dy = 3 \int (x-1)^2 dx,$$

then we obtain
$$\ln (1+y^2) = (x-1)^3 + C.$$

Plug the initial condition $y|_{x=0} = 0$ into it and we get $C = 1$. Thus the particular solution is
$$\ln (1+y^2) = (x-1)^3 + 1.$$

Example 3 For commodity's sales forecast, sales volume x is a function of time t. If sales volume x and time t satisfy the following differential equation
$$\frac{1}{x} \frac{dx}{dt} = k(B - x),$$
where k and B are constants greater than zero (B refers to the saturation level of market). Find the regularity that sales volume changes along with time.

Solution The equation is a separable equation. After separating variables, we obtain
$$\frac{dx}{x(B-x)} = k dt.$$

Rewrite the equality above into
$$\left[\frac{1}{x} + \frac{1}{(B-x)} \right] dx = Bk dt.$$

Integrate both sides, we obtain
$$\ln \left| \frac{x}{B-x} \right| = Bkt + C_1,$$

then
$$\frac{x}{B-x} = C_2 e^{Bkt} \quad (C_2 = \pm e^{C_1}).$$

Thus
$$x = \frac{BC_2 e^{Bkt}}{1+C_2 e^{Bkt}} = \frac{B}{1+Ce^{-Bkt}} \quad \left(C=\frac{1}{C_2}\right).$$

It can be seen that
$$\lim_{t \to +\infty} x(t) = \lim_{t \to +\infty} \frac{B}{1+Ce^{-Bkt}} = B.$$

That is to say, as time passing, sales volume will tend to the saturation level of market.

9.2.2 Homogeneous differential equation

If a first-order differential equation can be written into the form
$$\frac{dy}{dx} = f\left(\frac{y}{x}\right), \tag{4}$$

then the equation is called a homogeneous equation. Now we introduce the technique to solve homogeneous equations.

Let
$$u = \frac{y}{x},$$

then
$$y = ux, \quad \frac{dy}{dx} = x\frac{du}{dx} + u.$$

Plug these expressions into Equation (4), we get a separable equation
$$x\frac{du}{dx} = f(u) - u.$$

Integrate both sides after separating variables, we have
$$\int \frac{1}{f(u)-u} du = \int \frac{1}{x} dx.$$

After finding the indefinite integral, then substitute $\frac{y}{x}$ for u, we will get the general solution.

Example 4 Find the general solution of equation
$$\frac{dy}{dx} = \frac{y}{x} + \tan\frac{y}{x}.$$

Solution It's a homogeneous equation. Let $u = \frac{y}{x}$, then
$$y = ux, \quad \frac{dy}{dx} = x\frac{du}{dx} + u.$$

So original equation is written into
$$x\frac{du}{dx} + u = u + \tan u,$$

or
$$x\frac{du}{dx} = \tan u.$$

After separating variables, we have
$$\cot u\, du = \frac{1}{x} dx.$$

Integrate both sides and we obtain
$$\ln|\sin u| = \ln|x| + C_1,$$

or
$$\sin u = Cx \quad (C = \pm e^{C_1}).$$

Then substitute $\frac{y}{x}$ for u, we get the general solution

$$\sin \frac{y}{x} = Cx.$$

Example 5 Find the general solution of equation

$$(y^2 - 3x^2)\mathrm{d}x + 2xy\mathrm{d}y = 0.$$

Solution The equation can be written as

$$\frac{\mathrm{d}y}{\mathrm{d}x} = \frac{3x^2 - y^2}{2xy}.$$

Then divide numerator and denominator on the right side by x^2, we have

$$\frac{\mathrm{d}y}{\mathrm{d}x} = \frac{3 - \left(\frac{y}{x}\right)^2}{2\frac{y}{x}}.$$

Since the form is $y' = f\left(\frac{y}{x}\right)$, we find its general solution by substitution $u = \frac{y}{x}$. (The rest is left for readers to complete.)

The reason why regard the equation as homogeneous equation is that the order of x and y are same, which we can see in Example 4 and Example 5. Thus it can be written into the form $y' = f\left(\frac{y}{x}\right)$.

Example 6 The mirror of searchlight is a rotating surface generated by rotating a curve L on coordinate plane xOy about the x-axis. According to the performance requirements of searchlight, all lights from light source on its axis of rotation (x-axis) are parallel to the axis after reflection. Find the equation of curve L.

Solution Suppose the light source is the origin O of coordinates (Figure 9-2) and the curve L is in upper half plane ($y \geqslant 0$).

Assume that $M(x, y)$ is a point on L and after reflecting from M, the light from point O is a straight line MS, which is parallel to the x-axis. Then assume that the angle between M's tangent line AT and the x-axis is α, so $\angle SMT = \alpha$. On the other hand, $\angle OMA$

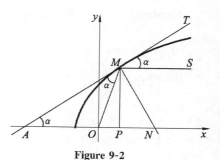

Figure 9-2

is the complementary angle of incident angle $\angle OMN$ and $\angle SMT$ is the complementary angle of reflected angle $\angle SMN$. According to reflection theorem of light,

$$\angle OMA = \angle SMT = \alpha,$$

so $AO = OM$. Because

$$AO = AP - OP = PM\cot \alpha - OP = \frac{y}{y'} - x,$$

$$OM = \sqrt{x^2 + y^2},$$

we obtain the differential equation

$$\frac{y}{y'} - x = \sqrt{x^2 + y^2}.$$

Regard x as dependent variable and y as independent variable, this equation is also written as

$$y\frac{\mathrm{d}x}{\mathrm{d}y} = x + \sqrt{x^2 + y^2}.$$

When $y > 0$, the formula above is

$$\frac{\mathrm{d}x}{\mathrm{d}y} = \frac{x}{y} + \sqrt{\left(\frac{x}{y}\right)^2 + 1}.$$

This is a homogeneous differential equation. Let $\frac{x}{y} = v$, then

$$x = yv, \quad \frac{\mathrm{d}x}{\mathrm{d}y} = y\frac{\mathrm{d}v}{\mathrm{d}y} + v.$$

Plug these formulas into the formula above, we obtain

$$y\frac{\mathrm{d}v}{\mathrm{d}y} = \sqrt{v^2 + 1}.$$

Separate variables, we have

$$\frac{\mathrm{d}v}{\sqrt{v^2 + 1}} = \frac{\mathrm{d}y}{y}.$$

Integrate both sides, we get

$$\ln(v + \sqrt{v^2 + 1}) = \ln y - \ln C$$

or

$$v + \sqrt{v^2 + 1} = \frac{y}{C}.$$

Then

$$\left(\frac{y}{C} - v\right)^2 = v^2 + 1,$$

so

$$\frac{y^2}{C^2} - \frac{2yv}{C} = 1.$$

Plug $yv = x$ into the formula above, then we obtain

$$y^2 = 2C\left(x + \frac{C}{2}\right).$$

Thus the curve L is a parabola with the x-axis as axis and the origin as the focus.

9.2.3 First-order linear differential equation

The differential equation

$$\frac{\mathrm{d}y}{\mathrm{d}x} + P(x)y = Q(x) \tag{5}$$

is called the first-order linear differential equation. If $Q(x) \equiv 0$, Equation (5) is called the first-order linear homogeneous differential equation and if $Q(x) \not\equiv 0$, Equation (5) is called the first-order linear inhomogeneous differential equation.

If Equation (5) is the first-order linear inhomogeneous differential equation, in order to find its solution, we should substitute 0 for $Q(x)$ firstly and we have

$$\frac{\mathrm{d}y}{\mathrm{d}x} + P(x)y = 0. \tag{6}$$

Equation (6) is called the corresponding homogeneous linear equation to inhomogeneous linear Equation (5).

Homogeneous linear Equation (6) is a separable differential equation. After separating variables, we obtain

$$\frac{\mathrm{d}y}{y} = -P(x)\mathrm{d}x.$$

Integrate both sides, we get

$$\ln|y| = -\int P(x)\mathrm{d}x + C_1,$$

where $\int P(x)\mathrm{d}x$ is an antiderivative of $P(x)$. Solve the equation above for y, we have

$$y = C\mathrm{e}^{-\int P(x)\mathrm{d}x} \quad (C = \pm \mathrm{e}^{C_1}).$$

Then notice that $y=0$ is also the solution of Equation (6), so the general solution of homogeneous linear Equation (6) is

$$y = C\mathrm{e}^{-\int P(x)\mathrm{d}x}, \tag{7}$$

where C is an arbitrary constant.

In order to find the general solution of inhomogeneous linear Equation (5), we use the so-called variation of parameters. Substitute a function $u(x)$ for the constant C in (7) and assume that the solution of inhomogeneous Equation (5) is

$$y = u\mathrm{e}^{-\int P(x)\mathrm{d}x}. \tag{8}$$

Then plug $y = u(x)\mathrm{e}^{-\int P(x)\mathrm{d}x}$ into the original Equation (5), we can find $u(x)$. Actually, if $y = u(x)\mathrm{e}^{-\int P(x)\mathrm{d}x}$, we have

$$\frac{\mathrm{d}y}{\mathrm{d}x} = u'(x)\mathrm{e}^{-\int P(x)\mathrm{d}x} - u(x)P(x)\mathrm{e}^{-\int P(x)\mathrm{d}x}.$$

Plug it into Equation (5), we obtain

$$u'(x)\mathrm{e}^{-\int P(x)\mathrm{d}x} - u(x)P(x)\mathrm{e}^{-\int P(x)\mathrm{d}x} + P(x)u(x)\mathrm{e}^{-\int P(x)\mathrm{d}x} = Q(x),$$

then

$$u'(x)\mathrm{e}^{-\int P(x)\mathrm{d}x} = Q(x),$$

or

$$u'(x) = Q(x)\mathrm{e}^{\int P(x)\mathrm{d}x}.$$

Thus

$$u(x) = \int Q(x)\mathrm{e}^{\int P(x)\mathrm{d}x}\mathrm{d}x + C.$$

Plug it into (8), we obtain the general solution of inhomogeneous linear Equation (5),

$$y = \mathrm{e}^{-\int P(x)\mathrm{d}x}\left(\int Q(x)\mathrm{e}^{\int P(x)\mathrm{d}x}\mathrm{d}x + C\right). \tag{9}$$

Then, we will analyze the structure of the general solution (9).

In the equality (9), when $C=0$, we obtain a particular solution of the linear inhomogeneous Equation (5) which is denoted by y^*, that is,

$$y^* = \mathrm{e}^{-\int P(x)\mathrm{d}x} \cdot \int Q(x)\mathrm{e}^{\int P(x)\mathrm{d}x}\mathrm{d}x.$$

And the equality (9) could also be written as

$$y = Ce^{-\int P(x)dx} + e^{-\int P(x)dx} \cdot \int Q(x)e^{\int P(x)dx} dx.$$

The left term is the general solution of the corresponding linear homogeneous equation and the right term is a particular solution y^* of the linear inhomogeneous equation. Therefore the general solution of the first-order linear inhomogeneous differential equation can be written into the form

$$y = \tilde{y} + y^*,$$

where $\tilde{y}$ is the general solution of corresponding homogeneous Equation (6) and y^* is a solution of inhomogeneous Equation (5).

According to the discussion above, for first-order linear inhomogeneous equation, we can not only use variation of parameters but also use the general solution Formula (9) of inhomogeneous equation directly to find the general solution.

Example 7 Find the general solution of

$$y' + \frac{1}{x}y = \frac{e^x}{x}.$$

Solution (Method 1) We use variation of parameters. The corresponding homogeneous equation is

$$y' + \frac{1}{x}y = 0.$$

Separate variables, we have

$$\frac{dy}{y} = -\frac{dx}{x}.$$

Then integrating both sides and we obtain

$$y = \frac{C}{x}.$$

Substitute $u(x)$ for C in the equality above and the solution of the original equation is $y = \frac{u}{x}$ and $y' = \frac{u'}{x} - \frac{u}{x^2}$. Plug it into the original equation, we obtain

$$\frac{u'}{x} - \frac{u}{x^2} + \frac{1}{x} \cdot \frac{u}{x} = \frac{e^x}{x},$$

and

$$u'(x) = e^x,$$

so

$$u(x) = e^x + C.$$

Thus the general solution of the original equation is

$$y = \frac{e^x + C}{x}.$$

(Method 2) We use the general solution formula. By (9), we obtain that

$$y = e^{-\int \frac{1}{x}dx} \left(\int \frac{e^x}{x} e^{\int \frac{1}{x}dx} dx + C \right)$$

$$= e^{-\ln x} \left(\int \frac{e^x}{x} \cdot e^{\ln x} dx + C \right)$$

$$= \frac{1}{x} \left(\int e^x dx + C \right)$$

$$=\frac{1}{x}(e^x+C).$$

Example 8 Find the general solution of
$$(x^2-1)y'+2xy-\cos x=0.$$

Solution Rewrite the equation into
$$y'+\frac{2x}{x^2-1}y=\frac{\cos x}{x^2-1},$$

therefore $P(x)=\frac{2x}{x^2-1}$, $Q(x)=\frac{\cos x}{x^2-1}$. Plug them into the Formula (9), we have

$$y=e^{-\int\frac{2x}{x^2-1}dx}\left(\int\frac{\cos x}{x^2-1}e^{\int\frac{2x}{x^2-1}dx}dx+C\right)$$
$$=\frac{1}{x^2-1}\left(\int\cos x dx+C\right)$$
$$=\frac{\sin x+C}{x^2-1}.$$

Example 9 Find the general solution of
$$y\ln y dx+(x-\ln y)dy=0.$$

Solution Rewrite the given equation into
$$\frac{dx}{dy}+\frac{1}{y\ln y}x=\frac{1}{y}.$$

Regard y as independent variable and x as dependent variable, then the equality above is a first-order linear equation with $P(y)=\frac{1}{y\ln y}$ and $Q(y)=\frac{1}{y}$. Therefore by the Formula (9) we have

$$x=e^{-\int\frac{1}{y\ln y}dy}\left(\int\frac{1}{y}e^{\int\frac{1}{y\ln y}dy}dy+C\right)$$
$$=e^{-\ln\ln y}\left(\int\frac{1}{y}e^{\ln\ln y}dy+C\right)$$
$$=\frac{1}{\ln y}\left(\int\frac{\ln y}{y}dy+C\right)$$
$$=\frac{1}{2}\ln y+\frac{C}{\ln y}.$$

Then we give another example, which we will use the differential equation to solve practical problems.

Example 10 Assume that the commodity's demand D and supply S are both functions of price p. Demand function is $D(p)=b-ap$ $(a,b>0)$ and supply function is $S(p)=cp-d$ $(c,d>0)$. Then we assume that price p is the function of time t and satisfies the equation
$$\frac{dp}{dt}=k[D(p)-S(p)].$$

Calculate the expression of price function $p(t)$ and $\lim\limits_{t\to+\infty}p(t)$.

Solution Plug $D(p)=b-ap$, $S(p)=cp-d$ into the given equation, we have
$$\frac{dp}{dt}=-k(a+c)p+k(b+d).$$

Then we use the general solution Formula (9) of the first-order linear inhomogeneous equation and obtain

$$p = e^{-\int k(a+c)dt}\left[\int k(b+d)e^{\int k(a+c)dt}dt + C\right]$$

$$= e^{-k(a+c)t}\left[\frac{b+d}{a+c}e^{k(a+c)t} + C\right]$$

$$= \frac{b+d}{a+c} + Ce^{-k(a+c)t}.$$

So we have

$$\lim_{t \to +\infty} p(t) = \frac{b+d}{a+c}.$$

The price which satisfies the balance between supply and demand is called equilibrium price of market. From $S(p) = D(p)$ or $cp - d = b - ap$, we can obtain that the equilibrium price denoted by p_e is

$$p_e = \frac{b+d}{a+c}.$$

Thus the analysis of this example illustrates that market price gradually tends to equilibrium price.

9.2.4 Bernoulli equation

The differential equation in the form

$$\frac{dy}{dx} + P(x)y = Q(x)y^n \tag{10}$$

is called the Bernoulli equation. If $n = 0$ or $n = 1$, it is a first-order linear differential equation. If $n \neq 0$ and $n \neq 1$, though it isn't linear, it could be rewritten into a first-order linear equation by variable substitution. First, divide both sides of Equation (10) by y^n. And then we obtain

$$y^{-n}\frac{dy}{dx} + P(x)y^{-n+1} = Q(x).$$

Let $z = y^{-n+1}$, since $\frac{dz}{dx} = (1-n)y^{-n}\frac{dy}{dx}$, the equation above could be written into

$$\frac{1}{1-n}\frac{dz}{dx} + P(x)z = Q(x),$$

or

$$\frac{dz}{dx} + (1-n)P(x)z = (1-n)Q(x).$$

This is a first-order linear differential equation. After finding the general solution of the equation, we can obtain the general solution of the Bernoulli equation by substituting y^{1-n} for z.

Example 11 Find the general solution of

$$\frac{dy}{dx} + \frac{1}{x}y = a(\ln x)y^2.$$

Solution Divide both sides of the equation by y^2, we have

$$y^{-2}\frac{dy}{dx} + \frac{1}{x}y^{-1} = a\ln x,$$

so
$$-\frac{d(y^{-1})}{dx}+\frac{1}{x}y^{-1}=a\ln x.$$

Let $z=y^{-1}$, then the equation above becomes
$$\frac{dz}{dx}-\frac{1}{x}z=-a\ln x.$$

This is a linear equation and the general solution is
$$z=x\left[C-\frac{a}{2}(\ln x)^2\right].$$

Substitute y^{-1} for z, the general solution of the equation is
$$yx\left[C-\frac{a}{2}(\ln x)^2\right]=1.$$

We always use variation substitution during the process of solving homogeneous equations introduced in 9.2.2 and Bernoulli equations introduced above. For some differential equations, they could be written into proper forms through suitable variation substitution and then find the solution by related methods.

Example 12 Find the general solution of
$$2y\frac{dy}{dx}+\frac{1}{x}y^2=x^2.$$

Solution This is a nonlinear differential equation.

However, notice that $\frac{dy^2}{dx}=2y\frac{dy}{dx}$, it could be written into
$$\frac{dy^2}{dx}+\frac{1}{x}y^2=x^2.$$

We introduce the substitution $z=y^2$, then the equation is rewritten into the first-order linear differential equation
$$\frac{dz}{dx}+\frac{1}{x}z=x^2,$$

and its general solution is
$$z=\frac{C}{x}+\frac{1}{4}x^3.$$

Thus the general solution of the original equation is
$$y^2=\frac{C}{x}+\frac{1}{4}x^3.$$

Example 13 Write $\frac{1}{y}y'-\frac{1}{x}\ln y=x^2$ into first-order linear equation.

Solution The original equation could be written into
$$(\ln y)'-\frac{1}{x}\ln y=x^2.$$

Let $z=\ln y$, we have
$$z'-\frac{1}{x}z=x^2,$$

which is a first-order linear differential equation.

Exercises 9-2

1. Find general solutions of the following differential equations or particular solutions that satisfy initial conditions:

   (1) $y' = 3x^2(1+y)^2$;

   (2) $y^2 dx + (x+1) dy = 0, y(0) = 1$;

   (3) $yy' + e^{y^2+3x} = 0$;

   (4) $dy = x(2y dx - x dy), y(1) = 4$.

2. Find general solutions of the following differential equations or particular solutions that satisfy initial conditions:

   (1) $\dfrac{dy}{dx} = \dfrac{y-x}{y+x}, y(1) = 0$;

   (2) $y' = \dfrac{x}{y} + \dfrac{y}{x}, y(1) = 0$;

   (3) $(x^2 - y^2) dx + xy dy = 0$;

   (4) $x(\ln y - \ln x) dy + y dx = 0$.

3. Find general solutions of the following differential equations or particular solutions that satisfy initial conditions:

   (1) $\dfrac{dy}{dx} - \dfrac{2y}{x+1} = (x+1)^3$;

   (2) $\dfrac{dy}{dx} + \dfrac{1-2x}{x^2} y - 1 = 0$;

   (3) $y' - 2xy = x - x^3, y(0) = 1$;

   (4) $x \dfrac{dy}{dx} + y = x^3, y(1) = \dfrac{5}{4}$;

   (5) $y' - y \tan x + y^2 \cos x = 0$;

   (6) $y' = \dfrac{y}{x+y^3}$.

4. Find the solution of differential equation $y = e^x + \displaystyle\int_0^x y(t) dt$.

5. Find a curve passing through the origin point with the tangent at any point (x, y) on the curve is $2x + y$:

6. Solve the equations by suitable variable substitution:

   (1) $\dfrac{dy}{dx} = (x+y)^2$;

   (2) $\dfrac{dy}{dx} = \dfrac{1}{(x+y)^2}$;

   (3) $\dfrac{dy}{dx} = \dfrac{x-y+5}{x-y-2}$.

9.3 Reducible high-order differential equation

Second-order differential equations and above are all called the high-order differential equations. For some high-order differential equations, they can be written into lower-order differentia equations by variable substitution so that we can use the corresponding methods of low-order differential equations to calculate them. For example, if a second-order differential equation can be written into a first-order differential equation, we apply the method in 9.2 to find the solution. This section mainly discusses the methods to solve the reduced high-order differential equations.

9.3.1 The form $y^{(n)} = f(x)$

For the differential equation
$$y^{(n)} = f(x), \tag{1}$$
the right side is just a function of independent variable x. Therefore, regard $y^{(n-1)}$ as a new

unknown function, then the equation (1) is the first-order differential equation of $y^{(n-1)}$. Integrate both sides and we get the general solution, which is

$$y^{(n-1)} = \int f(x)\,dx + C_1.$$

The equality above is a $(n-1)$ th-order differential equation, by the technique above, we get

$$y^{(n-2)} = \int \left[\int f(x)\,dx + C_1\right]dx + C_2.$$

Do this n times, the general solution of Equation (1) will includes n arbitrary constants.

Example 1 Find the general solution of $y^{(3)} = xe^x$.

Solution Integrate the given equation for three times, and we have

$$y'' = \int xe^x\,dx = (x-1)e^x + C_1,$$
$$y' = (x-2)e^x + C_1 x + C_2,$$
$$y = (x-3)e^x + Cx^2 + C_2 x + C_3 \quad \left(C = \frac{C_1}{2}\right).$$

9.3.2 The form $y'' = f(x, y')$

For the differential equation

$$y'' = f(x, y'), \tag{2}$$

we cannot find the unknown function y in it. Let $y' = p$, then $y'' = p'$, so the Equation (2) can be written into a first-order differential equation

$$p' = f(x, p).$$

Suppose $p = \varphi(x, C_1)$ is the solution, since $y' = p$, we get another first-order differential equation

$$y' = \varphi(x, C_1).$$

Integrate both sides and the general solution of the differential Equation (2) is

$$y = \int \varphi(x, C_1)\,dx + C_2.$$

Example 2 Find the solution of

$$(1-x^2)y'' - xy' = 0$$

which satisfies initial conditions $y|_{x=0} = 1$, $y'|_{x=0} = 3$.

Solution Let $y' = p$, we have

$$(1-x^2)p' - xp = 0.$$

Separate variables

$$\frac{dp}{p} = \frac{x}{1-x^2}\,dx.$$

Integrate both sides, we have

$$\ln|p| = -\frac{1}{2}\ln|1-x^2| + C,$$

so

$$p = \frac{C_1}{\sqrt{1-x^2}} \quad (C_1 = \pm e^C).$$

Since $y'|_{x=0}=3$, that is, $p|_{x=0}=3$, we get $C_1=3$. Thus we have
$$y'=\frac{3}{\sqrt{1-x^2}}.$$
Integrate both sides again and we get
$$y=3\arcsin x+C_2.$$
Then by $y|_{x=0}=1$, we get $C_2=1$. Therefore the solution is
$$y=3\arcsin x+1.$$

Example 3 Show that the curve whose curvature identically equals nonzero constants must be a circle.

Proof Assume that the curvilinear equation is $y=\varphi(x)$ and the curvature of any point identically equals to constant $\frac{1}{a}$ ($a>0$). Then according to curvature formula, we obtain
$$\frac{[1+(y')^2]^{3/2}}{|y''|}=a,$$
or
$$\frac{[1+(y')^2]^{3/2}}{y''}=\pm a.$$

This is a second-order equation without y. Let $y'=p$, the equation above becomes
$$(1+p^2)^{3/2}=\pm a p'.$$
Separate variables,
$$\pm a\frac{1}{(1+p^2)^{3/2}}dp=dx.$$
Integrate both sides, we have
$$\pm a\frac{p}{\sqrt{1+p^2}}=x+C_1.$$
Solve the formula for p and substitute y' for p, then
$$y'=\pm\frac{x+C_1}{\sqrt{a^2-(x+C_1)^2}}.$$
Integrate again and we obtain
$$y+C_2=\mp\sqrt{a^2-(x+C_1)^2},$$
therefore
$$(x+C_1)^2+(y+C_2)^2=a^2.$$
This is the equation of a circle with radius a.

9.3.3 The form $y''=f(y,y')$

For the differential equation
$$y''=f(y,y') \tag{3}$$
doesn't obviously include the independent variable x. Let $y'=p$, then
$$y''=\frac{dy'}{dx}=\frac{dp}{dx}=\frac{dp}{dy}\cdot\frac{dy}{dx}=\frac{dp}{dy}\cdot p,$$
therefore the Equation (3) can be written into
$$p\frac{dp}{dy}=f(y,p).$$

This is a first-order differential equation. Assume that its general solution is

then
$$p = \varphi(y, C_1),$$
$$y' = \varphi(y, C_1).$$

Separate variables and then integrate both sides, we have
$$\int \frac{dy}{\varphi(y, C_1)} = x + C_2.$$

Example 4 Find the general solution to the equation
$$y'' + \frac{2}{1-y} y'^2 = 0.$$

Solution This is a second-order differential equation that excludes independent variable x. Let $y' = p$ and then $y'' = p \dfrac{dp}{dx}$, so there is the equation
$$p \frac{dp}{dy} + \frac{2}{1-y} p^2 = 0.$$

By separating variables, integrating both sides and simplifying, we obtain
$$p = C_1 (1-y)^2,$$
or
$$y' = C_1 (1-y)^2.$$

Then separate variables and integrate again, the general solution is
$$\frac{1}{1-y} = C_1 x + C_2.$$

Exercises 9-3

1. Find the general solutions to the following differential equations:
   (1) $y'' = \ln x$; (2) $y'' = y' + x$;
   (3) $y^3 y'' - 1 = 0$; (4) $y y'' + (y')^2 = 0$.
2. Find the particular solutions to the following differential equations:
   (1) $y''' = e^{2x}, y(1) = y'(1) = y''(1) = 0$; (2) $(1+x^2) y'' - 2xy' = 0, y(0) = 1, y'(0) = 3$;
   (3) $y'' - a(y')^2 = 0, y(0) = 0, y'(0) = -1$; (4) $y^3 y'' + 1 = 0, y(1) = 1, y'(1) = 0$.
3. Find the integral curve of $y'' = x^2$ passing through the point $(1, 3)$ with tangent line $y = \dfrac{x}{2} + \dfrac{5}{2}$ at this point.

9.4 High-order linear differential equation

The equation
$$\frac{d^n y}{dx^n} + a_1(x) \frac{d^{n-1} y}{dx^{n-1}} + \cdots + a_n(x) y = f(x)$$

is called a n th-order linear differential equation. If $n = 1$, the equation is a first-order linear differential equation and if $n \geq 2$, the equation is called a high-order linear differential equation. If $f(x) \equiv 0$, the equation is homogeneous, and if $f(x)$ is not 0 identically, the equation is inhomogeneous. If $a_i(x)$ $(i = 1, 2, \cdots, n)$ are constants, the equation is called a linear constant-coefficient differential equation.

Example 1 Assume that a fixed string hangs an object with quality m. When the object is stationary, the gravity is equal to the elastic force in a contrary direction, so the resultant force is 0. This position is called the equilibrium position of the object. If pull the object away from this position straightly and then loosen it, the object will move up and down around the equilibrium position. Select the equilibrium point as the origin of coordinate. Constitute the coordinate system in Figure 9-3, and find the differential equation of the object's displacement function $x(t)$.

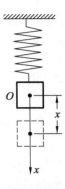

Figure 9-3

Solution Suppose the displacement that object moves away from equilibrium position is x (Figure 9-3), the string will produce elastic restoring force f to make the object back to the equilibrium position. Known from mechanics,

$$f = -cx,$$

where c is the elastic coefficient of string and the minus sign means the direction of restoring force is opposite to displacement.

In addition, the object also have resistance from damping medium (such as air, oil and etc.) when moving up and down. From the experiment, the direction of resistance R is opposite to the motion and its magnitude is directly proportional to the object's velocity. Assume that the proportional coefficient is μ, then

$$R = -\mu \frac{dx}{dt}.$$

According to the analysis of forces and the Newton's second law, we have

$$m \frac{d^2 x}{dt^2} = -cx - \mu \frac{dx}{dt}.$$

Let $2n = \frac{\mu}{m}$, $k^2 = \frac{c}{m}$, we get

$$\frac{d^2 x}{dt^2} + 2n \frac{dx}{dt} + k^2 x = 0.$$

This is the differential equation of its displacement function. It is a second-order linear homogeneous differential equation which is also called free vibration equation. If $n=0$ (that is, $\mu=0$), the equation is called non-damping free vibration equation and if $n \neq 0$ (that is, $\mu \neq 0$), the equation is called damping free vibration equation.

When the object vibrates, we also have the vertical force

$$F = H \sin pt,$$

so

$$\frac{d^2 x}{dt^2} + 2n \frac{dx}{dt} + k^2 x = h \sin pt,$$

where $h = \frac{H}{m}$. It is the forced vibration equation that satisfies string's vibration.

9.4.1 The property and structure of solution of second-order linear differential equation

In this part, we will take the second-order linear differential equation as example to discuss the property and structure of solution of the high-order linear differential equation. Similar to the first-order linear differential equation, the second-order linear differential equation also includes homogeneous equation and inhomogeneous equation. The form of second-order linear inhomogeneous differential equation is

$$y'' + P(x)y' + Q(x)y = f(x). \tag{1}$$

The form of second-order linear homogeneous differential equation is

$$y'' + P(x)y' + Q(x)y = 0. \tag{2}$$

If substitute 0 for the function $f(x)$ on the right side of (1), we will obtain its corresponding homogeneous equation as the form (2).

Theorem 1 Solutions of the linear differential Equations (1) and (2) have the following properties.

If $y_1(x)$ and $y_2(x)$ are two solutions of homogeneous Equation (2), then

$$y = C_1 y_1(x) + C_2 y_2(x) \tag{3}$$

is also the solution of homogeneous Equation (2), where C_1 and C_2 are arbitrary constants.

If $y_1(x)$ is a solution of homogeneous Equation (2) and $y_2(x)$ is a solution of inhomogeneous Equation (1), then

$$y = y_1(x) + y_2(x) \tag{4}$$

is also the solution of inhomogeneous Equation (1).

If $y_1(x)$ and $y_2(x)$ are two solutions of inhomogeneous Equation (1), then

$$y = y_1(x) - y_2(x) \tag{5}$$

is the solution of homogeneous Equation (2).

Proof Plug function (3) into the left of homogeneous Equation (2), and we obtain

$$[C_1 y_1'' + C_2 y_2''] + P(x)[C_1 y_1' + C_2 y_2'] + Q(x)[C_1 y_1 + C_2 y_2]$$
$$= C_1 [y_1'' + P(x)y_1' + Q(x)y_1] + C_2 [y_2'' + P(x)y_2' + Q(x)y_2].$$

Since $y_1(x)$ and $y_2(x)$ are solutions of homogeneous Equation (2) and the expressions in two square brackets on the right side both equal to 0, the whole equation equals 0 identically. Thus $y = y_1(x) + y_2(x)$ is the solution of Equation (2).

Plug function (4) into the left side of inhomogeneous Equation (1), then we have

$$(y_1'' + y_2'') + P(x)(y_1' + y_2') + Q(x)(y_1 + y_2) = [y_1'' + P(x)y_1' + Q(x)y_1] +$$
$$[y_2'' + P(x)y_2' + Q(x)y_2].$$

Since $y_1(x)$ and $y_2(x)$ are respectively solutions of homogeneous Equation (2) and inhomogeneous Equation (1), and the expressions in first and second square brackets on the right side equal to 0 and $f(x)$ respectively, the whole equation equals to $f(x)$ identically. Thus $y = y_1(x) + y_2(x)$ is the solution of inhomogeneous Equation (1).

Similarly, we can prove the property (iii), the process is left for readers to complete.

The property (i) can be used to discuss the structure of general solution of

homogeneous linear differential equation. It tells us that if we find two particular solutions $y_1(x)$ and $y_2(x)$ of homogeneous Equation (2), we can multiply them by arbitrary constants and then add them up to obtain the solution
$$y = C_1 y_1(x) + C_2 y_2(x).$$

And this solution includes two arbitrary constants C_1 and C_2 but it may not be the general solution. For example, assume that $y_1(x)$ is a solution of (2), we know $y_2(x) = 2y_1(x)$ is also a solution of (2) from the property. Then the solution $y = C_1 y_1(x) + C_2 y_2(x)$ may be rewritten as $y = C y_1(x)$ where $C = C_1 + 2C_2$. That is to say, under this condition, since there is actually a single constant C, $y = C_1 y_1(x) + C_2 y_2(x)$ is not the general solution of (2). When does (3) be the general solution of Equation (2)? To answer this question, we will introduce the concept of function's linear dependence and independence.

Assume that $y_1(x), y_2(x), \cdots, y_n(x)$ are functions defined on the interval I. If there is n nonzero constants $k_1, k_2, \cdots, k_n$ such that
$$k_1 y_1(x) + k_2 y_2(x) + \cdots + k_2 y_2(x) \equiv 0,$$
where $x \in I$, then these functions are linearly dependent on the interval I. Otherwise, they are linearly independent on the interval I.

For example, three functions $y_1 = 1, y_2 = x$ and $y_3 = 2 + 3x$ are linearly dependent on $(-\infty, +\infty)$ because there are $k_1 = -2, k_2 = -3$ and $k_3 = 1$ such that
$$k_1 y_1 + k_2 y_2 + k_3 y_3 \equiv 0,$$
on $(-\infty, +\infty)$. Then the functions 1, x and x^2 are linearly independent on any interval (a, b) because k_1, k_2, k_3 must be 0 such that $k_1 + k_2 x + k_3 x^2 = 0$.

We can easily know from the concept above that for two functions, their linear dependence or independence depend on their ratio value being constant or not. If it is constant, they are linearly dependent. Otherwise, they are linearly independent.

If two solutions $y_1(x)$ and $y_2(x)$ are linearly independent, then for the solution
$$y = C_1 y_1(x) + C_2 y_2(x),$$
C_1, C_2 cannot be written into one constant but just two independent arbitrary constants. Thus we have the following theorem about the structure of the general solution of the second-order homogeneous linear differential equation.

Theorem 2 If $y_1(x)$ and $y_2(x)$ are two linearly independent solutions of the second-order homogeneous linear differential Equation (2), then
$$y = C_1 y_1(x) + C_2 y_2(x) \ (C_1, C_2 \text{ are arbitrary constants})$$
is the general solution of Equation (2).

Example 2 Find the general solution of $y'' + y = 0$.

Solution Rewrite the equation as
$$y'' = -y.$$
It illustrates that the difference between the second derivative y'' and y is a negative

sign. Since $y_1 = \sin x$ and $y_2 = \cos x$ both satisfy this equation, they are particular solutions. Then because
$$\frac{y_1}{y_2} = \frac{\sin x}{\cos x} = \tan x \neq \text{constant},$$
y_1 and y_2 are linearly independent. Thus according to Theorem 2, the general solution is
$$y = C_1 \sin x + C_2 \cos x.$$

Example 3 Equation $(x-1)y'' - xy' + y = 0$ is also a second-order homogeneous linear differential equation. (It can be written into the standard form if divide both sides by $x-1$ and $p(x) = -\frac{x}{x-1}, Q(x) = \frac{1}{x-1}$.) It can easily prove that $y_1 = x$, $y_2 = e^x$ are two solutions of the given equation and
$$\frac{y_1}{y_2} = \frac{x}{e^x} \neq \text{constant},$$
so they are linearly independent. Thus the general solution of equation $(x-1)y'' - xy' + y = 0$ is
$$y = C_1 x + C_2 e^x.$$

Now, let us discuss the problem about the structure of inhomogeneous linear equation's solution.

As preciously mentioned, the corresponding homogeneous equation of inhomogeneous equation
$$y'' + P(x)y' + Q(x)y = f(x)$$
is
$$y'' + P(x)y' + Q(x)y = 0.$$

Suppose $Y(x)$ is the general solution of homogeneous Equation (2) and $y^*(x)$ is a particular solution of inhomogeneous Equation (1), then we know from the property of Theorem 1 that the function
$$y(x) = Y(x) + y^*(x) \tag{6}$$
is the solution of inhomogeneous Equation (1). And because $Y(x)$ is the general solution of homogeneous Equation (2), which includes two independent arbitrary constants, so function (3) also includes two independent arbitrary constants and then (3) is the general solution of inhomogeneous Equation (1). Thus we have the following theorem.

Theorem 3 If $y^*(x)$ is a particular solution of inhomogeneous linear differential Equation (1) and $Y(x)$ is the general solution of its corresponding homogeneous linear differential Equation (2), then
$$y(x) = Y(x) + y^*(x)$$
is the general solution of inhomogeneous linear differential Equation (1).

Example 4 Find the general solution of differential equation
$$y'' + y = 4x + 1.$$

Solution Known from Example 2, the general solution of its corresponding

homogeneous equation is
$$Y = C_1 \sin x + C_2 \cos x.$$
And we know that $y^* = 4x+1$ is a particular solution of the given equation. Thus
$$y = C_1 \sin x + C_2 \cos x + 4x + 1$$
is the general solution of the given equation.

To find the particular solution of the inhomogeneous differential Equation (1), it sometimes needs to use the following theorem.

Theorem 4 Assume that the function $f(x)$ on the right side of Equation (1) is the sum of two functions,
$$y'' + P(x)y' + Q(x)y = f_1(x) + f_2(x), \tag{7}$$
let $y_1^*(x)$ and $y_2^*(x)$ be particular solutions respectively, that is,
$$y'' + P(x)y' + Q(x)y = f_1(x)$$
and
$$y'' + P(x)y' + Q(x)y = f_2(x),$$
then $y_1^*(x) + y_2^*(x)$ is the particular solution of Equation (7).

The proof of this theorem is similar to Theorem 1, please complete the proof.

*9.4.2 The property and structure of solution of high-order linear differential equation

For the general n th-order linear differential equation
$$y^{(n)} + a_1(x)y^{(n-1)} + a_2(x)y^{(n-2)} + \cdots + a_n(x)y = 0, \tag{8}$$
$$y^{(n)} + a_1(x)y^{(n-1)} + a_2(x)y^{(n-2)} + \cdots + a_n(x)y = f(x) \tag{9}$$
and
$$y^{(n)} + a_1(x)y^{(n-1)} + a_2(x)y^{(n-2)} + \cdots + a_n(x)y = f_1(x) + f_2(x), \tag{10}$$
the properties and structures of solutions are similar to that of the second-order linear differential equation.

Theorem 1' Solutions of the n th-order linear differential Equation (8) and (9) have following properties.

(i) If $y_1(x), y_2(x), \cdots, y_n(x)$ are solutions of homogeneous Equation (8), then
$$y = C_1 y_1(x) + C_2 y_2(x) + \cdots + C_n y_n(x)$$
is the solution of homogeneous Equation (8) and $C_1, C_2, \cdots, C_n$ are arbitrary constants.

(ii) If $y_1(x)$ is a solution of homogeneous Equation (8) and $y_2(x)$ is a solution of inhomogeneous Equation (9), then
$$y = y_1(x) + y_2(x)$$
is the solution of inhomogeneous Equation (9).

(iii) If $y_1(x)$ and $y_2(x)$ are two solutions of inhomogeneous Equation (9), then
$$y = y_1(x) - y_2(x)$$
is the solution of homogeneous Equation (8).

Theorem 2′ If $y_1(x), y_2(x), \cdots, y_n(x)$ are linearly independent solutions of homogeneous Equation (8), then
$$y = C_1 y_1 + C_2 y_1 + \cdots + C_n y_n$$
is the general solution of homogeneous Equation (8).

Theorem 3′ If $y^*(x)$ is a particular solution of inhomogeneous Equation (9) and $Y(x)$ is the general solution of its corresponding homogeneous Equation (8), then
$$y(x) = Y(x) + y^*(x)$$
is the general solution of inhomogeneous Equation (9).

Theorem 4′ If $y_1^*(x)$ and $y_2^*(x)$ are respectively particular solutions of equations
$$y^{(n)} + a_1(x) y^{(n-1)} + \cdots + a_n(x) y = f_1(x)$$
and
$$y^{(n)} + a_1(x) y^{(n-1)} + \cdots + a_n(x) y = f_2(x),$$
then
$$y^*(x) = k_1 y_1^*(x) + k_2 y_2^*(x)$$
is the particular solution of Equation (10).

Proof of these theorems are similar to the second-order conditions.

Exercises 9-4

1. Determine whether the following function groups are linearly dependent or linearly independent:
   (1) x, x^2;
   (2) e^{2x}, e^{3x};
   (3) $5\sin 2x, \cos x \sin x$;
   (4) $\cos^2 x, 1 + \cos 2x$;
   (5) $e^x \cos x, e^x \sin x$.

2. Show that $y_1 = e^{x^2}$ and $y_2 = xe^{x^2}$ are solutions of homogeneous linear equation $y'' - 4xy' + (4x^2 - 2)y = 0$ and write out the general solution of this equation.

3. Show that $x_1 = \cos 2t, x_2 = \sin 2t$ are solutions of second-order homogeneous linear equation $\dfrac{d^2 x}{dt^2} + 4x = 0$ and write out the general solution of this equation. Then find the solution that satisfies the initial conditions $x(0) = 1, x'(0) = 1$.

4. Show that $y = C_1 x^2 + C_2 x^2 \ln x$ (C_1, C_2 are arbitrary constants) is the general solution of equation $x^2 y'' - 3xy' + 4y = 0$.

5. Show that $y = C_1 \cos 3x + C_2 \sin 3x + \dfrac{1}{32}(4x\cos x + \sin x)$ (C_1, C_2 are arbitrary constants) is the general solution of equation $y'' + 9y = x\cos x$ and calculate particular solution that satisfies the initial conditions $y(0) = 1, y'(0) = 1$.

9.5 Second-order linear differential equation with constant coefficients

9.5.1 Second-order homogeneous linear differential equation with constant coefficients

If p and q are constants, the equation
$$y'' + py' + qy = 0 \tag{1}$$
is called second-order homogeneous linear differential equation with constant coefficients. From the discussion above, if $y_1(x)$ and $y_2(x)$ are two linearly independent solutions of differential Equation (1), the general solution is $y = C_1 y_1(x) + C_2 y_2(x)$.

From the structure of Equation (1), the unknown function $y(x)$ and its derivatives $y'(x)$ and $y''(x)$ have the relationship of constant multiple. Since the difference between $y = e^{rx}$ and its all-order derivatives is only a constant factor, we can try to select a suitable constant $y = e^{rx}$ such that $y = e^{rx}$ satisfies Equation (1).

Plug $y = e^{rx}$, $y' = re^{rx}$, $y'' = r^2 e^{rx}$ into Equation (1), then we obtain
$$e^{rx}(r^2 + pr + q) = 0.$$
Since $e^{rx} \neq 0$, the equation above equals to
$$r^2 + pr + q = 0. \tag{2}$$

Thus we see that only if r is the root of algebraic Equation (2), $y = e^{rx}$ is the solution of differential Equation (1). We define algebraic Equation (2) as the characteristic equation of differential Equation (1).

Characteristic Equation (2) is quadratic algebraic equation whose r^2, r and constant term are the coefficients of y'', y', y in differential Equation (1) and its root can be expressed as
$$r_{1,2} = \frac{-p \pm \sqrt{p^2 - 4q}}{2}.$$

Then under the conditions $p^2 > 4q$, $p^2 = 4q$ or $p^2 < 4q$, we will discuss expressions of the general solution respectively of Equation (1).

① $p^2 > 4q$.

The characteristic Equation (2) has two unequal real roots r_1 and r_2. According to the discussion above, $y_1 = e^{r_1 x}$ and $y_2 = e^{r_2 x}$ are two solutions of Equation (1). Then because $\frac{y_1}{y_2} = e^{(r_1 - r_2)x}$ isn't a constant, they are two linearly independent solutions of Equation (1). Thus the general solution of differential Equation (1) can be expressed as
$$y = C_1 e^{r_1 x} + C_1 e^{r_2 x}.$$

② $p^2 = 4q$.

The characteristic Equation (2) has two equal real roots $r_1 = r_2 = r$, then we can obtain one solution
$$y_1 = e^{rx}.$$

Then find the other solution y_2 such that y_2 and y_1 are linearly independent. To ensure that y_2 and y_1 are linearly independent or $\dfrac{y_2}{y_1}$ isn't a constant, we assume that $\dfrac{y_2}{y_1} = u(x)$ or
$$y_2 = u(x) y_1 = u(x) e^{rx},$$
and $u(x)$ isn't a constant. Take the derivation of y_2, and obtain
$$y_2' = e^{rx}(u' + ru),$$
$$y_2'' = e^{rx}(u'' + 2ru' + r^2 u).$$

Plug y_2'', y_2' and y_2 into differential Equation (1), then we obtain
$$e^{rx}[(u'' + 2ru' + r^2 u) + p(u' + ru) + qu] = 0.$$

Because $e^{rx} \neq 0$, we have
$$u'' + (2r + p)u' + (r^2 + pr + q)u = 0.$$

Then r is the double root of characteristic Equation (2), we have $r^2 + pr + q = 0$ and $2r + p = 0$, so
$$u'' = 0.$$

Because we need a nonzero $u(x)$, we just let $u(x) = x$ and then obtain another solution of differential Equation (1)
$$y_2 = x e^{rx}.$$

Thus the general solution of differential Equation (1) can be expressed as
$$y = C_1 e^{rx} + C_2 x e^{rx},$$
or
$$y = (C_1 + C_2 x) e^{rx}.$$

③ $p^2 < 4q$.

The characteristic Equation (2) has a pair of conjugate complex roots, $r_1 = \alpha + i\beta$ and $r_2 = \alpha - i\beta$, then we have
$$\alpha = -\frac{p}{2},\ \beta = \frac{\sqrt{4q - p^2}}{2}.$$

Then we obtain two solutions of differential Equation (1),
$$\tilde{y}_1 = e^{(\alpha + i\beta)x},\ \tilde{y}_2 = e^{(\alpha - i\beta)x}.$$

By Euler's formula $e^{\pm i\beta x} = \cos \beta x \pm i \sin \beta x$, we rewrite $\tilde{y}_1, \tilde{y}_2$ into
$$\tilde{y}_1 = e^{(\alpha + i\beta)x} = e^{\alpha x} \cdot e^{i\beta x} = e^{\alpha x}(\cos \beta x + i \sin \beta x),$$
$$\tilde{y}_2 = e^{(\alpha - i\beta)x} = e^{\alpha x} \cdot e^{-i\beta x} = e^{\alpha x}(\cos \beta x - i \sin \beta x).$$

Because the two solutions include imaginary number i, they are not real-valued functions. By properties of differential equation's solution introduced in the above section,
$$y_1 = \frac{1}{2}(\tilde{y}_1 + \tilde{y}_2) = e^{\alpha x} \cos \beta x,$$
$$y_2 = \frac{1}{2i}(\tilde{y}_1 - \tilde{y}_2) = e^{\alpha x} \sin \beta x$$
are also solutions of differential Equation (1) and obviously linearly independent. Thus the general solution of differential Equation (1) could be expressed as
$$y = C_1 e^{\alpha x} \cos \beta x + C_2 e^{\alpha x} \sin \beta x,$$
or
$$y = e^{\alpha x}(C_1 \cos \beta x + C_2 \sin \beta x).$$

In conclusion, the procedures of calculating general solutions of second-order homogeneous linear differential equation with constant coefficients
$$y''+py'+qy=0$$
are as follows.

Step 1 Write out the characteristic equation of differential Equation (1)
$$r^2+pr+q=0.$$

Step 2 Solve equation above for r, then we have two roots r_1, r_2.

Step 3 Under different conditions of r_1 and r_2, we obtain the general solution of differential Equation (1) by the following table.

Two roots r_1, r_2 of characteristic equation $r^2+pr+q=0$	The general solution of differential equation $y''+py'+qy=0$
Real roots $r_1 \neq r_2$	$y=C_1 e^{r_1 x}+C_2 e^{r_2 x}$
Real roots $r_1=r_2=r$	$y=(C_1+C_2 x)e^{rx}$
A pair of conjugate complex roots $r_{1,2}=\alpha \pm i\beta$	$y=e^{\alpha x}(C_1 \cos \beta x + C_2 \sin \beta x)$

Example 1 Find the general solution of
$$y''-5y'+6y=0.$$
Solution The characteristic equation is
$$r^2-5r+6=0.$$
It has two unequal real roots $r_1=e$, $r_2=2$, so the general solution is
$$y=C_1 e^{3x}+C_2 e^{2x}.$$

Example 2 Find the general solution of
$$y''-4y'+4y=0.$$
Solution The characteristic equation is
$$r^2-4r+4=0.$$
It has two equal roots $r_1=r_2=2$, so the general solution is
$$y=(C_1+C_2 x)e^{2x}.$$

Example 3 Find the general solution of
$$y''+y'+y=0.$$
Solution The characteristic equation is
$$r^2+r+1=0.$$
It has a pair of conjugate complex roots
$$r_1=\frac{-1+\sqrt{3}i}{2},\ r_2=\frac{-1-\sqrt{3}i}{2}.$$
We have $\alpha=\frac{-1}{2}$, $\beta=\frac{\sqrt{3}}{2}$, so the general solution is
$$y=e^{-\frac{x}{2}}\left(C_1 \cos \frac{\sqrt{3}}{2}x + C_2 \sin \frac{\sqrt{3}}{2}x\right).$$

Finally, for general high-order homogeneous linear differential equation with constant coefficients
$$y^{(n)} + a_1 y^{(n-1)} + a_2 y^{(n-2)} + \cdots + a_{n-1} y' + a_n y = 0, \tag{3}$$
there are similar results to the second-order. We will not discuss it in details, and the related results are sketched as follows.

The characteristic equation of differential Equation (3) is
$$r^n + a_1 r^{n-1} + a_2 r^{n-2} + \cdots + a_{n-1} r + a_n = 0. \tag{4}$$

According to different conditions of roots of Equation (4), we give solutions of differential Equation (3).

① Each real simple root corresponds to one solution: e^{rx}.

② Each k multiple real root corresponds to k linearly independent solutions: $e^{rx}, xe^{rx}, x^2 e^{rx}, \cdots, x^{k-1} e^{rx}$.

③ Each pair of conjugate complex roots $\alpha \pm i\beta$ (Complex roots of algebraic equation must be conjugate in pair) corresponds to two linearly independent solutions:
$$e^{\alpha x} \cos \beta x, e^{\alpha x} \sin \beta x.$$

④ Each pair of k multiple conjugate complex roots $\alpha \pm i\beta$ corresponds to $2k$ linearly independent solutions:
$$e^{\alpha} \cos \beta x, xe^{\alpha x} \cos \beta x, x^2 e^{\alpha x} \cos \beta x, \cdots, x^{k-1} e^{\alpha x} \cos \beta x,$$
$$e^{\alpha} \sin \beta x, xe^{\alpha x} \sin \beta x, x^2 e^{\alpha x} \sin \beta x, \cdots, x^{k-1} e^{\alpha x} \sin \beta x.$$

From algebra, n-degree algebraic equation has n real roots and complex roots (Regard the number of k multiple complex roots as k), so it corresponds to n linearly independent solutions to differential Equation (3) according to the method above. By the n solutions, we can obtain the general solution of Equation (3).

Example 4 Find the general solution of
$$y^{(5)} + 2y''' + y' = 0.$$

Solution The characteristic equation is
$$r^5 + 2r^3 + r = 0,$$
or
$$r(r^2 + 1)^2 = 0.$$

It has a simple root $r = 0$ which corresponds to a solution
$$e^{0x} = 1.$$

It also has a pair of double complex roots $r = \pm i$ which corresponds to four solutions
$$\cos x, x\cos x, \sin x, x\sin x.$$

Thus the general solution is
$$y = C_1 + (C_2 + C_3 x) \cos x + (C_4 + C_5 x) \sin x.$$

9.5.2 Second-order inhomogeneous linear differential equation with constant coefficients

In this part, we will discuss how to find the solution of the second-order inhomogeneous linear differential equation with constant coefficients
$$y'' + py' + qy = f(x). \tag{5}$$

From the structure of linear differential equation's solution, to find the general solution of inhomogeneous Equation (5), it's required to find a particular solution of Equation (5) and the general solution of its corresponding homogeneous equation
$$y''+py'+qy=0. \tag{6}$$
In the previous part, there are the steps to find the general solution of linear homogeneous equation with constant coefficients, so now we just discuss the method of calculating the particular solution y^* of inhomogeneous linear differential Equation (5). Then we only introduce the method of calculating y^* for two common forms of $f(x)$. The character of this method is to calculate y^* by comparing the coefficients of polynomials, which is also called the method of undetermined coefficients. The two common forms of $f(x)$ are as follows.

① $f(x)=P_m(x)e^{\lambda x}$, where λ is a constant and $P_m(x)$ is a m-degree polynomial of x.

② $f(x)=e^{\alpha x}[P_l(x)\cos \beta x+P_n(x)\sin \beta x]$, where α, β are constants. $P_l(x), P_n(x)$ are respectively l-degree polynomial of x and n-degree polynomial of x while either $P_n(x)$ or $P_n(x)$ could be 0.

1) The form $f(x) = P_m(x)e^{\lambda x}$

Because the first-order and second-order derivatives of polynomial and exponential function's product are still the product of polynomial and exponential function, differential Equation (5) may have the following solution
$$y^* =Q(x)e^{\lambda x},$$
where $Q(x)$ is a polynomial. Then we analyze how to select suitable polynomial $Q(x)$ to make $y^* =Q(x)e^{\lambda x}$ be the solution of Equation (5). For this, plug
$$y^* =Q(x)e^{\lambda x},$$
$$y^{*\prime}=e^{\lambda x}[Q'(x)+\lambda Q(x)],$$
$$y^{*\prime\prime}=e^{\lambda x}[Q''(x)+2\lambda Q'(x)+\lambda^2 Q(x)]$$
into Equation (5) and divided by $e^{\lambda x}$, then we obtain
$$Q''(x)+(2\lambda+p)Q'(x)+(\lambda^2+p\lambda+q)Q(x)=P_m(x). \tag{7}$$

① If λ isn't the root of the characteristic equation
$$r^2+pr+q=0,$$
then $\lambda^2+p\lambda+q\neq 0$, we know from Formula (7) that the degree of polynomial $Q(x)$ equals to that of $P_m(x)$. Thus $Q(x)$ is a m-degree polynomial, then let
$$Q_m(x)=b_0 x^m+b_1 x^{m-1}+\cdots+b_m,$$
where $b_0, b_1, \cdots, b_m$ are undetermined coefficients. Plug $Q_m(x)$ into Equation (7) and compare the coefficients of the same x's power, then we will obtain a simultaneous group of $m+1$ equations of unknown numbers $b_0, b_1, \cdots, b_m$. Then solve these equations for coefficients $b_i (i=0,1,\cdots,m)$ and we obtain the particular solution $y^* =Q_m(x)e^{\lambda x}$.

② If λ is the simple root of characteristic equation $r^2+pr+q=0$, then $\lambda^2+p\lambda+q=0$ and $2\lambda+p\neq 0$. According to Equation (7), $Q'(x)$ must be a m-degree polynomial. Thus let

$$Q(x) = xQ_m(x),$$

and determine polynomial $Q_m(x)$ in the same way as ①.

③ If λ is the double root of characteristic equation $r^2 + pr + q = 0$, then $\lambda^2 + p\lambda + q = 0$ and $2\lambda + p = 0$. We know from Equation (7) that $Q''(x)$ must be a m-degree polynomial. Thus let

$$Q(x) = x^2 Q_m(x),$$

and determine polynomial $Q_m(x)$ in the same way as ①.

In conclusion, if $f(x) = P_m(x)e^{\lambda x}$, the second-order inhomogeneous linear differential equation with constant coefficients (5) has the solution

$$y^* = x^k Q_m(x) e^{\lambda x}$$

where $Q_m(x)$ is a polynomial, which has the same degree with $P_m(x)$. k is 0, 1 or 2 according to the conditions that λ isn't the root of characteristic equation, λ is the simple root of characteristic equation or λ is the double root of characteristic equation.

Example 5 Find the general solution of

$$y'' - 2y' - 3y = 3x + 1.$$

Solution The corresponding homogeneous equation is

$$y'' - 2y' - 3y = 0.$$

Its characteristic equation

$$r^2 - 2r - 3 = 0$$

has two unequal roots $r_1 = 3$, $r_2 = -1$, therefore the general solution of homogeneous equation is

$$y = C_1 e^{3x} + C_2 e^{-x}.$$

Because $f(x) = 3x + 1 = e^{0x}(3x + 1)$ is the form $P_m(x)e^{\lambda x}$, $P_m(x)$ is a linear polynomial and $\lambda = 0$ isn't the root of characteristic equation, assume that particular solution is

$$y^* = (b_0 x + b_1)e^{0x} = b_0 x + b_1.$$

Plug it into the given equation, then we obtain

$$-3b_0 x - 3b_1 - 2b_0 = 3x + 1.$$

Compare the coefficients of the same x's power on either side, then we obtain

$$\begin{cases} -3b_0 = 3, \\ -3b_1 - 2b_0 = 1. \end{cases}$$

And we have $b_0 = -1$, $b_1 = \dfrac{1}{3}$. Thus a particular solution is

$$y^* = -x + \dfrac{1}{3}.$$

Therefore, the general solution is

$$y = C_1 e^{3x} + C_2 e^{-x} - x + \dfrac{1}{3}.$$

Example 6 Find the general solution of

$$y'' - 5y' + 6y = x e^{2x}.$$

Chapter 9 Differential equations

Solution The characteristic equation of corresponding homogeneous equation is
$$r^2 - 5r + 6 = 0,$$
which has two unequal roots $r_1 = 2$ and $r_2 = 3$, so we obtain the general solution of homogeneous equation
$$y = C_1 e^{2x} + C_2 e^{3x}.$$

Because $P_m(x)$ in function $f(x)$ of the given inhomogeneous equation is a linear polynomial and $\lambda = 2$ is the simple root of characteristic equation, we assume the particular solution as
$$y^* = x(b_0 x + b_1) e^{2x}.$$

Plug it into the given equation, then we obtain
$$-2b_0 x + 2b_0 - b_1 = x.$$

Compare the coefficients of the same x's power on both sides, then we obtain
$$\begin{cases} -2b_0 = 1, \\ 2b_0 - b_1 = 0. \end{cases}$$

We get $b_0 = -\dfrac{1}{2}$, $b_1 = -1$. Thus a particular solution of the given equation is
$$y^* = x\left(-\dfrac{1}{2}x - 1\right) e^{2x}.$$

Therefore, the general solution is
$$y = C_1 e^{2x} + C_2 e^{3x} - \dfrac{1}{2}(x^2 + 2x) e^{2x}.$$

Example 7 Find the general solution
$$y'' + y' - 2y = (x-2) e^{5x} + (x^3 - 2x + 3) e^{-x}.$$

Solution Since the roots of the characteristic equation $r^2 + r - 2 = 0$ are $r_1 = -2$ and $r_2 = 1$, the general solution of the corresponding homogeneous equation is
$$y = C_1 e^{-2x} + C_2 e^{x}.$$

Then find the particular solution y^* of the given inhomogeneous equation. Because $f(x)$ on the right side is written as the sum of two functions $f_1(x)$, $f_2(x)$ and
$$f_1(x) = (x-2) e^{5x}, \quad f_2(x) = (x^3 - 2x + 3) e^{-x}.$$

Therefore we can use Theorem 4 in 9.4, then we have the following equations,
① $y'' + y' - 2y = (x-2) e^{5x}$;
② $y'' + y' - 2y = (x^3 - 2x + 3) e^{-x}$.

Firstly find the particular solution of the Equation ①. Because the polynomial of the function on the right side is linear and $\lambda = 5$ isn't the root of characteristic equation, assume that the form of its solution y_1^* is
$$y_1^* = (b_0 x + b_1) e^{5x}.$$

Plug it into ① and we obtain
$$11 b_0 + 28(b_0 x + b_1) = x - 2.$$

Solve it, and we have $b_0 = \dfrac{1}{28}$, $b_1 = -\dfrac{67}{784}$, therefore Equation ① has the particular solution

$$y_1^* = \left(\frac{1}{28}x - \frac{67}{784}\right)e^{5x}.$$

Then find the particular solution of Equation ②. Because the polynomial on the right side is cubical and $\lambda = -1$ isn't the root of characteristic equation, assume that the form of its solution y_2^* is

$$y_2^* = (d_0 x^3 + d_1 x^2 + d_2 x + d_3)e^{-x}.$$

Plug it into ② and we obtain

$$6d_0 x + 2d_1 - (3d_0 x^2 + 2d_1 x + d_2) - 2(d_0 x^3 + d_1 x^2 + d_2 x + d_3) = x^3 - 2x + 3.$$

Compare the coefficients of polynomials on both sides and then obtain $d_0 = -\frac{1}{2}$, $d_1 = \frac{3}{4}$, $d_2 = -\frac{5}{4}$, $d_3 = -\frac{1}{8}$. Thus a particular solution of Equation ② is

$$y_2^* = \left(-\frac{1}{2}x^3 + \frac{3}{4}x^2 - \frac{5}{4}x - \frac{1}{8}\right)e^{-x}.$$

Then add y_1^* and y_2^* together to obtain the particular solution of original equation

$$y^* = y_1^* + y_2^* = \left(\frac{1}{28}x - \frac{67}{784}\right)e^{5x} + \left(-\frac{1}{2}x^3 + \frac{3}{4}x^2 - \frac{5}{4}x - \frac{1}{8}\right)e^{-x},$$

therefore the general solution of the given equation is

$$y = C_1 e^{-2x} + C_2 e^x + \left(\frac{1}{28}x - \frac{67}{784}\right)e^{5x} + \left(-\frac{1}{2}x^3 + \frac{3}{4}x^2 - \frac{5}{4}x - \frac{1}{8}\right)e^{-x}.$$

2) The form $f(x) = e^{\alpha x}[P_l(x)\cos \beta x + P_n(x)\sin \beta x]$

The form of particular solution of second-order inhomogeneous linear differential equation with constant coefficients is

$$y^* = x^k \cdot e^{\alpha x}[Q_m^{(1)}(x)\cos \beta x + Q_m^{(2)}(x)\sin \beta x],$$

where $Q_m^{(1)}(x)$ and $Q_m^{(2)}(x)$ are both m-degree polynomials of undetermined coefficients. The degree m equals to the larger one of polynomial's degree l and n in $f(x)$, that is, $m = \max\{l, n\}$ and k is selected respectively as 1 or 0 according to the conditions that $\alpha \pm i\beta$ is the root of characteristic equation or not.

After determining the form of inhomogeneous equation's particular solution, plug it into the equation and determine polynomials $Q_m^{(1)}(x)$ and $Q_m^{(2)}(x)$ by comparing coefficients. Thus we can obtain the particular solution of the inhomogeneous equation.

Leave out the deduction of the particular solution's form, please readers complete the proof in details by yourselves or refer to related references.

Example 8 Find the general solution of

$$y'' + 3y' + 2y = e^{-x}\sin x.$$

Solution The characteristic equation of the corresponding homogeneous equation is

$$r^2 + 3r + 2 = 0,$$

whose roots are $r_1 = -1$ and $r_2 = -2$, so the general solution of the corresponding homogeneous equation is

$$y = C_1 e^{-x} + C_2 e^{-2x}.$$

The free term of the given inhomogeneous equation is
$$f(x)=e^{-x}\sin x=e^{-x}(0\cdot\cos x+1\cdot\sin x),$$
$m=\max\{0,0\}=0$ and $\alpha\pm i\beta=-1\pm i$ aren't the characteristic roots, so assume that
$$y^*=(a\cos x+b\sin x)e^{-x}.$$
Then
$$y^{*\prime}=e^{-x}[(b-a)\cos x-(a+b)\sin x],$$
$$y^{*\prime\prime}=e^{-x}(-2b\cos x+2a\sin x).$$
Plug into the original differential equation and we obtain
$$(b-a)\cos x-(a+b+1)\sin x=0,$$
so
$$b-a=0,\ a+b+1=0.$$
We have $a=b=-\dfrac{1}{2}$, therefore a particular solution of the given equation is
$$y^*=-\frac{1}{2}e^{-x}(\cos x+\sin x).$$
Thus the general solution is
$$y=C_1 e^{-x}+C_2 e^{-2x}-\frac{1}{2}e^{-x}(\cos x+\sin x).$$

*9.5.3 Vibration equation

As an example, we will discuss the differential equation in 9.4 which satisfies the vibration of spring (see Example 1 in 9.4) undamped free vibration equation, damped free vibration equation and forced vibration equation. They are all linear differential equations with constant coefficients, so we can use the methods introduced in this part to find their solutions and then analyze the laws of object's vibration.

1) Undamped free vibration equation

$$\frac{d^2 x}{dt^2}+k^2 x=0. \tag{8}$$

This is a second-order linear homogeneous differential equation with constant coefficients. Its characteristic equation $r^2+k^2=0$ has a pair of conjugate complex roots $r=\pm ki$, so its general solution is

$$x=C_1\cos kt+C_2\sin kt. \tag{9}$$

Let

$$\sqrt{C_1^2+C_2^2}=A,\ \frac{C_1}{\sqrt{C_1^2+C_2^2}}=\sin\delta,\ \frac{C_2}{\sqrt{C_1^2+C_2^2}}=\cos\delta,$$

then Equation (9) can also be written into

$$x=A\sin(kt+\delta). \tag{10}$$

We see that under the condition without resistance, the displacement of object which leaves away from equilibrium point makes periodic variation by sinusoidal function of time t. This motion is simple harmonic vibration. The amplitude of this vibration is A and the angular frequency is k. Because $k=\sqrt{\dfrac{c}{m}}$ (see Example 1 in 9.4), and it is only related to object's quality m and spring's modulus of elasticity c, it is completely determined by vibration system itself. Thus k is also called inherent frequency of system.

2) Damped free vibration equation

$$\frac{d^2 x}{dt^2} + 2n \frac{dx}{dt} + k^2 x = 0. \tag{11}$$

This equation is still a homogeneous linear equation with constant coefficients and the two roots of its characteristic equation $r^2 + 2nr + k^2 = 0$ are

$$r_1 = -n + \sqrt{n^2 - k^2}, \quad r_2 = -n - \sqrt{n^2 - k^2}.$$

(1) Small damping: $n < k$.

The roots of characteristic equation are a pair of conjugate complex roots

$$r_1 = -n + i\omega, \quad r_2 = -n - i\omega,$$

where $\omega = \sqrt{k^2 - n^2}$, so the general solution of differential Equation (11) is

$$x = e^{-nt}(C_1 \cos \omega t + C_2 \sin \omega t). \tag{12}$$

Let

$$\sqrt{C_1^2 + C_2^2} = A, \quad \frac{C_1}{\sqrt{C_1^2 + C_2^2}} = \sin \delta, \quad \frac{C_2}{\sqrt{C_1^2 + C_2^2}} = \cos \delta,$$

then (12) can be written into

$$x = A e^{-nt} \sin(\omega t + \delta). \tag{13}$$

The graph of function (13) is shown in Figure 9-4. It illustrates that when the resistance from medium is small, the object also makes alternate vibration. However, because amplitude $Ae^{-\beta t}$ decreases gradually with time t, the object gradually comes back to the equilibrium position as time passing.

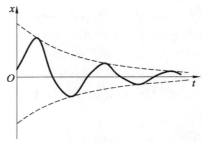

Figure 9-4

(2) Large damping: $n > k$.

The roots of characteristic equation are two unequal negative roots

$$r_1 = -n + \sqrt{n^2 - k^2}, \quad r_2 = -n - \sqrt{n^2 - k^2},$$

so the general solution of differential Equation (11) is

$$x = C_1 e^{-r_1 t} + C_2 e^{-r_2 t}. \tag{14}$$

From Equation (14), we can just find at most one value of t that makes $x = 0$. Since as $t \to +\infty$, we have $x \to 0$, which means the object gradually comes back to the equilibrium position with time increasing. That is to say, the motion of object will never have vibration phenomenon. The figure of function (14) can be seen in Figure 9-5a or 9-5b. (Figures are related to equation's initial conditions, so readers can analyze them with different initial conditions by yourselves.)

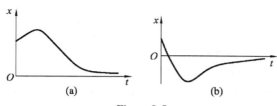

Figure 9-5

(3) Critical damping: $n=k$.

The roots of characteristic equation are two equal real roots $r_1 = r_2 = -n$, so the general solution of differential equation is

$$x = e^{-nt}(C_1 + C_2 t). \tag{15}$$

From the formula above, when the damping is critical, there is at most one value of t that makes $x=0$, so the system also has no vibration phenomenon. But as $t \to +\infty$, we have $e^{-nt}(C_1 + C_2 t) \to 0$, so the object also gradually comes back to the equilibrium position.

Please readers draw out the graph of function (14) by yourselves.

3) Forced vibration equation

$$\frac{d^2 x}{dt^2} + 2n \frac{dx}{dt} + k^2 x = h \sin pt. \tag{16}$$

For simplicity, we only discuss undamped forced vibration. Assume that $n=0$, then discuss the following equation

$$\frac{d^2 x}{dt^2} + k^2 x = h \sin pt. \tag{17}$$

This is a second-order inhomogeneous linear differential equation with constant coefficients, and its corresponding homogeneous equation is undamped free vibration Equation (8) whose characteristic equation's roots are a pair of conjugate complex roots $r = \pm ki$ and the general solution is

$$x = C_1 \cos kt + C_2 \sin kt = A \sin(kt + \delta).$$

Then find a particular solution of inhomogeneous Equation (17). Compare the function

$$f(t) = h \sin pt$$

in Equation (17) with

$$f(t) = e^{\alpha t}[P_l(t) \cos \beta t + P_n(t) \sin \beta t],$$

there are $\alpha = 0, \beta = p, P_l(t) = 0, P_n(t) = h$.

① If $p \neq k$, then $\alpha \pm i\beta$ aren't the roots of characteristic equation. Thus assume that

$$x^* = a \cos pt + b \sin pt.$$

Plug it into Equation (17) and obtain $a = 0, b = \dfrac{h}{k^2 - p^2}$, so

$$x^* = \frac{h}{k^2 - p^2} \sin pt.$$

Thus if $p \neq k$, the general solution of differential Equation (17) is

$$x = A\sin(kt+\delta) + \frac{h}{k^2-p^2}\sin pt. \tag{18}$$

The formula above illustrates the motion of object consists of two parts that are both simple harmonic vibration. The left term of Equation (18) refers to free vibration and the right term is called forced vibration. Forced vibration is caused by external force and its angular frequency is same to that of external force p. When the angular frequency of external force p is close to system's inherent frequency, its amplitude $\left|\frac{h}{k^2-p^2}\right|$ will be very large.

② If $p=k$, $\alpha \pm i\beta = \pm ip$ are roots of characteristic equation. Thus we assume that
$$x^* = t(a\cos kt + b\sin kt).$$
Plug it into Equation (17) and obtain $a = -\frac{h}{2k}, b=0$, so
$$x^* = -\frac{h}{2k}t\cos kt.$$
So when $p=k$, the general solution of differential Equation (17) is
$$x = A\sin(kt+\delta) - \frac{h}{2k}t\cos kt. \tag{19}$$

The second term on the right side illustrates that the amplitude of forced vibration $\frac{h}{2k}t$ increases infinitely as time t increasing. This is the so-called resonance phenomenon.

According to the results of ① and ② discussed above, it's supposed to make angular frequency of external force p away from system's inherent frequency in order to avoid resonance phenomenon, whereas, for resonance phenomenon, it should be $p=k$ or get p as close as possible to k.

Exercises 9-5

1. Find general solutions of the following differential equations:
   (1) $y'' + 6y' + 5y = 0$; (2) $y'' - 4y = 0$;
   (3) $y'' + 4y' + 4y = 0$; (4) $y'' + 2y' + 5y = 0$;
   (5) $y'' + 9y = 0$; (6) $y^{(4)} - 5y'' + 4y = 0$.

2. Find the following differential equations' particular solutions that satisfy initial conditions:
   (1) $y'' - 6y' + 8y = 0, y(0) = 1, y'(0) = 6$; (2) $y'' - 6y' + 9y = 0, y(0) = 1, y'(0) = 2$;
   (3) $y'' + 6y' + 13y = 0, y(0) = 3, y'(0) = -1$.

3. An integral curve of equation $y'' + 9y = 0$ passes through the point $(\pi, -1)$ and the tangent at this point is the straight line $y + 1 = x - \pi$. Please find this curve.

4. Write out particular solutions of the following differential equations:
   (1) $y'' + 3y' + 2y = (x+1)e^x$; (2) $y'' + 3y' - 4y = (2x^2+1)e^x$;
   (3) $y'' - 4y' + 4y = (3x+1)e^{2x}$; (4) $y'' - 2y' + 4y = e^{2x}(x\cos x + \sin x)$;
   (5) $y'' - 2y' + 5y = xe^x \sin 2x$.

5. Find the general solutions or particular solutions that satisfy initial conditions of the following differential equations:

(1) $y''-2y'=x+2, y(0)=0, y'(0)=1$;   (2) $y''+3y'+2y=3xe^{-x}$;
(3) $y''-3y'+2y=5, y(0)=1, y'(0)=2$;   (4) $y''+y+\sin 2x=0, y(\pi)=y'(\pi)=1$;
(5) $y''+y=4\sin x$;   (6) $y''+y=e^x+\cos x$.

6. Given that $f(0)=0$ and $f'(x) = 1 + \int_0^x [3e^t - f(t)]dt$, find function $f(x)$.

7. Assume that the function $\varphi(x)$ is continuous and satisfies $\varphi(x) = e^x + \int_0^x t\varphi(t)dt - x\int_0^x \varphi(t)dt$, find $\varphi(x)$.

8. Given that the equation $y''-5y'+6y=f(x)$. When $f(x)=1, x, e^x$, it respectively has particular solutions $\frac{1}{6}, \frac{1}{6}x+\frac{5}{36}, \frac{1}{2}e^x$. Find the general solution of $y''-5y'+6y=2-12x+6e^x$.

Summary

1. Main contents

This chapter mainly introduces basic concepts of differential equation, some main kinds of differential equations and their solutions.

(1) Basic concepts of differential equation.

Differential equation, order of differential equation, solution of differential equation, particular solution of differential equation, general solution of differential equation, initial conditions of differential equation.

(2) Some common first-order differential equation and their solutions.

① Separable equation. After separating variables, it is $g(y)dy = f(x)dx$, then integrate both sides to find the general solution.

② Homogeneous equation $\frac{dy}{dx} = f\left(\frac{y}{x}\right)$. Through the variable substitution $u = \frac{y}{x}$, it can be written into separable equation.

③ First-order linear equation. For first-order linear equation $\frac{dy}{dx}+P(x)y=0$, solve it by separating variables. For first-order linear inhomogeneous differential equation $\frac{dy}{dx}+P(x)y=Q(x)$, we can find the general solution by variation of parameters or general solution formula

$$y = e^{-\int P(x)dx}\left(\int Q(x)e^{\int P(x)dx}dx + C\right).$$

④ Bernoulli equation $\frac{dy}{dx} + P(x)y = Q(x)y^n$ $(n \neq 0, 1)$. Through the variable substitution $z = y^{1-n}$, it can be written into first-order linear equation.

(3) Three kinds of high-order differential equation.

$y^{(n)} = f(x)$, $y'' = f(x, y')$ and $y'' = f(y, y')$. Use variable substitution to rewrite them into first-order differential equations.

(4) Properties and structures of high-order linear differential equation's solution.

The general solution of second-order homogeneous linear differential equation can be expressed as $y = C_1 y_1(x) + C_2 y_2(x)$ where $y_1(x)$ and $y_2(x)$ are two linearly independent particular solutions. The general solution of inhomogeneous linear equation can be expressed as $y(x) = Y(x) + y^*(x)$. $Y(x)$ is the general solution of its corresponding homogeneous equation and $y^*(x)$ is the particular solution of inhomogeneous equation.

(5) Methods of calculating the general solution of second-order homogenous linear differential equation with constant coefficients.

① Write out related characteristic equation of differential equation
$$r^2 + pr + q = 0.$$

② Find two roots of characteristic equation r_1 and r_2.

③ According to different conditions of the two roots, write out the general solution of differential equation by the following table.

Two roots r_1, r_2 of characteristic equation $r^2 + pr + q = 0$	The general solution of differential equation $y'' + py' + qy = 0$
Real roots $r_1 \neq r_2$	$y = C_1 e^{r_1 x} + C_2 e^{r_2 x}$
Real roots $r_1 = r_2 = r$	$y = (C_1 + C_2 x) e^{r x}$
A pair of conjugate complex roots $r_{1,2} = \alpha \pm i\beta$	$y = e^{\alpha x}(C_1 \cos \beta x + C_2 \sin \beta x)$

(6) If the free term is one of the two common forms, the method of calculating general solution of second-order inhomogeneous linear differential equation with constant coefficients is as follows.

① Find the general solution $Y(x)$ of corresponding homogeneous differential equation.

② Find a particular solution $y^*(x)$ of inhomogeneous equation.

If the free term on the right side is $f(x) = P_m(x) e^{\lambda x}$, the particular solution is
$$y^* = x^k Q_m(x) e^{\lambda x},$$
where $Q_m(x)$ is a polynomial which has the same degree with $P_m(x)$, k is respectively selected as 0, 1 or 2 according to conditions that λ isn't the root of characteristic equation, λ is the simple root of characteristic equation or λ is the double root of characteristic equation.

If $f(x) = e^{\alpha x}[P_l(x) \cos \beta x + P_n(x) \sin \beta x]$, the particular solution is
$$y^* = x^k \cdot e^{\alpha x}[Q_m^{(1)}(x) \cos \beta x + Q_m^{(2)}(x) \sin \beta x],$$
where $Q_m^{(1)}(x)$ and $Q_m^{(2)}(x)$ are both m-degree polynomials, $m = \max\{l, n\}$, k is respectively selected as 1, 0 according to conditions that $\alpha \pm i\beta$ are the roots of characteristic equation or not.

③ The general solution of inhomogeneous equation is $y = Y(x) + y^*(x)$.

*(7) The example about the application of high-order linear differential equation: vibration equation.

2. Basic requirements

(1) Understand these concepts including differential equation, order of differential equation, solution of differential equation, general solution of differential equation, particular solution of differential equation and initial conditions of differential equation.

(2) Master separable equation and methods of solving first-order linear equation.

(3) Be able to solve homogeneous equation and Bernoulli equation, then understand the thought of variable substitution.

(4) Be able to solve the following equations by decreasing order, $y^{(n)} = f(x)$, $y'' = f(x, y')$ and $y'' = f(y, y')$.

(5) Understand the structure of second-order linear differential equation's solution.

(6) Master methods of solving second-order homogeneous linear differential equation with constant coefficients and know the way to calculate high-order homogeneous linear differential equation with constant coefficients.

(7) Master methods of solving the particular solution of second-order inhomogeneous linear differential equation with constant coefficients whose free term is $P_m(x) e^{\lambda x}$ or $e^{\alpha x} [P_l(x) \cos \beta x + P_n(x) \sin \beta x]$.

(8) Be able to use differential equations to solve some simple geometric and physical problems.

Quiz

1. Find the general solutions or particular solutions that satisfy initial conditions:

   (1) $\sqrt{1-y^2} \, dx + y\sqrt{1-x^2} \, dy = 0$;

   (2) $\dfrac{dy}{dx} = \dfrac{y}{x - \sqrt{xy}}$, $y(1) = 1$;

   (3) $x \dfrac{dy}{dx} + (1+x) y = 3x^2 e^{-x}$, $y(1) = \dfrac{2}{e}$;

   (4) $\dfrac{dy}{dx} = \dfrac{y}{x} + \dfrac{y^2}{x^3}$;

   (5) $y' = xy'' + (y'')^2$;

   (6) $y'' = 2yy'$, $y(0) = 1$, $y'(0) = 2$.

2. Given that a differentiable function $f(x)$ satisfies the equation
$$\int_1^x \frac{f(t)}{f^2(t) + t} dt = f(x) - 1,$$
find the expression of function $f(x)$.

3. Find the general solutions or particular solutions that satisfy initial conditions:

   (1) $y'' - y' - 6y = 0$, $y(0) = 1$, $y'(0) = -1$;
   (2) $y'' + 6y' + 9y = 0$;
   (3) $y'' - a^2 y = x + 1$;
   (4) $y'' - 2y' + y = xe^x$, $y(0) = y'(0) = 0$;
   (5) $y'' + 4y = \cos 2x$;
   (6) $y'' - 2y' + 2y = 4e^x \cos x$.

Exercises

1. Gap filling.

   (1) $3xy''' + 2x(y')^2 + 5x^4 y = x^4 + \sin x$ is _____-order differential equation.

(2) Assume that first-order linear differential equation is $y'+P(x)y=Q(x)$.

① When $Q(x)=0$, its general solution is _____ ;

② When $Q(x)\neq 0$, its general solution is _____ .

(3) Assume that second-order linear differential equation $y''+P(x)y'+Q(x)y=f(x)$ ($f(x)$ doesn't equal to 0 identically) has two solutions $y_1(x)$, $y_2(x)$. If $\alpha_1 y_1(x)+\alpha_2 y_2(x)$ is also the solution to this equation, then $\alpha_1+\alpha_2=$ _____ .

(4) Given that $y=1$, $y=x$, $y=x^2$ are three solutions of a second-order inhomogeneous linear differential equation, then the general solution of this equation is _____ .

2. Find general solutions of the following differential equations:

(1) $ydx-xdy=x^2 ydy$;

(2) $xy'+y=2\sqrt{xy}$;

(3) $(e^{x+y}-e^x)dx+(e^{x+y}+e^y)dy=0$;

(4) $y\sin x+\cos x \dfrac{dy}{dx}=1$;

(5) $(x-2)\dfrac{dy}{dx}-y=2(x-2)^3$;

(6) $y'+f'(x)y=f(x)f'(x)$;

(7) $xy'+y=y(\ln x+\ln y)$;

(8) $\dfrac{dy}{dx}+xy-x^3 y^3=0$;

(9) $yy''-(y')^2-y^2 y'=0$;

(10) $y'''+y''-2y'=x(e^x+4)$;

(11) $y''+2y'+5y=\sin 2x$;

(12) $y''-4y'+4y=e^x+e^{2x}+1$.

3. Find the particular solutions that satisfy initial conditions:

(1) $y^3 dx+2(x^2-xy^2)dy=0$, $y(1)=1$;

(2) $y''-3y'+2y=e^{3t}$, $y(0)=1$, $y'(0)=0$;

(3) $2y''-\sin 2y=0$, $x=0$, $y(0)=\dfrac{\pi}{2}$, $y'(0)=1$;

(4) $y''+2y'+y=\cos x$, $x=0$, $y(0)=0$, $y'(0)=\dfrac{3}{2}$.

4. Assumed that differentiable function $f(x)$ satisfies

$$f(x)\cos x + 2\int_0^x f(t)\sin t\, dt = x+1,$$

then find the expression of $f(x)$.

5. Assume that one object whose quality is m starts to fall down in the air. If the air resistance is $R=kv$ (k is constant and v refers to velocity of the object), find the functional relation between distance s and time t.

6. Assume that $y_1(x)$, $y_2(x)$ are two solutions of second-order homogeneous linear equation $y''+p(x)y'+q(x)y=0$ and let

$$W(x)=y_1(x)y_2'(x)-y_1'(x)y_2(x).$$

Prove: (1) $W(x)$ satisfies equation $W'+p(x)W=0$; (2) $W(x)=W(x_0)e^{-\int_{x_0}^x p(x)dx}$.

Chapter 10 Vectors and analytic geometry of space

Analytic geometry studies the relationship of figures and numbers. In high school, we learned analytic geometry of plane, which is constituted on the basis of orthogonal plane coordinate system and point in the plane is corresponding to a pair of real numbers. Based on this, we get the straight lines and curves of the plane corresponding to equations. Thus we can use algebraic methods to study geometric problems, and also use geometric figures to explain some phenomenon arisen in the research of quantitative relations. It can be known from the study of previous chapters that the knowledge of analytic geometry of plane is necessary for studying calculus of unary function. Similarly, in order to study calculus of multivariate function, we should master the knowledge of analytic geometry of space which constitutes the relationship of three-dimensional figure and quantity.

The analytic geometry of plane uses quantity algebra as tool to study geometric problems, while the analytic geometry of space uses vectors as tool to constitute the equation of plane and spatial line when studying the relationship of their figures and quantities.

This chapter firstly introduces the space orthogonal coordinate system and vectors, then use vectors to discuss planes and spatial lines.

10.1 Space right angle coordinate system

10.1.1 The space right angle coordinate system

In high school, we learned one-dimensional coordinate system (Number line) and plane right angle coordinate system to research geometric problems on the plane. In order to use algebraic methods to study three-dimensional figures, firstly we introduce space right angle coordinate system. Space right angle coordinate system is the generalization of plane right angle coordinate system.

In space, we draw three axes Ox, Oy and Oz which intersect at point O and are mutually perpendicular, and they all have the same unit of length. The intersection point O

is called the origin of coordinates and the three axes, whose joint name is coordinate axis, which are respectively called x-axis (abscissa axis), y-axis (ordinate axis) and z-axis (vertical axis) (Figure 10-1). The direction of z-axis is determined by the right-hand rule that use right hand to grasp z-axis, and the direction of right thumb is the direction of z-axis when other four fingers turn $\frac{\pi}{2}$ from the direction of the x-axis to that of the y-axis (Figure 10-2). These three axes form a space right angle coordinate system which is denoted by $O\text{-}xyz$.

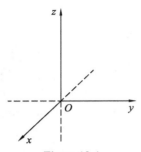

Figure 10-1

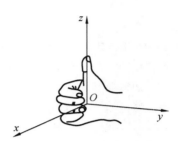

Figure 10-2

In the space right angle coordinate system, the plane determined by any two coordinate axes is called coordinate plane. Obviously there are three coordinate planes that are respectively called plane xOy, plane yOz and plane xOz. These three coordinate planes divide space into eight parts and each part is called an octant. The eight octants are denoted by letters I, II, III, IV, V, VI, VII, VIII respectively (Figure 10-3) and the octants I, II, III, IV are above the plane xOy. The octant which includes positive part of x-axis, y-axis and z-axis is called octant I and the others are determined anticlockwise. The octants V, VI, VII, VIII are below the plane xOy which are named similar to the four octants above in turn.

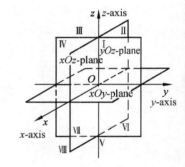

Figure 10-3

10.1.2 Right angle coordinate of spatial point

Assume that P is some point of space right angle coordinate system.

Passing through the point P we draw three planes which are perpendicular to x-axis, y-axis and z-axis respectively, and the intersection points in three coordinate axes are denoted by P_x, P_y and P_z respectively (Figure 10-4). The coordinates of the three intersection points on x-axis, y-axis and z-axis are respectively x, y and z. Therefore, spatial point P uniquely determines a group of orderly real numbers x, y and z.

Conversely, for a group of orderly real numbers x, y and z, we can find the corresponding points P_x, P_y and P_z on the axes. Then passing through these three points we draw the planes which are

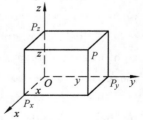

Figure 10-4

perpendicular to x-axis, y-axis and z-axis. These three planes will intersect at the unique point P (Figure 10-4).

In the right angle coordinate system, we have the one-to-one correspondence between spatial point P and the orderly array x, y and z. This group of orderly real numbers is called the right angle coordinates of spatial point P or the coordinates of P, which is denoted by $P(x, y, z)$. x, y and z are respectively called the abscissa axis, ordinate axis and vertical axis of the point P.

Now we get the relationship of the spatial figure "point" and the "numbers" (x,y,z), that is, for any spatial point P, there is (x,y,z). For example, the coordinate of the origin is $O(0,0,0)$, then the positive unit points on three coordinate axes are $(1,0,0)$, $(0,1,0)$ and $(0,0,1)$ respectively. We also get the following properties. The point on coordinate axis has two coordinates zero. For example, the coordinate of the point on x-axis is $(x,0,0)$ where $y=z=0$. The point on coordinate plane has one coordinate zero. For example, the coordinate of point on the plane yOz is $(0,y,z)$ where $x=0$. Whereas, the point with two coordinates zero must be on the coordinate axis and the point with one coordinate zero must be on the coordinate plane. In addition, in eight octants, signs of point's coordinates are also different but we have a rule to follow which is shown in the table.

Octant	I	II	III	IV	V	VI	VII	VIII
Sign	$(+,+,+)$	$(-,+,+)$	$(-,-,+)$	$(+,-,+)$	$(+,+,-)$	$(-,+,-)$	$(-,-,-)$	$(+,-,-)$

10.1.3 Distance between two points in space

After constituting a space right angle coordinate system and determining the coordinates of points, it infer the formula of distance between two points in space.

Assume that $P_1(x_1, y_1, z_1)$, $P_2(x_2, y_2, z_2)$ are two points in space and the distance between them is denoted by $d=|P_1P_2|$.

Passing through the points P_1, P_2 we draw six planes perpendicular to three coordinate axes respectively, then the six planes create a cuboid with diagonal line P_1P_2 (Figure 10-5). According to the Pythagorean theorem, we find the length of cuboid's diagonal line.

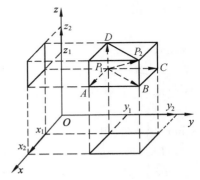

Figure 10-5

Known from Figure 10-5,
$|P_1A|=|x_2-x_1|$, $|AB|=|y_2-y_1|$, $|BP_2|=|z_2-z_1|$.
For the $\text{Rt}\triangle P_1AB$,
$$|P_1B|^2=|P_1A|^2+|AB|^2,$$
and in the $\text{Rt}\triangle P_1BP_2$,
$$|P_1P_2|^2=|P_1B|^2+|BP_2|^2,$$
then
$$d^2=|P_1P_2|^2=|P_1A|^2+|AB|^2+|BP_2|^2$$
$$=(x_2-x_1)^2+(y_2-y_1)^2+(z_2-z_1)^2,$$

therefore
$$d = |P_1 P_2| = \sqrt{(x_2-x_1)^2+(y_2-y_1)^2+(z_2+z_1)^2}. \quad (1)$$

This is the formula of distance between two points in space and it is the generalization of distance in the plane.

In particular, the distance between an arbitrary point $P(x,y,z)$ and the origin point $O(0,0,0)$ is
$$d = |OP| = \sqrt{x^2+y^2+z^2}. \quad (2)$$

Example 1 Show that the triangle with vertexes $A(4,1,9)$, $B(10,-1,6)$, $C(2,4,3)$ is isosceles right triangle.

Proof
$$|AB|^2 = (10-4)^2+(-1-1)^2+(6-9)^2 = 49,$$
$$|AC|^2 = (2-4)^2+(4-1)^2+(3-9)^2 = 49,$$
$$|BC|^2 = (2-10)^2+(4+1)^2+(3-6)^2 = 98.$$

And since
$$|AB|^2+|AC|^2 = |BC|^2, \quad |AB| = |AC|,$$
the $\triangle ABC$ is an isosceles right triangle.

Example 2 Find the distances from point $P(1,2,4)$ to three coordinates axes.

Solution Passing through the point P we draw lines perpendicular to the coordinate axes and the intersection points are $A(1,0,0), B(0,2,0)$ and $C(0,0,4)$ respectively.

Thus the distances from P to three coordinate axes are respectively
$$d_x = |PA| = \sqrt{0^2+2^2+4^2} = 2\sqrt{5},$$
$$d_y = |PB| = \sqrt{1^2+0^2+4^2} = \sqrt{17},$$
$$d_z = |PC| = \sqrt{1^2+2^2+0^2} = \sqrt{5}.$$

Example 3 Suppose a moving point $P(x,y,z)$ keeps equivalent distances away from two fixed points $O(0,0,0)$ and $M(a,b,c)$. Find the equation of the trajectory of P.

Solution Assume that $P(x,y,z)$ is an arbitrary point and there is $|PO|=|PM|$.

Because
$$|PO| = \sqrt{x^2+y^2+z^2},$$
$$|PM| = \sqrt{(x-a)^2+(y-b)^2+(z-c)^2},$$
we have
$$2ax+2by+2cz = a^2+b^2+c^2.$$

We can also write it in short
$$Ax+By+Cz = D,$$
where $A=2a, B=2b, C=2c$ and $D=a^2+b^2+c^2$.

This is a plane which is called perpendicular bisected plane of the line segment OM.

Example 4 Suppose a moving point $P(x,y,z)$, the distance R from P to a fixed point $P_0(x_0, y_0, z_0)$ always remains unchanged and the trajectory of point P is called a spherical surface. Find the equation of the spherical surface.

Solution Assume that $P(x,y,z)$ is an arbitrary point on the spherical surface, we have
$$|P_0 P| = R,$$

or
$$\sqrt{(x-x_0)^2+(y-y_0)^2+(z-z_0)^2}=R.$$

Squared on both sides of the equation, we obtain the equation of spherical surface
$$(x-x_0)^2+(y-y_0)^2+(z-z_0)^2=R^2, \tag{3}$$
and $P_0(x_0, y_0, z_0)$ is called the center of spherical surface and R is called the spherical radius.

In particular, the spherical equation with center $O(0,0,0)$ is
$$x^2+y^2+z^2=R^2. \tag{4}$$

Expand the spherical Equation (3), we have
$$x^2+y^2+z^2-2x_0 x-2y_0 y-2z_0 z+(x_0^2+y_0^2+z_0^2-R^2)=0.$$

Thus the spherical equation is a ternary quadratic equation with same coefficients of all quadratic terms excluding cross terms. Whereas, if such a ternary quadratic equation can be rewritten into the form of Equation (3), its figure is a spherical surface.

Exercises 10-1

1. Point out the characteristics of the following points' coordinates.

(1) P is on the coordinate axis;

(2) P is on the coordinate plane;

(3) P is on the plane which is parallel to plane xOz and the distance of the two planes is 3;

(4) P is on the plane which is perpendicular to z-axis and the distance from the plane to origin is 5.

2. Point out the characteristics of the following points' position, $A(3,0,1)$, $B(0,1,2)$, $C(0,0,1)$ and $D(0,-2,0)$.

3. In space right angle coordinate system, which octant does the following point belong to? $A(1,-2,3), B(2,3,-4), C(2,-3,-4), D(-2,-3,1)$.

4. In space right angle coordinate system, find the symmetry points of the two points, $P(2,-3,-1)$ and $M(a,b,c)$, with respect to (1) coordinate planes; (2) coordinate axes; (3) origin.

5. Show that the triangle with vertexes $A(4,3,1)$, $B(7,1,2)$, $C(5,2,3)$ is an isosceles triangle.

6. On the plane yOz, find the point which has same distance to the points $A(3,1,2)$, $B(4,-2,-2)$ and $C(0,5,1)$.

7. Find a spherical equation passing through the origin with center $(1,3,-2)$.

8. Find the centre and spherical radius of $x^2+y^2+z^2+2x-4y-4=0$.

10.2 Vector algebra

10.2.1 Concepts of vector

In dynamics, physics and our daily life, there are many kinds of quantities. Except that those quantities like temperature, time, length, area, volume and so on which are called scalar just having magnitude but no direction, there are also some complex quantities like force, displacement, velocity and so on which have magnitude and direction, which are called vector.

> **Definition 1** The quantity which has magnitude and direction is called a vector.

We always use a directed line segment to represent a vector, the length of the line segment represents the magnitude of the vector and the direction from starting point P_1 to terminal point P_2 represents the direction of vector, so the vector is always denoted by $\overrightarrow{P_1P_2}$. This kind of representation is called geometric representation of vector (Figure 10-6). In the textbook, we also use the bold-type letters or a letter with an arrow to represent the vector, such as a, b, x or $\vec{a}$, $\vec{b}$, $\vec{x}$ and so on. In some problems, we just consider the magnitude and direction of a vector and ignore its starting position. This kind of vector is called free vector. That is to say, the free vector will be the same one after parallel moving and the equal vectors can be regarded as the same free vector. Due to the arbitration of free vector's starting point, we could select a point as the common starting point of some vectors for researching. In this chapter, unless specifically specified, vectors discussed are all free vectors.

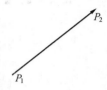

Figure 10-6

The magnitude of vector is called the norm of vector, denoted by $|\overrightarrow{P_1P_2}|$ or $|a|$. The vector with norm 1 is called an unit vector. The vector with norm 0 is called a null vector, denoted by **0** and the direction of **0** is arbitrary. If the norms and directions of two vectors a and b are same, then the two vectors are called equal vectors, denoted by $a=b$. Assume that a is a vector, the vector which has the same magnitude of a and in the opposite direction is called the negative vector (or reverse vector) of a, denoted by $-a$. If two nonzero vectors have the same or opposite directions, then the two vectors are parallel or collinear, denoted by $a \parallel b$. A group of vectors parallel to the same plane is called coplanar vectors. Obviously, null vector is coplanar with arbitrary group of coplanar vector. A group of collinear vectors must be coplanar vector. If two of three vectors are collinear, then three vectors are also coplanar.

10.2.2 Linear operations of vector

1) Addition and subtraction of vectors

(1) Parallelogram law and triangle law of vector's addition.

There are two vectors a and b. Choose a fixed point O and let $\overrightarrow{OA}=a$, $\overrightarrow{OB}=b$. Then let $\overrightarrow{OA}$, $\overrightarrow{OB}$ be adjacent sides, we draw the parallelogram $OACB$ (Figure 10-7). Diagonal vector $\overrightarrow{OC}=c$ is called the sum of vectors a and b, denoted by

$$c=a+b.$$

The method of obtaining the sum of two vectors is called the parallelogram law of vector's addition.

Chapter 10　Vectors and analytic geometry of space

Given that two vectors a and b. Select a fixed point O and let $\overrightarrow{OA}=a$. Regard the terminus A of $\overrightarrow{OA}$ as starting point to draw $\overrightarrow{OA}=b$. Then we can obtain $a+b=c=\overrightarrow{OC}$ by connecting O and C (Figure 10-8). This method is called the triangle law of vector's addition.

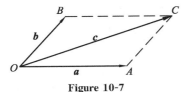

Figure 10-7

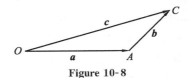

Figure 10-8

Vector's addition satisfies the following rules.

① Commutative law　$a+b=b+a$.

② Associative law　$(a+b)+c=a+(b+c)$.

(2) Polygonal law of the sum of several vectors.

Because vector's addition satisfies commutative law and associative law, triangle law infers the sum of finite vectors $a_1, a_2, \cdots, a_n$. Start from arbitrary point O and draw in turn
$$\overrightarrow{OA_1}=a_1, \overrightarrow{A_1A_2}=a_2, \cdots, \overrightarrow{A_{n-1}A_n}=a_n,$$
then we obtain a broken line $OA_1A_2\cdots A_n$ (Figure 10-9). Thus the vector $\overrightarrow{OA_n}=a$ is the sum of n vectors

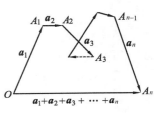

Figure 10-9

$$a=a_1+a_2+\cdots+a_n.$$

This method is also called the polygonal law of finding the sum of several vectors.

(3) Vector's subtraction.

By negative vector, we could stipulate the subtraction of two vectors.

If $b+c=a$, then
$$c=a-b=a+(-b).$$

Then, we obtain two useful conclusions.

① Give vector $\overrightarrow{AB}$ and point O at random, there will be (Figure 10-10)
$$\overrightarrow{AB}=\overrightarrow{AO}+\overrightarrow{OB}=\overrightarrow{OB}-\overrightarrow{OA}.$$

② If use a and b as adjacent sides to draw parallelogram, then $a+b$ and $a-b$ are two diagonal vectors (Figure 10-11).

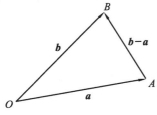

Figure 10-10

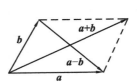

Figure 10-11

Because of the principle that the sum of two triangle's sides is longer than the third one, there is
$$|a+b|\leqslant|a|+|b|, \quad |a-b|\leqslant|a|+|b|,$$

93

and equal mark stands up if and only if a and b are collinear.

2) Scalar multiplication of vectors

Assume that k is a quantity, then the product ka of vector a and quantity k is also a vector. Its norm is $|ka|=|k||a|$ and if $k>0$, ka and a have the same direction, while if $k<0$, ka and a have the opposite directions, if $k=0$ or $a=0$, then $ka=0$.

The multiplication of quantity and vector satisfies the following rules.

(1) Associative law $\quad \lambda(\mu a) = \mu(\lambda a) = (\lambda \mu) a$.

It can be known from the stipulation of multiplication of quantity and vector that the vectors $\lambda(\mu a), \mu(\lambda a)$ and $(\lambda \mu) a$ have the same direction and their norms

$$|\lambda(\mu a)| = |\mu(\lambda a)| = |(\lambda \mu) a| = |\lambda \mu||a|,$$

so $\lambda(\mu a) = \mu(\lambda a) = (\lambda \mu) a$.

(2) Distributive law $\quad (\lambda + \mu) a = \lambda a + \mu a$;

$$\lambda(a+b) = \lambda a + \lambda b.$$

According to the definition of scalar multiplication of vectors, ka is a vector parallel to a. Then we can obtain the following conclusions.

① Assume that $a°$ is an unit vector which has the same direction of a, then

$$a = |a| a° \left(\text{or } a° = \frac{a}{|a|} \right).$$

② Assume that $a \neq 0$, then $b /\!/ a$ if and only if there is an unique real number k such that

$$b = ka.$$

Proof The sufficient condition is obvious, so we just prove the necessary condition here.

Assume that $b /\!/ a$. Select $|k| = \frac{|b|}{|a|}$, and k is positive if a and b have the same direction while k is negative if a and b have the opposite directions, so the vectors b and ka have the same direction and

$$|ka| = |k||a| = \frac{|b|}{|a|}|a| = |b|,$$

then $b = ka$.

Then prove the uniqueness of k. Assume that $b = ka$ and $b = \lambda a$, we have

$$(\lambda - k) a = 0 \quad \text{or} \quad |\lambda - k||a| = 0.$$

Because $|a| \neq 0$, we have $|\lambda - k| = 0$. So $\lambda = k$.

Addition and subtraction of vectors and scalar multiplication of vectors are together called the linear operations of vectors, such as $2a + 3b - 4c, k_1 a + k_2 b$, and so on.

Example 1 Assume that $a = 2e_1 + 3e_2 + 5e_3, b = -e_1 - e_3, c = 4e_2 - 2e_3$, find $2a + 3b - 2c$.

Solution $2a + 3b - 2c = 2(2e_1 + 3e_2 + 5e_3) + 3(-e_1 - e_3) - 2(4e_2 - 2e_3)$
$= e_1 - 2e_2 + 11e_3.$

Example 2 Given that $a = e_1 + e_2 + 2e_3, b = -e_1 + e_3, c = -2e_1 - e_2 - e_3$, show that a,

b, c constitute a triangle.

Solution Because
$$b-a=(-e_1+e_3)-(e_1+e_2+2e_3)=-2e_1-e_2-e_3=c \text{ (or } a+c=b).$$

According to the triangle law of vector's addition and subtraction, we know that a, b, c constitute a triangle.

10.2.3 Coordinates of vector

For the algebraization of vector's operations, it needs to constitute algebraic expression of vector. Thus firstly we introduce the concept of vector a's coordinates, then use the coordinates of vector to obtain the quantitative expression of vector's norm and direction. At first, introduce the concept of vector's projection on the axis.

1) Vector's projection on the axis

Given that an axis u and vector $\overrightarrow{AB}$, passing through the points A and B we draw two planes perpendicular to the axis u. The intersection points of the two planes and the axis u, A' and B' are called projections of points A and B on axis u (foot of a perpendicular) respectively (Figure 10-12) and $\overrightarrow{A'B'}$ is called the projected vector on axis u.

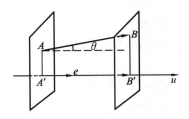

Figure 10-12

If select an unit vector e which has the same direction with axis u, then there are
$$\overrightarrow{A'B'} /\!/ e, \quad \overrightarrow{A'B'}=xe.$$

The quantity x is called vector $\overrightarrow{AB}$'s projection on axis u and denoted by
$$\text{Prj}_u \overrightarrow{AB}=x. \tag{1}$$

If $\overrightarrow{A'B'}$ has the same direction with axis u, then $x>0$, and if $\overrightarrow{A'B'}$ has the opposite direction with axis u, then $x<0$.

As for vector's projection on axis, we can obtain the following two properties.

Property 1 Vector $\overrightarrow{AB}$'s projection on axis u equals to the product of vector's norm $\overrightarrow{AB}$ and the cosine of angle φ between vector $\overrightarrow{AB}$ and axis u's positive direction (Figure 10-12),
$$\text{Prj}_u \overrightarrow{AB}=|\overrightarrow{AB}| \cdot \cos \varphi. \tag{2}$$

Property 2 Vector's projection on the axis keeps linear operations (Figure 10-13),
$$\text{Prj}_u(a+b)=\text{Prj}_u a+\text{Prj}_u b, \tag{3}$$
$$\text{Prj}_u(\lambda a)=\lambda \cdot \text{Prj}_u a. \tag{4}$$

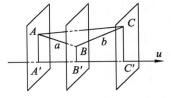

Figure 10-13

In particular, if the point O of vector $\overrightarrow{OA}$ is on axis u, to find the projection of $\overrightarrow{OA}$, it just needs to find the projection of point A on axis u.

2) Coordinates of vector

In space right angle coordinate system O-xyz, select unit vectors $\boldsymbol{i}$, $\boldsymbol{j}$, $\boldsymbol{k}$ on three coordinate axes respectively. Put vector $\boldsymbol{a}$'s starting point on the origin point O and assume that vector's terminus is P, that is, $\boldsymbol{a}=\overrightarrow{OP}$.

Assume that point P's projections (foot of a perpendicular) on three coordinate axes are P_x, P_y, P_z respectively and its projections (foot of a perpendicular) on plane xOy is P_0 (Figure 10-14), then there is
$$\boldsymbol{a}=\overrightarrow{OP}=\overrightarrow{OP_x}+\overrightarrow{P_xP_0}+\overrightarrow{P_0P}=\overrightarrow{OP_x}+\overrightarrow{OP_y}+\overrightarrow{OP_z}.$$

Because $\overrightarrow{OP_x}$ is parallel to $\boldsymbol{i}$, there exists an unique x such that $\overrightarrow{OP_x}=x\boldsymbol{i}$. Similarly, we can obtain the unique y and z such that $\overrightarrow{OP_y}=y\boldsymbol{j}$, $\overrightarrow{OP_z}=z\boldsymbol{k}$. Thus
$$\boldsymbol{a}=x\boldsymbol{i}+y\boldsymbol{j}+z\boldsymbol{k},$$

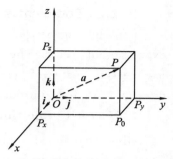

Figure 10-14

and ordered array x, y, z are called the components or coordinates of vector $\boldsymbol{a}=\overrightarrow{OP}$, which is denoted by $\boldsymbol{a}=x\boldsymbol{i}+y\boldsymbol{j}+z\boldsymbol{k}=(x,y,z)$.

According to the definition of vector's coordinate, it's easy to know that $\overrightarrow{OP_x}$, $\overrightarrow{OP_y}$, $\overrightarrow{OP_z}$ are projected vectors of $\boldsymbol{a}=\overrightarrow{OP}$ on three coordinate axes and x, y, z are vector $\boldsymbol{a}$'s projections on three coordinate axes. This is the geometric meaning of vector's three coordinates in right angle coordinate system. We can know from the definition of spatial point's right angle coordinates in 10.1 that x, y and z are also the coordinates of spatial point P, $P(x,y,z)$. So, there exists the one-to-one correspondence among spatial point P, vector $\overrightarrow{OP}$ and coordinates x,y,z.

3) Use coordinates into the vector's linear operations

Assume that $\boldsymbol{a}=(x_1,y_1,z_1)$ and $\boldsymbol{b}=(x_2,y_2,z_2)$, we obtain the following rules

(1) $\boldsymbol{a}+\boldsymbol{b}=(x_1+x_2)\boldsymbol{i}+(y_1+y_2)\boldsymbol{j}+(z_1+z_2)\boldsymbol{k}=(x_1+x_2,y_1+y_2,z_1+z_2)$;

(2) $\boldsymbol{a}-\boldsymbol{b}=(x_1-x_2)\boldsymbol{i}+(y_1-y_2)\boldsymbol{j}+(z_1-z_2)\boldsymbol{k}=(x_1-x_2,y_1-y_2,z_1-z_2)$;

(3) $\lambda\boldsymbol{a}=\lambda x_1\boldsymbol{i}+\lambda y_1\boldsymbol{j}+\lambda z_1\boldsymbol{k}=(\lambda x_1,\lambda y_1,\lambda z_1)$;

(4) The necessary and sufficient condition of $\boldsymbol{a}/\!/\boldsymbol{b}$ is $\dfrac{x_1}{x_2}=\dfrac{y_1}{y_2}=\dfrac{z_1}{z_2}$.

Note In rule (4), if one of x_2,y_2,z_2 is 0, for example $x_2=0$, it means that $x_1=0$ and $\dfrac{y_1}{y_2}=\dfrac{z_1}{z_2}$. If two of x_2,y_2,z_2 are 0, for example $x_2=y_2=0$ and $z_2\neq 0$, it means that $x_1=0$ and $y_1=0$.

This shows that it only needs to do related quantitative operations on vector's coordinates when do vector's addition, subtraction and scalar-multiplication's linear operations.

Example 3 Assume that $\boldsymbol{a}=(3,5,-1)$, $\boldsymbol{b}=(2,2,3)$, $\boldsymbol{c}=(2,-1,-3)$ calculate
$$2\boldsymbol{a}-3\boldsymbol{b}+4\boldsymbol{c}.$$

Solution $2\boldsymbol{a}-3\boldsymbol{b}+4\boldsymbol{c}=2(3,5,-1)-3(2,2,3)+4(2,-1,-3)=(8,0,-23)$.

Example 4 Given that $P_1(x_1,y_1,z_1), P_2(x_2,y_2,z_2)$, calculate $\overrightarrow{P_1P_2}$.

Solution Because $\overrightarrow{P_1P_2}=\overrightarrow{OP_2}-\overrightarrow{OP_1}$ (Figure 10-15), and $\overrightarrow{OP_2}=(x_2,y_2,z_2), \overrightarrow{OP_1}=(x_1,y_1,z_1)$, by vector's subtraction, we obtains

$$\overrightarrow{P_1P_2}=(x_2-x_1,y_2-y_1,z_2-z_1). \tag{5}$$

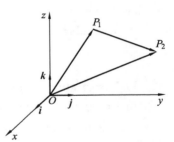

Figure 10-15

This example illustrates that coordinates of vector $\overrightarrow{P_1P_2}$ equals to the difference of coordinates of end point and the coordinates of start point.

Known from Example 4, if $\boldsymbol{a}=\overrightarrow{P_1P_2}$, then $\boldsymbol{a}$'s three projections are $x_2-x_1, y_2-y_1, z_2-z_1$. For convenience, write them respectively as $a_x=x_2-x_1, a_y=y_2-y_1, a_z=z_2-z_1$.

Thus $\boldsymbol{a}$ can be also expressed as $\boldsymbol{a}=(a_x,a_y,a_z)$ and a_x, a_y, a_z are called vector $\boldsymbol{a}$'s three coordinates.

Example 5 Given two points $A(0,1,-4)$ and $B(2,3,0)$, use coordinates to express the vectors $\overrightarrow{AB}$ and $-2\overrightarrow{AB}$.

Solution By Example 4, we obtain
$$\overrightarrow{AB}=(2-0,3-1,0+4)=(2,2,4),$$
$$-2\overrightarrow{AB}=(-4,-4,-8).$$

4) Definite proportion and division point formula

Suppose the points $P_1(x_1,y_1,z_1)$ and $P_2(x_2,y_2,z_2)$. If point P on straight line P_1P_2 satisfies $\overrightarrow{P_1P}=\lambda\overrightarrow{PP_2}(\lambda\neq -1)$, point P is called the point dividing $\overrightarrow{P_1P_2}$ in definite proportion λ. Calculate division point P's coordinates.

As shown in Figure 10-16, because
$$\overrightarrow{P_1P}=\overrightarrow{OP}-\overrightarrow{OP_1},$$
$$\overrightarrow{PP_2}=\overrightarrow{OP_2}-\overrightarrow{OP},$$

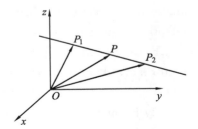

Figure 10-16

there is
$$\overrightarrow{OP}-\overrightarrow{OP_1}=\lambda(\overrightarrow{OP_2}-\overrightarrow{OP}).$$

Then solve it, we have
$$\overrightarrow{OP}=\frac{1}{1+\lambda}(\overrightarrow{OP_1}+\lambda\overrightarrow{OP_2}).$$

Substitute coordinates for vectors, we obtain
$$(x,y,z)=\frac{1}{1+\lambda}[(x_1,y_1,z_1)+\lambda(x_2,y_2,z_2)]$$
$$=\frac{1}{1+\lambda}(x_1+\lambda x_2, y_1+\lambda y_2, z_1+\lambda z_2),$$

therefore the coordinates of point P are
$$x=\frac{x_1+\lambda x_2}{1+\lambda},\quad y=\frac{y_1+\lambda y_2}{1+\lambda},\quad z=\frac{z_1+\lambda z_2}{1+\lambda}. \tag{6}$$

The formula is called spatial formula of definite proportion and division point which is similar to the formula on the plane.

In particular, if $\lambda=1$, P is the midpoint of $\overrightarrow{P_1P_2}$ and the coordinates are

$$x=\frac{x_1+x_2}{2},\ y=\frac{y_1+y_2}{2},\ z=\frac{z_1+z_2}{2}. \tag{7}$$

5) Norm and direction cosine formula

According to the definition of vector's coordinates, assume that $r=(x,y,z)$, then $|r|=|\overrightarrow{OP}|$ is the length of the diagonal line of cuboid, and $|x|=|\overrightarrow{OP_x}|, |y|=|\overrightarrow{OP_y}|, |z|=|\overrightarrow{OP_z}|$. Thus we obtain

$$|r|^2=|\overrightarrow{OP}|^2=|\overrightarrow{OP_x}|^2+|\overrightarrow{OP_y}|^2+|\overrightarrow{OP_z}|^2=x^2+y^2+z^2,$$

$$|r|=\sqrt{x^2+y^2+z^2}. \tag{8}$$

If the coordinates are given, we can find the vector's norm by the Formula (8).

Then, introduce the concept of included angle of two vectors.

For two nonzero vectors a and b, we select a point O in space to let $\overrightarrow{OA}=a$ and $\overrightarrow{OB}=b$, then $\angle AOB$ is called the angle φ between the vectors a and b, denoted by $(\widehat{a,b})$ or $(\widehat{b,a})$, where $0\leqslant\varphi\leqslant\pi$. If one of the vectors a and b is null vector, the angle φ could be any value from 0 to π.

Similarly, we can define the angle of a vector and an axis or the angle of two axes in space.

The angles α, β, γ of a vector a and three coordinate axes are called the direction angles of the vector a, and the cosines of direction angles $\cos\alpha$, $\cos\beta$, $\cos\gamma$ are called the direction cosines of the vector a.

Known from the property of vector's projection on the axis and the definition of vector's coordinates, if $a=(x,y,z)$,

$$x=|a|\cos\alpha,\ y=|a|\cos\beta,\ z=|a|\cos\gamma.$$

By Formula (8), we obtain the direction cosines formulas

$$\begin{cases}\cos\alpha=\dfrac{x}{|a|}=\dfrac{x}{\sqrt{x^2+y^2+z^2}},\\ \cos\beta=\dfrac{y}{|a|}=\dfrac{y}{\sqrt{x^2+y^2+z^2}},\\ \cos\gamma=\dfrac{z}{|a|}=\dfrac{z}{\sqrt{x^2+y^2+z^2}}.\end{cases} \tag{9}$$

Then, it's easy to obtain

$$\cos^2\alpha+\cos^2\beta+\cos^2\gamma=1, \tag{10}$$

and

$$a^\circ=\frac{a}{|a|}=\frac{1}{\sqrt{x^2+y^2+z^2}}(x,y,z)=(\cos\alpha,\cos\beta,\cos\gamma). \tag{11}$$

Example 6 Find the unit vector which is parallel to the vector $a=(6,7,-6)$.

Solution Since $|a|=\sqrt{6^2+7^2+(-6)^2}=11$, the unit vector a° which parallels to the

vector a is
$$a° = \pm \frac{a}{|a|} = \pm \frac{1}{11}(6,7,-6),$$
where $\frac{1}{11}(6,7,-6)$ has the same direction with a and $-\frac{1}{11}(6,7,-6)$ has the opposite direction.

10.2.4 Dot product of two vectors

In physics, the displacement that one object moves straightly under constant force F is s, then the work made by force F is
$$W = |F||s|\cos\theta,$$
where θ is the angle between F and s (Figure 10-17). The work W is a quantity determined by F and s in the formula above.

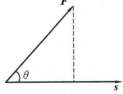

Figure 10-17

Definition 2 The product of the norms of two vectors a, b and the cosine of their angle is called the dot product of vectors a and b (also called inner product or scalar product), which is denoted by $a \cdot b$, that is,
$$a \cdot b = |a||b|\cos(\widehat{a,b}). \tag{12}$$

The dot product of two vectors is a quantity. According to the definition of dot product, it can infer that

(1) $a \cdot b = |a| \cdot \text{Prj}_a b = |b| \cdot \text{Prj}_b a$. In particular, if e is unit vector, $a \cdot e = \text{Prj}_e a$.

(2) $a \cdot a = a^2 = |a|^2$.

Because the angle between the same vector is $\theta = 0$, we have
$$a \cdot a = |a||a|\cos 0 = |a|^2.$$

(3) Two nonzero vectors a and b are called perpendicular if the angle between them is $\theta = \frac{\pi}{2}$. Then a and b are perpendicular if and only if $a \cdot b = 0$.

If $a \cdot b$ and $|a| \neq 0$, $|b| \neq 0$, we have $\cos\theta = 0$ or $\theta = \frac{\pi}{2}$, that is, $a \perp b$. Whereas, if $a \perp b$, we have $\theta = \frac{\pi}{2}$ or $\cos\theta = 0$, so $a \cdot b = |a||b|\cos\theta = 0$.

By this, it can infer that
$$i \cdot j = j \cdot k = k \cdot i = 0, \quad i \cdot i = j \cdot j = k \cdot k = 1.$$

The dot product of two vectors satisfies the following rules.

(1) Commutative law $\quad a \cdot b = b \cdot a$.

According to the definition, there are
$$a \cdot b = |a||b|\cos(\widehat{a,b}), \quad b \cdot a = |b||a|\cos(\widehat{b,a}),$$
and
$$|a||b| = |b||a|, \quad \cos(\widehat{a,b}) = \cos(\widehat{b,a}),$$
so
$$a \cdot b = b \cdot a.$$

(2) Distributive law $\quad (a+b) \cdot c = a \cdot c + b \cdot c$.

If $c = 0$, the formula above obviously stands up.

If $c \neq 0$,
$$(a+b) \cdot c = |c| \, \text{Prj}_c(a+b).$$

From the property of projection, we know that
$$\text{Prj}_c(a+b) = \text{Prj}_c a + \text{Prj}_c b.$$

Thus
$$(a+b) \cdot c = |c| \text{Prj}_c(a+b) = |c| \text{Prj}_c a + |c| \text{Prj}_c b = a \cdot c + b \cdot c.$$

(3) Associative law $(\lambda a) \cdot b = a \cdot (\lambda b) = \lambda(a \cdot b)$.

If $b = 0$, the formula above obviously stands up.

If $b \neq 0$, from the property of projection, we have
$$(\lambda a) \cdot b = |b| \, \text{Prj}_b(\lambda a) = |b| \lambda \text{Prj}_b a = \lambda |b| \text{Prj}_b a = \lambda (a \cdot b).$$

In the right angle coordinate system, we also have another expression of two vector's dot product. Suppose
$$a = (a_x, a_y, a_z) = a_x i + a_y j + a_z k,$$
$$b = (b_x, b_y, b_z) = b_x i + b_y j + b_z k,$$

we obtain
$$\begin{aligned} a \cdot b &= (a_x i + a_y j + a_z k) \cdot (b_x i + b_y j + b_z k) \\ &= a_x b_x i \cdot i + a_x b_y i \cdot j + a_x b_z i \cdot k + a_y b_x j \cdot i + a_y b_y j \cdot j + a_y b_z j \cdot k + a_z b_x k \cdot i + \\ &\quad a_z b_y k \cdot j + a_z b_z k \cdot k \\ &= a_x b_x + a_y b_y + a_z b_z. \end{aligned} \tag{13}$$

By the definition of dot product $a \cdot b = |a| |b| \cos(\widehat{a,b})$, we get the angle formula of two nonzero vectors
$$\cos(\widehat{a,b}) = \frac{a \cdot b}{|a||b|} = \frac{a_x b_x + a_y b_y + a_z b_z}{\sqrt{a_x^2 + a_y^2 + a_z^2} \sqrt{b_x^2 + b_y^2 + b_z^2}}. \tag{14}$$

From the formula above, a and b are perpendicular if and only if
$$a_x b_x + a_y b_y + a_z b_z = 0.$$

Example 7 Given that $|a| = \sqrt{3}, |b| = 1, (\widehat{a,b}) = \frac{\pi}{6}$, find $|a+b|, |a-b|$.

Solution $|a+b| = \sqrt{(a+b) \cdot (a+b)} = \sqrt{a \cdot a + 2a \cdot b + b \cdot b}$
$$= \sqrt{|a|^2 + 2|a||b|\cos(\widehat{a,b}) + |b|^2} = \sqrt{3 + 2\sqrt{3} \cdot \frac{\sqrt{3}}{2} + 1} = \sqrt{7},$$
$|a-b| = \sqrt{(a-b) \cdot (a-b)} = \sqrt{a \cdot a - 2a \cdot b + b \cdot b}$
$$= \sqrt{|a|^2 - 2|a||b|\cos(\widehat{a,b}) + |b|^2} = \sqrt{3 - 2\sqrt{3} \cdot \frac{\sqrt{3}}{2} + 1} = 1.$$

Example 8 Given that $A(-1,2,3)$, $B(1,1,1)$, $C(0,0,5)$, show that $\triangle ABC$ is orthogonal triangle and find $\angle B$.

Proof As shown in Figure 10-18, we have
$$\overrightarrow{BA} = (-2, 1, 2),$$

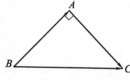

Figure 10-18

$$\vec{BC}=(-1,-1,4), \vec{AC}=(1,-2,2).$$

(1) Because
$$\vec{BA} \cdot \vec{AC}=-2-2+4=0,$$
we have $\vec{BA} \perp \vec{AC}$, so $\triangle ABC$ is orthogonal triangle.

(2) Because $\cos\angle B=\dfrac{\vec{BA}\cdot\vec{BC}}{|\vec{BA}||\vec{BC}|}=\dfrac{2-1+8}{3\sqrt{18}}=\dfrac{1}{\sqrt{2}}$, we have $\angle B=\dfrac{\pi}{4}$.

10.2.5 Vector product of two vectors

Definition 3 Vector product(also called exterior product or cross product) of two vectors $\boldsymbol{a}$ and $\boldsymbol{b}$ is a vector, denoted by $\boldsymbol{a}\times\boldsymbol{b}$.

(1) Norm of $\boldsymbol{a}\times\boldsymbol{b}$, $|\boldsymbol{a}\times\boldsymbol{b}|=|\boldsymbol{a}||\boldsymbol{b}|\sin(\widehat{\boldsymbol{a},\boldsymbol{b}})$;

(2) Direction of $\boldsymbol{a}\times\boldsymbol{b}$ is perpendicular to $\boldsymbol{a}$, $\boldsymbol{b}$, and $\boldsymbol{a}$, $\boldsymbol{b}$, $\boldsymbol{a}\times\boldsymbol{b}$ constitute right-handed system (Figure 10-19).

From the definition of two vectors' cross product, we obtain:

(1) $\boldsymbol{a}\times\boldsymbol{a}=\boldsymbol{0}$.

Because the angle $\theta=0$, we have
$$|\boldsymbol{a}\times\boldsymbol{a}|=|\boldsymbol{a}|^2\sin 0=0.$$

(2) $\boldsymbol{a}\parallel\boldsymbol{b}$ if and only if $\boldsymbol{a}\times\boldsymbol{b}=\boldsymbol{0}$.

If $\boldsymbol{a}\times\boldsymbol{b}=\boldsymbol{0}$, since $|\boldsymbol{a}|\neq 0, |\boldsymbol{b}|\neq 0$, we have $\sin\theta=0$, then

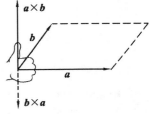

Figure 10-19

$\theta=0$ or π, so $\boldsymbol{a}\parallel\boldsymbol{b}$. Whereas, if $\boldsymbol{a}\parallel\boldsymbol{b}$, we have $\theta=0$ or π, then $\sin\theta=0$ and $|\boldsymbol{a}\times\boldsymbol{b}|=0$, so $\boldsymbol{a}\times\boldsymbol{b}=\boldsymbol{0}$.

(3) The geometric meaning of the norm of the vector product $|\boldsymbol{a}\times\boldsymbol{b}|$ is that $|\boldsymbol{a}\times\boldsymbol{b}|$ equals to the area of parallelogram with adjacent sides $\boldsymbol{a}$, $\boldsymbol{b}$, that is $|\boldsymbol{a}\times\boldsymbol{b}|=S_{a\times b}$.

The vector product of two vectors satisfies the following rules.

(1) Anti-commutative law $\boldsymbol{a}\times\boldsymbol{b}=-\boldsymbol{b}\times\boldsymbol{a}$.

According to the right-handed rule, the direction from $\boldsymbol{b}$ to $\boldsymbol{a}$ is opposite to that from $\boldsymbol{a}$ to $\boldsymbol{b}$. This means that commutative law cannot stand up in vector product.

(2) Distributive law $(\boldsymbol{a}+\boldsymbol{b})\times\boldsymbol{c}=\boldsymbol{a}\times\boldsymbol{c}+\boldsymbol{b}\times\boldsymbol{c}$.

(3) Associative law $(\lambda\boldsymbol{a})\times\boldsymbol{b}=\boldsymbol{a}\times(\lambda\boldsymbol{b})=\lambda(\boldsymbol{a}\times\boldsymbol{b})$.

In particular, there are
$$\boldsymbol{i}\times\boldsymbol{i}=\boldsymbol{j}\times\boldsymbol{j}=\boldsymbol{k}\times\boldsymbol{k}=\boldsymbol{0},\ \boldsymbol{i}\times\boldsymbol{j}=-\boldsymbol{j}\times\boldsymbol{i}=\boldsymbol{k},$$
$$\boldsymbol{j}\times\boldsymbol{k}=-\boldsymbol{k}\times\boldsymbol{j}=\boldsymbol{i},\ \boldsymbol{k}\times\boldsymbol{i}=-\boldsymbol{i}\times\boldsymbol{k}=\boldsymbol{j}.$$

Then in right angle coordinate system, we find the coordinate expression of two vectors' cross product. Suppose
$$\boldsymbol{a}=(a_x,a_y,a_z)=a_x\boldsymbol{i}+a_y\boldsymbol{j}+a_z\boldsymbol{k},$$
$$\boldsymbol{b}=(b_x,b_y,b_z)=b_x\boldsymbol{i}+b_y\boldsymbol{j}+b_z\boldsymbol{k},$$
we can obtain

$$\begin{aligned}
\boldsymbol{a} \times \boldsymbol{b} &= (a_x \boldsymbol{i} + a_y \boldsymbol{j} + a_z \boldsymbol{k}) \times (b_x \boldsymbol{i} + b_y \boldsymbol{j} + b_z \boldsymbol{k}) \\
&= a_x b_x \boldsymbol{i} \times \boldsymbol{i} + a_x b_y \boldsymbol{i} \times \boldsymbol{j} + a_x b_z \boldsymbol{i} \times \boldsymbol{k} + a_y b_x \boldsymbol{j} \times \boldsymbol{i} + a_y b_y \boldsymbol{j} \times \boldsymbol{j} + a_y b_z \boldsymbol{j} \times \boldsymbol{k} + \\
&\quad a_z b_x \boldsymbol{k} \times \boldsymbol{i} + a_z b_y \boldsymbol{k} \times \boldsymbol{j} + a_z b_z \boldsymbol{k} \times \boldsymbol{k} \\
&= (a_y b_z - b_y a_z) \boldsymbol{i} + (b_x a_z - a_x b_z) \boldsymbol{j} + (a_x b_y - b_x a_y) \boldsymbol{k}.
\end{aligned}$$

Use third-order determinant, the formula above is often written in the form

$$\boldsymbol{a} \times \boldsymbol{b} = \begin{vmatrix} \boldsymbol{i} & \boldsymbol{j} & \boldsymbol{k} \\ a_x & a_y & a_z \\ b_x & b_y & b_z \end{vmatrix}. \tag{15}$$

Example 9 Given that $\boldsymbol{a} = (2,2,1)$, $\boldsymbol{b} = (4,5,3)$, find $\boldsymbol{a} \times \boldsymbol{b}$, $|\boldsymbol{a} \times \boldsymbol{b}|$ and unit vector $(\boldsymbol{a} \times \boldsymbol{b})°$.

Solution

$$\boldsymbol{a} \times \boldsymbol{b} = \begin{vmatrix} \boldsymbol{i} & \boldsymbol{j} & \boldsymbol{k} \\ 2 & 2 & 1 \\ 4 & 5 & 3 \end{vmatrix} = (2 \times 3 - 5 \times 1)\boldsymbol{i} + (1 \times 4 - 2 \times 3)\boldsymbol{j} + (2 \times 5 - 2 \times 4)\boldsymbol{k}$$

$$= (1, -2, 2),$$

$$|\boldsymbol{a} \times \boldsymbol{b}| = \sqrt{1^2 + (-2)^2 + 2^2} = 3,$$

$$(\boldsymbol{a} \times \boldsymbol{b})° = \frac{1}{3}(1, -2, 2) = \left(\frac{1}{3}, -\frac{2}{3}, \frac{2}{3}\right).$$

Example 10 Given that three vertexes of triangle are $A(1,2,3)$, $B(2,-1,5)$, $C(3,2,-5)$, find: (1) the area of $\triangle ABC$; (2) the height of side AB in $\triangle ABC$.

Solution (1) $S_{\triangle ABC} = \frac{1}{2} S_{\square ABCD} = \frac{1}{2} |\overrightarrow{AB} \times \overrightarrow{AC}|$ (Figure 10-20).

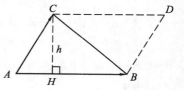

Figure 10-20

$$\overrightarrow{AB} = (1, -3, 2), \quad \overrightarrow{AC} = (2, 0, -8),$$

$$\overrightarrow{AB} \times \overrightarrow{AC} = \begin{vmatrix} \boldsymbol{i} & \boldsymbol{j} & \boldsymbol{k} \\ 1 & -3 & 2 \\ 2 & 0 & -8 \end{vmatrix} = 24\boldsymbol{i} + 12\boldsymbol{j} + 6\boldsymbol{k},$$

then $$|\overrightarrow{AB} \times \overrightarrow{AC}| = \sqrt{24^2 + 12^2 + 6^2} = 6\sqrt{21},$$

so $$S_{\triangle ABC} = \frac{1}{2} |\overrightarrow{AB} \times \overrightarrow{AC}| = 3\sqrt{21}.$$

(2) Because the height CH on the side AB in $\triangle ABC$ is a height of $\square ABCD$, we have

$$|\overrightarrow{CH}| = \frac{S_{\square ABCD}}{|\overrightarrow{AB}|} = \frac{|\overrightarrow{AB} \times \overrightarrow{AC}|}{|\overrightarrow{AB}|}.$$

Then $$|\overrightarrow{AB}| = \sqrt{(1)^2 + (-3)^2 + 2^2} = \sqrt{14},$$

so $$|\overrightarrow{CH}| = \frac{6\sqrt{21}}{\sqrt{14}} = 3\sqrt{6}.$$

Exercises 10-2

1. Given that $\boldsymbol{a} = \boldsymbol{e}_1 + 2\boldsymbol{e}_2 - \boldsymbol{e}_3$, $\boldsymbol{b} = 3\boldsymbol{e}_1 - 2\boldsymbol{e}_2 + 2\boldsymbol{e}_3$, find $\boldsymbol{a} + \boldsymbol{b}$, $\boldsymbol{a} - \boldsymbol{b}$ and $3\boldsymbol{a} - 2\boldsymbol{b}$.

2. Given that $\overrightarrow{AB}=a+5b, \overrightarrow{BC}=-2a+8b, \overrightarrow{CD}=3(a-b)$, show that the three points A, B, D are collinear.

3. Vector $\overrightarrow{AB}=(-3,2,1)$ and point $A(1,2,-4)$, find the coordinates of point B.

4. Given that two points $P_1(1,2,3)$ and $P_2(-1,0,1)$, use coordinates to express the vectors $\overrightarrow{P_1P_2}$ and $5\overrightarrow{P_1P_2}$.

5. Find the norms of vectors $a=i+j+k, b=2i-3j+5k, c=-2i-j+2k$ and the unit vectors $a°, b°, c°$, then use $a°, b°, c°$ to express vectors a, b, c respectively.

6. Given that a line segment AB is trisected by $C(2,0,2)$ and $D(5,-2,0)$, find the coordinates of the line segment's endpoints A and B.

7. Assume that $a=3i-j-2k, b=i+2j-k$, find:
(1) $a \cdot b$ and $a \times b$;   (2) $(-2a) \cdot 3b$ and $a \times 2b$;
(3) $\cos(\widehat{a,b}), \sin(\widehat{a,b})$ and $\tan(\widehat{a,b})$.

8. Which is the value of l that makes vectors $a=6i-3j+3k$ and $b=4i+lj+2k$ to be:
(1) perpendicular;   (2) parallel.

9. Given that $a=2i-3j+k, b=i-j+3k, c=i-2j$, find:
(1) $(a \cdot b)c-(b \cdot c)b$;   (2) $(a+b) \times (b+c)$.

10. Given that $a=(2,3,1), b=(5,6,4)$. Find:
(1) Area of parallelogram whose adjacent sides are a, b.
(2) Heights on the two sides a, b in parallelogram.

10.3 Plane or space straight line

Geometric figures in space such as plane, straight line, curved surface and curve all can be regarded as loci generated by a point moving by some laws, so the moving point $P(x,y,z)$ on the figure can be regarded as the set of points which have some characteristic properties. This characteristic (the law of moving point's motion) reflected by quantitative relationship is a constraint condition that could be expressed by $F(x,y,z)=0$, which is an equation of the three coordinates of the moving point $P(x,y,z)$. It is also the quantitative expression of the geometric figure.

Characteristics of points on geometric figure includes two following aspects. ① The coordinates of arbitrary point $P(x,y,z)$ on geometric figure satisfy the equation $F(x,y,z)=0$. ② All points $P(x,y,z)$ whose coordinates satisfy the equation $F(x,y,z)=0$ are on the geometric figure.

10.3.1 Plane and its equation

Space plane is the easiest figure in space and there are plenty of methods to determine it. For instance, we can determine a plane by three fixed points which aren't collinear or by a straight line and a point not in the line. For example, the equation of plane in Example 3 and the equation of sphere in Example 4 in 10.1.3, which are the constraint conditions that points on a plane or a sphere should satisfy

$$Ax+By+Cz=D \quad \text{and} \quad (x-x_0)^2+(y-y_0)^2+(z-z_0)^2=R^2$$

by the laws "distances from one moving point to two fixed points are same" and "distance from one moving point to one fixed point is constant".

Given a fixed point P_0 and a nonzero vector $\boldsymbol{n}$ in space, the plane passing through the point P_0 and perpendicular to the vector $\boldsymbol{n}$ can be determined uniquely. Now we regard the vector $\boldsymbol{n}$ which is perpendicular to the plane as a normal vector. Obviously, an arbitrary vector on the plane is perpendicular to this normal vector $\boldsymbol{n}$.

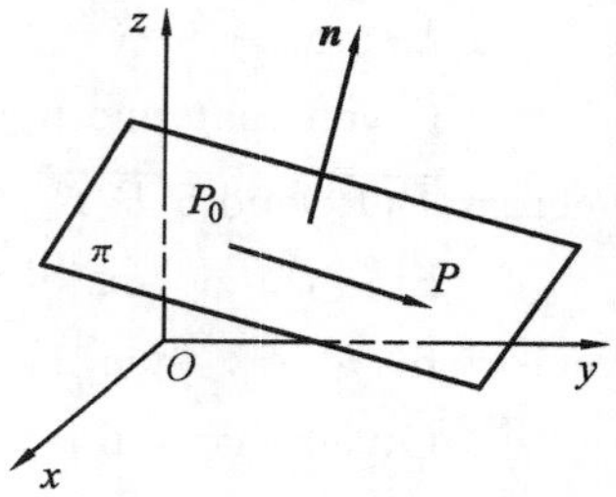

Assume that the plane π passes through a given point $P_0(x_0, y_0, z_0)$ with the normal vector $\boldsymbol{n} = (A, B, C)$ (A, B, C aren't all 0) (Figure 10-21).

Figure 10-21

Suppose $P(x, y, z)$ is an arbitrary point on the plane π, we know that the point P is on plane π if and only if the vector $\overrightarrow{P_0P}$ is perpendicular to the normal vector $\boldsymbol{n}$, that is, $\boldsymbol{n} \cdot \overrightarrow{P_0P} = 0$.

Since $\boldsymbol{n} = (A, B, C)$ and $\overrightarrow{P_0P} = (x - x_0, y - y_0, z - z_0)$, we have
$$A(x - x_0) + B(y - y_0) + C(z - z_0) = 0. \tag{1}$$

This is a ternary linear equation of x, y and z. Obviously, the coordinates of an arbitrary point on plane π must satisfy the Equation (1) and the coordinates of the points not on the plane π don't satisfy the Equation (1). Thus Equation (1) is the plane equation determined by the point P_0 and the normal vector $\boldsymbol{n}$, which is the equation determined by point and normal vector.

Example 1 Find the equation of the plane passing through the point $(-1, 2, 0)$ with normal vector $\boldsymbol{n} = (1, 3, 1)$.

Solution By the Equation (1), we obtain the equation
$$(x + 1) + 3(y - 2) + (z - 0) = 0,$$
or
$$x + 3y + z - 5 = 0.$$

Example 2 Find the equation of the plane passing through the points $A(0, 1, -1)$, $B(1, 0, 3)$ and $C(-1, 2, 0)$.

Solution Firstly, find the normal vector $\boldsymbol{n}$. Because $\boldsymbol{n} \perp \overrightarrow{AB}$, $\boldsymbol{n} \perp \overrightarrow{AC}$, we have $\boldsymbol{n} /\!/ \overrightarrow{AB} \times \overrightarrow{AC}$ and we could select $\boldsymbol{n} = k \overrightarrow{AB} \times \overrightarrow{AC}$. Since
$$\overrightarrow{AB} \times \overrightarrow{AC} = \begin{vmatrix} \boldsymbol{i} & \boldsymbol{j} & \boldsymbol{k} \\ 1 & -1 & 4 \\ -1 & 1 & 1 \end{vmatrix} = (-5, -5, 0) = -5(1, 1, 0),$$
we select $\boldsymbol{n} = (1, 1, 0)$ and plug it into the Equation (1), the equation is
$$(x - 0) + (y - 1) + 0(z + 1) = 0,$$
or
$$x + y - 1 = 0.$$

In Equation (1), we let $D = -(Ax_0 + By_0 + Cz_0)$, then Equation (1) will become
$$Ax + By + Cz + D = 0. \tag{2}$$

Regard Equation (2) as the common equation of plane π which is a ternary linear

Chapter 10 Vectors and analytic geometry of space

equation of x, y, z.

In right angle coordinate system, the coefficients A, B, C in Equation (2) are three coordinates of the normal vector $\boldsymbol{n}$ of the plane π.

If one or more coefficients A, B, C, D in Equation (2) equals to 0, the corresponding plane has some special positions.

(1) $D=0$. The Equation (2) becomes $Ax+By+Cz=0$. Obviously, the origin $(0,0,0)$ satisfies the equation, so this plane passes the origin. Whereas, if the plane passes the origin, there is $D=0$.

(2) One of A, B, C is 0. For example, if $C=0$, Equation (2) becomes $Ax+By+D=0$ and the normal vector $\boldsymbol{n}=(A,B,0)$ is perpendicular to the z-axis, so the plane is parallel to the z-axis or perpendicular to the coordinate plane xOy. In particular, if $C=D=0$, the plane passes through the z-axis. Similarly we can obtain that if $A=0$, the plane is parallel to the x-axis; if $B=0$, the plane is parallel to the y-axis.

(3) Two of A, B, C are 0. For example, if $A=B=0$, the equation becomes $Cz+D=0$ or $z=-\dfrac{D}{C}$. The plane is parallel to the x-axis and the y-axis simultaneously or parallel to the plane xOy. Similarly we can obtain that if $B=C=0$ or $A=C=0$, the plane is parallel to the plane yOz or the plane xOz.

In particular, $x=0$, $y=0$, $z=0$ refer to the three coordinate planes respectively.

Example 3 Find the plane equation which passes $M_1(1,-5,1)$ and $M_2(3,2,-2)$ and is parallel to the y-axis.

Solution Because the plane is parallel to the y-axis, we assume that the plane equation is
$$Ax+Cz+D=0.$$
Since the plane passes the points $M_1(1,-5,1)$, $M_2(3,2,-2)$, there are
$$\begin{cases} A+C+D=0, \\ 3A-2C+D=0. \end{cases}$$
Solve them and obtain $A=\dfrac{3}{2}C, D=\dfrac{5}{2}C$. Plug them into the equation above and divided the both sides by $C(C\neq 0)$, then the equation is
$$3x+2z-5=0.$$
This example can also find out the normal vector firstly
$$\boldsymbol{n}=\boldsymbol{j}\times\overrightarrow{M_1M_2}=(0,1,0)\times(2,7,-3)=\begin{vmatrix} \boldsymbol{i} & \boldsymbol{j} & \boldsymbol{k} \\ 0 & 1 & 0 \\ 2 & 7 & -3 \end{vmatrix}=-3\boldsymbol{i}-2\boldsymbol{k}.$$

Then use the dot method to find the plane equation.

Example 4 Assume that the plane passes through three points $P_1(a,0,0), P_2(0,b,0), P_3(0,0,c)$ $(abc\neq 0)$ (Figure 10-22), find the plane equation.

Solution Assume that the plane equation is
$$Ax+By+Cz+D=0.$$

The plane passes three points P_1, P_2, P_3, so
$$\begin{cases} aA+D=0, \\ bB+D=0, \\ cC+D=0. \end{cases}$$

Solve it and then obtain $A=-\dfrac{D}{a}, B=-\dfrac{D}{b}, C=-\dfrac{D}{c}$.

Substitute them into the equation above and divided the both sides by $D(D\neq 0)$, so the plane equation is

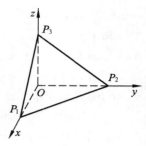

Figure 10-22

$$\frac{x}{a}+\frac{y}{b}+\frac{z}{c}=1. \tag{3}$$

Equation (3) is called the intercept equation and a, b, c are the intercepts of the plane on three coordinate axes respectively.

10.3.2 Included angle between two planes

Assume that two planes π_1 and π_2

$$\pi_1: A_1 x+B_1 y+C_1 z+D_1=0,$$
$$\pi_2: A_2 x+B_2 y+C_2 z+D_2=0,$$

then their normal vectors are

$$\boldsymbol{n}_1=(A_1,B_1,C_1), \boldsymbol{n}_2=(A_2,B_2,C_2).$$

Assume that the angle between the two planes π_1 and π_2 is θ (Figure 10-23) and $0\leqslant\theta\leqslant\dfrac{\pi}{2}$. Actually, θ also equals to the angle between the normal vectors $\boldsymbol{n}_1$ and $\boldsymbol{n}_2$, that is, $\theta=(\widehat{\boldsymbol{n}_1,\boldsymbol{n}_2})$, or is complementary to the angle between the normal vectors $\boldsymbol{n}_1$ and $\boldsymbol{n}_2$, that is,

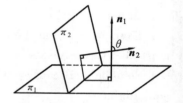

Figure 10-23

$$\theta=\pi-(\widehat{\boldsymbol{n}_1,\boldsymbol{n}_2}).$$

By the angle formula of two vectors, we obtain

$$\cos\theta=|\cos(\widehat{\boldsymbol{n}_1,\boldsymbol{n}_2})|=\frac{|\boldsymbol{n}_1\cdot\boldsymbol{n}_2|}{|\boldsymbol{n}_1||\boldsymbol{n}_2|}=\frac{|A_1 A_2+B_1 B_2+C_1 C_2|}{\sqrt{A_1^2+B_1^2+C_1^2}\sqrt{A_2^2+B_2^2+C_2^2}}. \tag{4}$$

Formula (4) is called the angle formula of two planes.

Through the conditions of two vectors being mutually perpendicular or parallel, we can obtain the following conclusions.

(1) Two planes π_1 and π_2 are mutually perpendicular if and only if
$$A_1 A_2+B_1 B_2+C_1 C_2=0.$$

(2) Two planes π_1 and π_2 are mutually parallel if and only if
$$\frac{A_1}{A_2}=\frac{B_1}{B_2}=\frac{C_1}{C_2}.$$

Example 5 A plane passes the x-axis and the angle between it and the plane $x+y=0$ is $\dfrac{\pi}{3}$, find its equation.

Solution Assume that the plane equation is $By+Cz=0$, then

$$\cos\frac{\pi}{3}=\frac{|1\times 0+1\times B+0\times C|}{\sqrt{1^2+1^2+0^2}\sqrt{0^2+B^2+C^2}}=\frac{1}{2},$$

or $|B|=\frac{\sqrt{2}}{2}\sqrt{B^2+C^2}$. Solve it and we obtain that $B^2=C^2$, $C=\pm B$.

Thus the plane equation is $y\pm z=0$.

Example 6 Find the plane which passes the point $A(1,1,-1)$ and is perpendicular to the planes $x-y+z-7=0$ and $3x+2y-12z+5=0$.

Solution Assume that the normal vector is $n=(A,B,C)$, and the vectors of the given planes are $n_1=(1,-1,1), n_2=(3,2,-12)$. Since $n\perp n_1, n\perp n_2$, we have $n/\!/n_1\times n_2$, so we can select $n=kn_1\times n_2$. Now

$$n_1\times n_2=\begin{vmatrix} i & j & k \\ 1 & -1 & 1 \\ 3 & 2 & -12 \end{vmatrix}=(10,15,5)=5(2,3,1),$$

so $n=(2,3,1)$. Plug it into the Equation (1) and the equation is
$$2(x-1)+3(y-1)+(z+1)=0,$$
or
$$2x+3y+z-4=0.$$

10.3.3 Distance from point to plane

As shown in Figure 10-24, the distance from point P_0 (x_0,y_0,z_0) to plane $Ax+By+Cz+D=0$ is

$$d=\frac{|Ax_0+By_0+Cz_0+D|}{\sqrt{A^2+B^2+C^2}}. \tag{5}$$

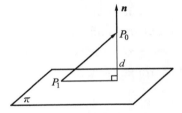

Figure 10-24

Example 7 The distance from the point $(1,2,-3)$ to the plane $2x-y+2z+3=0$ is

$$d=\frac{|2\times 1-1\times 2+2\times(-3)+3|}{\sqrt{2^2+(-1)^2+2^2}}=1.$$

10.3.4 Space line and its equation

We will use the vector as a tool to find the equation of space line.

Assume that a space line l is regarded as the intersection line of two planes π_1 and π_2 (Figure 10-25). If the equations of two intersecting planes π_1 and π_2 are $A_1x+B_1y+C_1z+D_1=0$ and $A_2x+B_2y+C_2z+D_2=0$ respectively, the point on straight

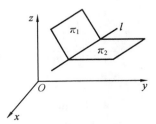

Figure 10-25

line l is on both two planes, so its coordinates must satisfy the equations of two planes or the equation set

$$\begin{cases} A_1x+B_1y+C_1z+D_1=0, \\ A_2x+B_2y+C_2z+D_2=0. \end{cases} \tag{6}$$

Whereas, the point whose coordinates satisfy the equation set (6) is on two planes simultaneously, so it must be on the intersection line of the two planes. Thus the equation set is the equation of the straight line l, which is called the common equation.

There are infinite numbers of planes passing through the space line l, we can get the equation set just by combining equations of two planes selected from these planes arbitrarily.

In space, given one fixed point $P_0(x_0, y_0, z_0)$ and a nonzero vector $s=(m,n,p)$, the straight line l passing through the point P_0 and parallel to vector s can be determined uniquely (Figure 10-26), and the vector s is called the direction vector of the straight line l. (Obviously, any nonzero vector which is parallel to straight line l is the direction vector.)

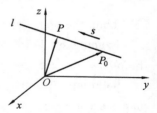

Figure 10-26

Assume that $P(x,y,z)$ is an arbitrary point on the straight line l, then the point P is on the straight line l if and only if $\overrightarrow{P_0P}$ and $s(\neq 0)$ are collinear, that is, $\overrightarrow{P_0P} /\!/ s$. We obtain

$$\frac{x-x_0}{m} = \frac{y-y_0}{n} = \frac{z-z_0}{p}. \tag{7}$$

Equation (7) is called the symmetric equation or point direction form equation or standard form equation.

The coordinates m, n, p of the direction vector s are called a group of direction numbers, and the direction cosine of s is called the line's direction cosine.

From the symmetric equation, it's easy to obtain the direction vector s and the coordinates of the fixed point P_0.

Let

$$\frac{x-x_0}{m} = \frac{y-y_0}{n} = \frac{z-z_0}{p} = t,$$

and we obtain

$$\begin{cases} x = x_0 + mt, \\ y = y_0 + nt, \\ z = z_0 + pt. \end{cases} \tag{8}$$

Equation set (8) it called the parametric equation and t is the parameter.

Note (1) In symmetric Equation (7), m, n, p are the coordinates of the direction vector s. If one of m, n, p is 0, for instance, $n=0$, the equation set is same to $\frac{x-x_0}{m} = \frac{z-z_0}{p}$ and $y-y_0=0$. If two of m, n, p are 0, for instance, $n=p=0$, the equation set is same to $y-y_0=0$ and $z-z_0=0$.

(2) The coefficients m,n,p in the parametric equation are the coordinates of direction vector.

Example 8 Find the symmetric equation of the straight line passing through the point $(1,3,-2)$ and perpendicular to the plane $2x+y-3z+1=0$.

Solution The normal vector of the given plane is $n=(2,1,-3)$. Because the straight line is perpendicular to the plane, n is parallel to the straight line, we can select n as the

direction vector **s**. Plug it into the Equation (7), we obtain the symmetric equation
$$\frac{x-1}{2}=\frac{y-3}{1}=\frac{z+2}{-3}.$$

Example 9 Write the common equation
$$\begin{cases} 2x-3y+z-5=0, \\ 3x+y-2z-2=0 \end{cases}$$
into symmetric equation and parametric equation.

Solution Firstly, we find a point $P_0(x_0,y_0,z_0)$ on the straight line. Let $x_0=1$, then plug it into the equation set and we obtain
$$\begin{cases} -3y+z=3, \\ y-2z=-1. \end{cases}$$
Solve it, we have $y_0=-1, z_0=0$. Thus $(1,-1,0)$ is a point on the straight line.

Then we will find the direction vector **s** of the straight line. Because $s \perp n_1, s \perp n_2$, we have $s // n_1 \times n_2$. And $n_1=(2,-3,1)$, $n_2=(3,1,-2)$, then
$$n_1 \times n_2 = \begin{vmatrix} i & j & k \\ 2 & -3 & 1 \\ 3 & 1 & -2 \end{vmatrix} = (5,7,11).$$

So, we get $s=(5,7,11)$, and the symmetric equation is
$$\frac{x-1}{5}=\frac{y+1}{7}=\frac{z-0}{11},$$
the parametric equation is
$$\begin{cases} x=1+5t, \\ y=-1+7t, \\ z=11t. \end{cases}$$

10.3.5 Included angle between two straight lines

The angle θ between two space straight lines l_1 and l_2 is defined by the angle of their direction vectors and $0 \leqslant \theta \leqslant \frac{\pi}{2}$. Thus $\theta=(\widehat{s_1,s_2})$ or $\theta=\pi-(\widehat{s_1,s_2})$.

Assume that the equations of two straight lines l_1, l_2 are
$$l_1: \frac{x-x_1}{m_1}=\frac{y-y_1}{n_1}=\frac{z-z_1}{p_1},$$
$$l_2: \frac{x-x_2}{m_2}=\frac{y-y_2}{n_2}=\frac{z-z_2}{p_2}.$$

According to the angle formula of two vectors, we obtain
$$\cos\theta=|\cos(\widehat{s_1,s_2})|=\frac{|s_1 \cdot s_2|}{|s_1||s_2|}=\frac{|m_1m_2+n_1n_2+p_1p_2|}{\sqrt{m_1^2+n_1^2+p_1^2}\sqrt{m_2^2+n_2^2+p_2^2}}. \tag{9}$$

Through the conditions that two vectors are mutually perpendicular or parallel, it's easy to obtain that:

(1) the straight lines l_1 and l_2 are mutually perpendicular if and only if
$$m_1m_2+n_1n_2+p_1p_2=0;$$

(2) the straight lines l_1 and l_2 are mutually parallel or overlapped if and only if
$$\frac{m_1}{m_2}=\frac{n_1}{n_2}=\frac{p_1}{p_2}.$$

Example 10 Find the angle between two straight lines
$$l_1: \frac{x+3}{4}=\frac{y-2}{3}=\frac{z-5}{1} \text{ and } l_2: \frac{x}{1}=\frac{y-2}{-1}=\frac{z-5}{2}.$$

Solution Since $s_1=(4,3,1), s_2=(1,-1,2)$, by the Formula (9), we obtain
$$\cos\theta=\frac{|4\times1+3\times(-1)+1\times2|}{\sqrt{4^2+3^2+1^2}\sqrt{1^2+(-1)^2+2^2}}=\frac{3}{2\sqrt{39}}=\frac{\sqrt{39}}{26}.$$

Thus the angle between the two straight lines is $\theta=\arccos\frac{\sqrt{39}}{26}$.

10.3.6 Angle between straight line and plane

If the straight line l isn't perpendicular to the plane π, the angle $\left(0\leqslant\varphi<\frac{\pi}{2}\right)$ between the straight line and its projected straight line l_0 is the angle between straight line and plane (Figure 10-27). If the straight line is perpendicular to the plane, the angle φ between straight line and plane is right-angle.

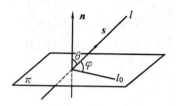

Figure 10-27

The angle φ between the straight line l and plane π can be determined by the direction vector s of the line l and the normal vector n of the plane π (Figure 10-27). If the angle between s and n is $(\widehat{n,s})=\theta(0\leqslant\theta<\pi)$, since $\varphi=\left|\frac{\pi}{2}-\theta\right|$, we have $\sin\varphi=|\cos\theta|$.

Assume that the equation of plane π is
$$Ax+By+Cz+D=0,$$
and the equation of straight line l is
$$\frac{x-x_0}{m}=\frac{y-y_0}{n}=\frac{z-z_0}{p}.$$

We can obtain
$$\sin\varphi=|\cos(\widehat{n,s})|=\frac{|n\cdot s|}{|n||s|}=\frac{|Am+Bn+Cp|}{\sqrt{A^2+B^2+C^2}\sqrt{m^2+n^2+p^2}}. \quad (10)$$

From the straight line being parallel to plane, we have that the direction vector s is perpendicular to the normal vector n, so the straight line is parallel to the plane if and only if
$$Am+Bn+Cp=0.$$
Similarly, the straight line is perpendicular to the plane if and only if
$$\frac{A}{m}=\frac{B}{n}=\frac{C}{p}.$$

Example 11 Find the angle between the straight line $\frac{x-1}{1}=\frac{y}{-4}=\frac{z+3}{1}$ and the plane $\pi: 2x-2y-z+3=0$.

Solution Given that the direction vector of the straight line is $s = i - 4j + k$, and the normal vector of the plane π is $n = (2, -2, -1)$, we have
$$\sin \varphi = \frac{|1 \times 2 + (-4) \times (-2) + 1 \times (-1)|}{\sqrt{1^2 + (-4)^2 + 1^2} \sqrt{2^2 + (-2)^2 + (-1)^2}} = \frac{9}{\sqrt{18} \times 3} = \frac{1}{\sqrt{2}},$$

so
$$\varphi = \frac{\pi}{4}.$$

Example 12 Find the equation of a plane which passes through the straight line l: $\frac{x-2}{5} = \frac{y+1}{2} = \frac{z-2}{4}$ and is perpendicular to the plane π_0: $x + 4y - 3z + 7 = 0$.

Solution Assume that the normal vector of the unknown plane is $n = (A, B, C)$. Since the plane passes through the straight line l, we have $n \perp s_0$. Then the unknown plane is perpendicular to the given plane, so $n \perp n_0$. Thus we have
$$n = n_0 \times s_0 = \begin{vmatrix} i & j & k \\ 1 & 4 & -3 \\ 5 & 2 & 4 \end{vmatrix} = 22i - 19j - 18k.$$

Because the plane passes through the point $P_0(2, -1, 2)$, the plane equation is
$$22(x-2) - 19(y+1) - 18(z-2) = 0,$$
or
$$22x - 19y - 18z - 27 = 0.$$

Exercises 10-3

1. Find the plane equation which passes through the point $(3, 0, -1)$ and is parallel to the plane $3x - 7y + 5z - 12 = 0$.

2. Given the points $P_1(0, 4, -5), P_2(-1, -2, 2)$ and $P_3(4, 2, 1)$, find the equation of the plane passing through the three points.

3. Find the planes which pass through the points $P_1(2, -1, 1), P_2(3, -2, 1)$, and are parallel to three coordinate axes respectively.

4. Point out the specific positions of the following planes:
   (1) $x - y + 1 = 0$;
   (2) $4x - 4y + 7z = 0$;
   (3) $x + 2 = 0$;
   (4) $x + 5z = 0$.

5. Find the unit normal vectors and their direction cosines of the following planes:
   (1) $2x + 3y + 6z - 35 = 0$;
   (2) $x - 2y + 2z + 21 = 0$.

6. Find the angles between the following planes:
   (1) $x + y - 11 = 0, 3x + 8 = 0$;
   (2) $2x - 3y + 6z - 12 = 0, x + 2y + 2z - 7 = 0$.

7. Find the values of l and m making two planes $2x + my + 3z - 5 = 0$ and $lx - 6y - 6z + 2 = 0$ to be: (1) mutually perpendicular; (2) mutually parallel.

8. Find the distance from the point to the plane:
   (1) $P(-2, 4, 3), \pi: 2x - y + 2z + 3 = 0$;
   (2) $P(1, 2, -3), \pi: 5x - 3y + z + 4 = 0$;
   (3) $P(3, -5, -2), \pi: 2x - y + 3z + 11 = 0$.

9. Find equation of line which satisfies the following conditions.
   (1) Pass through the origin and be parallel to $s=(1,-1,1)$.
   (2) Pass through two points $(2,5,8),(-1,0,3)$.
   (3) Pass through the point $(2,-8,3)$ and be perpendicular to plane $x+2y-3z-2=0$.
   (4) Pass through the point $P(1,0,-2)$ and be perpendicular to two straight lines $\frac{x-1}{1}=\frac{y}{1}=\frac{z+1}{-1}$ and $\frac{x}{1}=\frac{y-1}{-1}=\frac{z+1}{0}$.

10. Write the common equation of a straight line $\begin{cases} x-y+2z-6=0, \\ 2x+y+z-5=0 \end{cases}$ into symmetric equation.

11. Find the angle between two straight lines $\frac{x-1}{3}=\frac{y+2}{6}=\frac{z-5}{2}$ and $\frac{x}{2}=\frac{y-3}{9}=\frac{z+1}{6}$.

12. Find the equation of a line which passes through the point $P(1,0,-2)$, and is parallel to the plane $3x-2y+2z-1=0$ and perpendicular to straight line $\frac{x-1}{4}=\frac{y-3}{-2}=\frac{z}{1}$.

13. Find the plane equation which passes the point $(2,0,-3)$ and is perpendicular to the straight line $\begin{cases} x-2y+4z-7=0, \\ 3x+5y-2z+1=0. \end{cases}$

14. Find the angle between the straight line $l: \frac{x}{-1}=\frac{y-1}{1}=\frac{z-1}{2}$ and the plane $\pi: 2x+y-z-3=0$.

15. Find the plane equation which passes the point $P(4,0,-1)$ and the straight line $\frac{x-4}{5}=\frac{y+3}{2}=\frac{z}{1}$.

16. Find the plane equation which passes the point $(1,0,-1)$ and is parallel to two straight lines $\frac{x-1}{2}=\frac{y-1}{1}=\frac{z+1}{1}$ and $\frac{x-2}{1}=\frac{y+1}{1}=\frac{z-3}{0}$.

10.4 Curved surface and space curve

In the previous section, we use vectors as a tool to discuss plane and space linear. In this section, we will discuss the curved surface and the space curve. By the relations between the curved surface or the space curve and their equations, we will attribute the geometric problems of studying curved surface and space curve to the algebraic problems of studying their equations.

10.4.1 Equation of spatial curved surface

Similar to regarding any plane curve as the geometric loci of a moving point by some laws, we could regard the spatial curved surface as the geometric loci of a moving point by some laws in space analytic geometry.

Chapter 10 Vectors and analytic geometry of space

1) Common equation of spatial curved surface

Since curved surface S is regarded as the locus of a moving point by some laws, S can be expressed as a ternary equation of the coordinates x, y, z of the moving point, that is,

$$F(x, y, z) = 0. \tag{1}$$

If the coordinates of an arbitrary point on curved surface S satisfy the Equation (1), and the points whose coordinates don't satisfy the Equation (1) aren't on the curved surface S, then the Equation (1) is called the equation of the curved surface S and the curved surface S is called the figure of the Equation (1) (Figure 10-28).

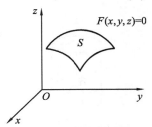

Figure 10-28

From the Example 3 and 4 of 10.1.3, we know plane and sphere are loci of a moving point and the equations of these loci are found. Now, we will introduce the common curved surface, cylinder and rotating surface.

2) Cylinder

Cylinder is the locus generated by a straight line L making parallel moving along with a curve C. The curve C is called the directrix of the cylinder and the straight line L is called the generatrix of the cylinder.

The directrix of a cylinder in this book is always a curve on the coordinate plane and the generatrix is always a straight line parallel to the coordinate axis. For example, find the equation of a cylinder whose directrix is the curve y on the coordinate plane xOy and generatrix is a straight line parallel to the z-axis. As shown in Figure 10-29, assume that $P(x, y, z)$ is any point on cylinder, then its projection $P_1(x_1, y_1)$ on the coordinate plane xOy must be on the curve L, that is, there exists a function F

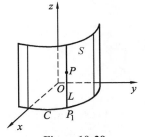

Figure 10-29

such that $F(x_1, y_1) = 0$ (Figure 10-29). And because every point on the curve L is on the curved surface S, in addition, for some point P on the generatrix L, assume that $|P_1 P| = z_1$, then the coordinate of the point P is (x_1, y_1, z_1). It's easy to know that (x_1, y_1, z_1) also satisfies the equation $F(x, y) = 0$. On the other hand, if $P'(x'_1, y'_1, z)$ satisfies $F(x', y') = 0$, the point P' must be on the cylinder.

Therefore, the space equation $F(x, y) = 0$ is the equation of the cylinder whose generatrix parallels to the z-axis and directrix is the curve $F(x, y) = 0$ on the plane xOy.

From this, we know that $F(x, y) = 0$ is a curve on the plane xOy or is a curved surface-cylinder in space.

In the same way, equations $F(y, z) = 0$ and $F(x, z) = 0$ are both the cylinders in space and their generatrixes are parallel to the x-axis and the y-axis respectively.

For example, the equation $x^2 + y^2 = R^2$ is the cylinder whose directrix is the circle $x^2 + y^2 = R^2$ on the plane xOy and the generatrix is parallel to the z-axis (Figure 10-30). Similarly, equations

$$\frac{x^2}{a^2}+\frac{y^2}{b^2}=1, \quad \frac{x^2}{a^2}-\frac{y^2}{b^2}=1, \quad x^2=2py, \quad x-y=0$$

are elliptic cylinder (Figure 10-31), hyperbolic cylinder (Figure 10-32), parabolic cylinder (Figure 10-33) and plane (Figure 10-34) whose generatrixes are all parallel to the z-axis.

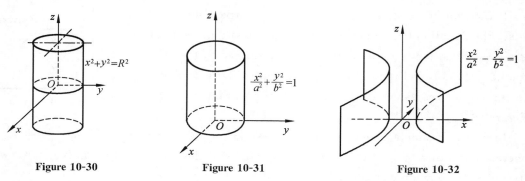

Figure 10-30 **Figure 10-31** **Figure 10-32**

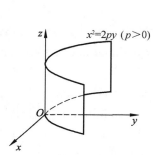 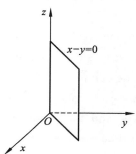

Figure 10-33 **Figure 10-34**

3) Surface of revolution

The curved surface generated by rotating a plane curve C around a fixed straight line on this plane is called the surface of revolution. The fixed straight line is called the axis of the surface of revolution and the curve C is called the generatrix of the surface of revolution. In the following discussion, the generatrix is a curve on the coordinate plane and the axis is some coordinate axis.

Assume that there is a curve L on the coordinate plane yOz and its equation is $f(y,z)=0$. The curved surface S is generated by rotating the curve L around the z-axis (Figure 10-35). Find the equation of the surface of revolution.

When curve L rotates around the z-axis, a point $P_1(0,y_1,z_1)$ on curve L will go to the position of the point $P(x,y,z)$. Because the point P_1 is on the curve L, its coordinates satisfy the equation

$$f(y_1,z_1)=0. \tag{2}$$

Meanwhile, the coordinates of P_1 and P satisfy the following relation, $z_1=z$, and

$$|y_1|=\sqrt{x^2+y^2} \quad \text{or} \quad y_1=\pm\sqrt{x^2+y^2}.$$

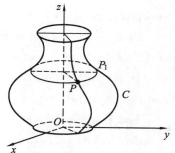

Figure 10-35

Plug y_1, z_1 into the Equation (2) and we obtain
$$f(\pm\sqrt{x^2+y^2},z)=0. \tag{3}$$
Equation (3) is satisfield only when $P(x,y,z)$ is on the surface of revolution, so it is the equation of the surface of revolution.

By the same way, the equation of the surface of revolution generated by rotating curve L around the y-axis is
$$f(y,\pm\sqrt{x^2+z^2})=0.$$

Generally, if the curve L on the coordinate plane rotates around a coordinate axis on this coordinate plane, to find the equation of the surface, it's just required to keep the variable same to the name of the axis and substitute the other variable by the square root of the quadratic sum of the other two variables.

Example 1 Let the following curves on plane yOz rotate around the given coordinate axes, find the equation of the surface of revolution.

(1) $\dfrac{y^2}{a^2}+\dfrac{z^2}{b^2}=1$ rotates around the y-axis;

(2) $\dfrac{y^2}{a^2}-\dfrac{z^2}{b^2}=1$ rotates around the y-axis;

(3) $y^2=2pz$ $(p>0)$ rotates around the z-axis;

(4) $z=ky(k>0)$ rotates around the z-axis.

Solution (1) By the technique above, we have
$$\frac{y^2}{a^2}+\frac{x^2+z^2}{b^2}=1,$$
which is called the ellipsoid of revolution.

(2) The equation is
$$\frac{y^2}{a^2}-\frac{x^2+z^2}{b^2}=1,$$
which is called the hyperboloid of revolution.

(3) The equation is
$$x^2+y^2=2pz(p>0),$$
which is called the paraboloid of revolution.

(4) The equation is
$$z=\pm k\sqrt{x^2+y^2} \text{ 或 } z^2-k^2(x^2+y^2)=0,$$
which is called the circular conical surface. The vertex of this circular conical surface is the origin, and regard the z-axis as symmetry axis.

10.4.2 Equation of spatial curve

1) Common equation

Any space curve L can be regarded as the intersection line of two curved surfaces passing through this curve. Assume that equations of the two curved surfaces are $F_1(x,y,z)=0$ and $F_2(x,y,z)=0$ respectively, and the intersection curve is L.

Then any point on curve L is also on the two curved surfaces, so for any point (x,y,z) on the curve L satisfy the equation set

$$\begin{cases} F_1(x,y,z)=0, \\ F_2(x,y,z)=0. \end{cases} \quad (4)$$

On the other hand, for any point whose coordinates satisfy the equation set (4) is on the two curved surfaces or on the intersection line L of the two curved surfaces (Figure 10-36). Therefore, equation set (4) is the equation of the space curve L, which is called the common equation.

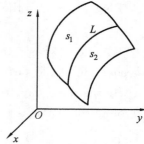

Figure 10-36

There are infinite curved surfaces passing through the space curve L, so the common equation of the curve L isn't unique.

Example 2 Find the equation of the z-axis.

Solution The z-axis can be regarded as the intersection line of the plane yOz and the plane xOz (Figure 10-37), so its equation is

$$\begin{cases} x=0, \\ y=0. \end{cases}$$

This equation set above has the same solution as the equation set

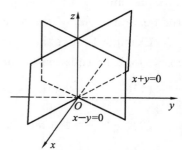

Figure 10-37

$$\begin{cases} x+y=0, \\ x-y=0, \end{cases}$$

so the equation of the z axis can also be the second equation set.

Example 3 Which curve is expressed by the equation set $\begin{cases} x^2+y^2+z^2=R^2, \\ z=0? \end{cases}$ And find another two equations of this curve.

Solution $x^2+y^2+z^2=R^2$ is the equation of the spherical surface with radius R and center at the origin. $z=0$ is the plane xOy. So the equation set expresses the intersection curve of them, or the circle on the plane xOy with center at the origin and radius R. This curve can also be expressed by the following equation sets

$$\begin{cases} x^2+y^2=R^2, \\ z=0 \end{cases} \quad \text{or} \quad \begin{cases} x^2+y^2+z^2=R^2, \\ x^2+y^2=R^2. \end{cases}$$

2) Parametric equation

Similar to the plane curve, space curve can also be expressed by a parametric equation. When regard the space curve as the movement locus of a particle, we often use the parametric representation.

In plane analytic geometry, the parametric equation of a plane curve is

$$\begin{cases} x=x(t), \\ y=y(t), \end{cases} \quad t \text{ is a parameter.}$$

Similarly, we also can use the function of the parameter t to express the space curve

$$\begin{cases} x=x(t), \\ y=y(t), \quad t \text{ is a parameter.} \\ z=z(t), \end{cases} \qquad (5)$$

Equation (5) is called the parametric equation of space curve.

Example 4 Assume that a point P on cylindrical surface goes along on the cylindrical surface $x^2+y^2=a^2$ and rotates around the z-axis with constant angular velocity ω. Then rise along in the positive direction which parallels to the z-axis with linear velocity v (ω, v are both constants). And the movement locus of moving the point P is called the cylindrical spiral or helix. Find the parametric equation of the cylindrical spiral.

Solution Establish the coordinate system as Figure 10-38 and let time t be parameter. Assume that when $t=0$, the point P is on $A(a, 0, 0)$. Then the point moves to the point $P(x,y,z)$ and the projection of the point P on the plane xOy is $P'(x,y,0)$. Obviously, point P' is on the base circle and the moving point rotates around the z-axis with angular velocity ω, so $\angle AOP'=\omega t$ and

$$x=|OP'|\cos\angle AOP'=a\cos\omega t,$$
$$y=|OP'|\sin\angle AOP'=a\sin\omega t.$$

The moving point rises along in the positive direction which parallels to the z-axis with linear velocity v, so

$$z=P'P=vt.$$

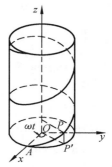

Figure 10-38

Therefore, the parametric equation of cylindrical spiral is

$$\begin{cases} x=a\cos\omega t, \\ y=a\sin\omega t, \quad 0<t<+\infty. \\ z=vt, \end{cases}$$

If regard $\theta=\omega t$ as a parameter and make $b=\dfrac{v}{\omega}$, the parametric equation of cylindrical spiral can also be written into

$$\begin{cases} x=a\cos\theta, \\ y=a\sin\theta, \quad 0<\theta<\infty. \\ z=b\theta, \end{cases}$$

3) Projection of space curve on coordinate plane

Assume that the equation of space curve L is

$$\begin{cases} F_1(x,y,z)=0, \\ F_2(x,y,z)=0. \end{cases}$$

Eliminate z from this equation set and then we obtain a cylindrical surface which passes through the curve L

$$F(x,y)=0. \qquad (6)$$

The coordinates of every point on the space curve L all satisfy the Equation (6). We know from the concept of cylindrical surface that this is a cylindrical equation whose directrix is curve $F(x,y)=0$ on the plane xOy and generatrix is parallel to the z-axis. Because the cylindrical surface includes curve L, this cylindrical surface is called the projected cylindrical surface of the curve L on the plane xOy. The intersection curve of the projected cylindrical surface and the plane xOy is

$$\begin{cases} F(x,y)=0, \\ z=0, \end{cases} \quad (7)$$

which is called the projected curve of the curve L on the plane xOy.

By the same way, eliminate x or y from the equation set, we will obtain $G(y,z)=0$ or $H(x,z)=0$, which are the projected cylindrical surfaces on the plane yOz or on the plane xOz. The projected curves on the plane yOz and the plane xOz are

$$\begin{cases} G(y,z)=0, \\ x=0, \end{cases} \quad \text{and} \quad \begin{cases} H(x,z)=0, \\ y=0. \end{cases}$$

Example 5 Find the projected curves of space curve

$$L: \begin{cases} 2x^2+y^2+z^2=16, \\ x^2-y^2+z^2=0 \end{cases}$$

on the three coordinate planes.

Solution To get the projected cylindrical surface on the plane xOy, eliminate z from L's equation and we obtain

$$x^2+2y^2=16.$$

Therefore the projected curve on the plane xOy is

$$\begin{cases} x^2+2y^2=16, \\ z=0. \end{cases}$$

Similarly, we find the projected curves on the plane yOz and the plane xOz are

$$\begin{cases} 3y^2-z^2=16, \\ x=0 \end{cases} \quad \text{and} \quad \begin{cases} 3x^2+2z^2=16, \\ y=0. \end{cases}$$

In the later chapter, we will introduce the technique to find the projected curve of a space curve on the coordinate plane by multiple integral.

10.4.3 Quadric surface

The curved surface determined by a ternary quadric equation is called the quadric surface such as spherical surface, cylindrical surface, paraboloid of revolution and so on. Now we will introduce some common quadric surfaces and their equations, then use the standard equation of quadric surfaces to discuss the figure of quadric surface.

1) Ellipsoidal surface

In space right angle coordinate system, the curved surface expressed by

$$\frac{x^2}{a^2}+\frac{y^2}{b^2}+\frac{z^2}{c^2}=1 \ (a,b,c>0) \quad (8)$$

is called the ellipsoidal surface.

If $a=b$ or $b=c$, it is called the ellipsoidal surface of revolution. In particular, if $a=b=c$, the Equation (8) becomes $x^2+y^2+z^2=a^2$, which is a spherical surface. That is to say, the ellipsoidal surface of revolution and the spherical surface are both particular cases of the ellipsoidal surfaces.

Known from the Equation (8), ellipsoidal surface is symmetric about three coordinate planes, three coordinate axes and the origin. There are also

$$\frac{x^2}{a^2} \leqslant 1, \ \frac{y^2}{b^2} \leqslant 1, \ \frac{z^2}{c^2} \leqslant 1, \tag{9}$$

or
$$|x| \leqslant a, \ |y| \leqslant b, \ |z| \leqslant c.$$

This means that ellipsoidal surface is in a cuboid surrounded by planes $x = \pm a, y = \pm b, z = \pm c$ and the curved surface is bounded. a, b, c are called the semiaxis of the ellipsoidal surface.

According to the discussion above, we know that the ellipsoidal surface is the figure in Figure 10-39.

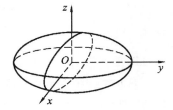

Figure 10-39

2) Hyperbolic surface

(1) In space right angle coordinate system, the curved surface expressed by

$$\frac{x^2}{a^2} + \frac{y^2}{b^2} - \frac{z^2}{c^2} = 1 \ (a,b,c > 0) \tag{10}$$

is called the uniparted hyperboloid.

Known from the Equation (10), uniparted hyperboloid is symmetric about three coordinate planes, three coordinate axes and the origin, and its figure can be seen in Figure 10-40.

(2) In space right angle coordinate system, the curved surface of the equation

$$\frac{x^2}{a^2} + \frac{y^2}{b^2} - \frac{z^2}{c^2} = -1 \ (a,b,c > 0) \tag{11}$$

is called the biparted hyperboloid.

Known from the Equation (11), biparted hyperboloid is symmetric about three coordinate planes, three coordinate axes and the origin. And the curved surface doesn't exist if $-c < z < c$, but it is a point at $z = \pm c$. The curved surface is above $z = c$ or below $z = -c$, so it is biparted and its figure can be seen in Figure 10-41.

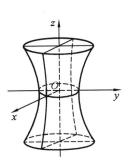   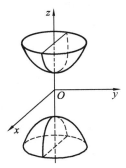

Figure 10-40   Figure 10-41

3) Parabolic surface

(1) In space right angle coordinate system, the curved surface

$$\frac{x^2}{a^2}+\frac{y^2}{b^2}=2z \quad (a,b>0) \tag{12}$$

is called the elliptical paraboloid.

Known from the Equation (12), the elliptical paraboloid is symmetric about the plane xOz, the plane yOz and the z-axis, but not symmetric about the plane xOy, the x-axis, the y-axis and the origin. The curved surface has no symmetric center and is above the plane xOy, $z \geqslant 0$. The figure can be seen in Figure 10-42.

(2) In space right angle coordinate system, the curved surface

$$\frac{x^2}{a^2}-\frac{y^2}{b^2}=2z \quad (a,b>0) \tag{13}$$

is called the hyperbolic paraboloid.

Known from the Equation (13), the hyperbolic paraboloid is symmetric about the plane xOz, the plane yOz and the z-axis, but not symmetric to the plane xOy, the y-axis and the origin. The curved surface has no symmetric center and its figure can be seen in Figure 10-43. Because the figure is like a saddle, it is also called the saddle surface (Figure 10-43).

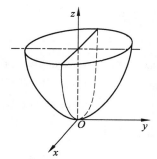

Figure 10-42

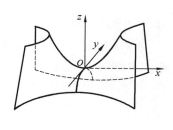

Figure 10-43

Exercises 10-4

1. Assume that there are two points $A(2,3,1)$ and $B(6,-4,2)$, find the locus of the moving point P which satisfies the condition $2|\overrightarrow{PA}|=|\overrightarrow{PB}|$.

2. Describe the curved surfaces expressed by the following equations:
  (1) $3x^2+4y^2=25$;
  (2) $y=2x^2$;
  (3) $z^2-x^2=1$;
  (4) $y=x+1$.

3. Find the rotating curved surfaces generated by the following plane curves rotating around the given axes:
  (1) $\frac{x^2}{4}+\frac{y^2}{9}=1$ rotates around the x-axis;
  (2) $x^2-z^2=1$ rotates around the z-axis;
  (3) $z^2=5x$ rotates around the x-axis;
  (4) $4x^2-9y^2=36$ rotates around the y-axis.

4. Describe the figures expressed by the following simultaneous equations:
  (1) $\begin{cases} x+y+z=3, \\ x+2y=1; \end{cases}$
  (2) $\begin{cases} \dfrac{x^2}{25}-\dfrac{y^2}{16}=1, \\ z=3. \end{cases}$

5. Write the following parametric equations of curves into the common equations:

(1) $\begin{cases} x=6t+1, \\ y=(t+1)^2, \quad -\infty<t<+\infty; \\ z=2t, \end{cases}$

(2) $\begin{cases} x=3\sin t, \\ y=5\sin t, \quad 0\leqslant t<2\pi. \\ z=4\cos t, \end{cases}$

6. Write the following curves' common equations into the parametric equations:

(1) $\begin{cases} x^2+y^2+z^2=9, \\ y=x; \end{cases}$

(2) $\begin{cases} (x-1)^2+y^2+(z-1)^2=4, \\ z=0. \end{cases}$

7. Find the equations of the projective cylinders and the projective curves of the space curve

$$\begin{cases} x^2+z^2-3yz-2x+3z-3=0, \\ x+y+z=1 \end{cases}$$

on three coordinate planes.

8. Describe the curved surfaces expressed by the following equations:

(1) $x^2+y^2+4z^2-1=0$;

(2) $-x^2+\dfrac{y^2}{2}+\dfrac{z^2}{3}=1$;

(3) $x^2+\dfrac{y^2}{4}-\dfrac{z^2}{6}=-1$;

(4) $x^2+y^2-z=0$;

(5) $x^2-y^2-z^2-1=0$;

(6) $3x^2-4y^2+12z=0$.

9. Describe the curves expressed by the following equations:

(1) $\begin{cases} \dfrac{x^2}{25}+\dfrac{y^2}{16}+\dfrac{z^2}{9}=1, \\ x=3; \end{cases}$

(2) $\begin{cases} x^2+y^2+z^2=16, \\ y=2; \end{cases}$

(3) $\begin{cases} \dfrac{x^2}{25}-\dfrac{z^2}{16}=1, \\ y=4; \end{cases}$

(4) $\begin{cases} \dfrac{x^2}{16}+\dfrac{y^2}{16}-\dfrac{z^2}{9}=1, \\ z=4. \end{cases}$

10. Draw the solid figures surrounded by the following curved surfaces.

(1) The solid figure surrounded by $\dfrac{x^2}{25}+\dfrac{y^2}{9}+\dfrac{z^2}{16}=1$ in the first octant.

(2) $y=0, z=0, 3x+y=6, 3x+2y=12, x+y+z=6$.

Summary

Analytic geometry uses algebraic methods to study geometric problems. In order to introduce the algebraic methods, it must use the algebraization of geometric construction in space, which is also the basic of analytic geometry. On the basis of establishing the space right angle coordinate system, we get the one-to-one correspondence between vectors and groups of ordered real numbers(coordinates or components), points and groups of ordered real numbers (coordinates) by introducing concepts of vector and coordinate, then the geometric construction in space is quantified. Thus operations of vectors are also transformed into the operations of numbers that brings great convenience to calculation.

Use vector and its coordinates to discuss space line and plane, and transform the study of geometric problems into the discussion of algebraic equations.

1. Main contents

(1) Space right angle coordinate system.

The method of establishing the space right angle coordinate system is similar to that of the plane right angle coordinate system. Passing through a point O in space to elicit three coordinate axes which are mutually perpendicular and then a space right angle coordinate system $O\text{-}xyz$ is constituted. In space right angle coordinate system, point P in space has one-to-one correspondence to a group of ordered real numbers (x, y, z) which is the relation of points in geometry and quantities in algebra. Then use the coordinates to get the basic formula in space analytic geometry—distance formula between two points

$$|P_1P_2| = \sqrt{(x_2-x_1)^2+(y_2-y_1)^2+(z_2-z_1)^2}.$$

(2) Concept and operation of vector.

A vector is expressed by directed line segment, and size, direction are two factors of the vector. The size of a vector is called the vector's norm, which is expressed by the length of the directed line segment and the direction is from the starting point to the terminal point. To get the algebraization of vector's operations, we introduce the algebraic representation of vector in the right angle coordinate system

$$\boldsymbol{a} = a_x\boldsymbol{i} + a_y\boldsymbol{j} + a_z\boldsymbol{k} = (a_x, a_y, a_z).$$

Use vector's coordinates, we can transform vector's various operations into coordinate operations.

Assume that $\boldsymbol{a} = a_x\boldsymbol{i} + a_y\boldsymbol{j} + a_z\boldsymbol{k}, \boldsymbol{b} = b_x\boldsymbol{i} + b_y\boldsymbol{j} + b_z\boldsymbol{k}, \boldsymbol{c} = c_x\boldsymbol{i} + c_y\boldsymbol{j} + c_z\boldsymbol{k}$, then

$$\boldsymbol{a} \pm \boldsymbol{b} = (a_x \pm b_x)\boldsymbol{i} + (a_y \pm b_y)\boldsymbol{j} + (a_z \pm b_z)\boldsymbol{k};$$

$$\lambda\boldsymbol{a} = (\lambda a_x)\boldsymbol{i} + (\lambda a_y)\boldsymbol{j} + (\lambda a_z)\boldsymbol{k};$$

$$\boldsymbol{a} \cdot \boldsymbol{b} = a_xb_x + a_yb_y + a_zb_z;$$

$$\boldsymbol{a} \times \boldsymbol{b} = \begin{vmatrix} \boldsymbol{i} & \boldsymbol{j} & \boldsymbol{k} \\ a_x & a_y & a_z \\ b_x & b_y & b_z \end{vmatrix}.$$

Meanwhile, we can use vector's coordinates to express vector's norm and the conditions of the vectors being parallel or perpendicular.

$$|\boldsymbol{a}| = \sqrt{a_x^2+a_y^2+a_z^2}; \quad \boldsymbol{a}^\circ = \frac{\boldsymbol{a}}{|\boldsymbol{a}|} = \frac{1}{\sqrt{a_x^2+a_y^2+a_z^2}}(a_x, a_y, a_z);$$

$$\boldsymbol{a} /\!/ \boldsymbol{b} \Leftrightarrow \boldsymbol{a} \times \boldsymbol{b} = \boldsymbol{0} \Leftrightarrow \frac{a_x}{b_x} = \frac{a_y}{b_y} = \frac{a_z}{b_z};$$

$$\boldsymbol{a} \perp \boldsymbol{b} \Leftrightarrow \boldsymbol{a} \cdot \boldsymbol{b} = 0 \Leftrightarrow a_xb_x + a_yb_y + a_zb_z = 0;$$

$$\cos(\widehat{\boldsymbol{a}, \boldsymbol{b}}) = \frac{\boldsymbol{a} \cdot \boldsymbol{b}}{|\boldsymbol{a}||\boldsymbol{b}|} = \frac{a_xb_x + a_yb_y + a_zb_z}{\sqrt{a_x^2+a_y^2+a_z^2}\sqrt{b_x^2+b_y^2+b_z^2}}.$$

(3) Space linear and plane equation.

In analytic geometry, determining a plane and a straight line in space means determining its equation. Equations of plane are mainly the point normal form and the general form.

Point normal form $A(x-x_0)+B(y-y_0)+C(z-z_0)=0$.

General form $Ax+By+Cz+D=0$.

Equations of straight line mainly include the symmetric form, the parametric form and the general form.

Symmetric form $\dfrac{x-x_0}{m}=\dfrac{y-y_0}{n}=\dfrac{z-z_0}{p}$.

Parametric form $\begin{cases} x=x_0+mt, \\ y=y_0+nt, \\ z=z_0+pt, \end{cases}$ t is a parameter.

General form $\begin{cases} A_1x+B_1y+C_1z+D_1=0, \\ A_2x+B_2y+C_2z+D_2=0. \end{cases}$

The key to find the plane equation is to determine a given point $P_0(x_0,y_0,z_0)$ on plane and the normal vector $\boldsymbol{n}=(A,B,C)$. The key to find the linear equation is to determine a fixed point $P_0(x_0,y_0,z_0)$ on straight line and its direction vector $\boldsymbol{s}=(m,n,p)$.

When discuss relations of planes, relations of straight lines or relations of plane and straight line, it is boil down to the relation of the plane's normal vector and the straight line's direction vector.

(4) Space curve and curved surface.

In space, a ternary equation $F(x,y,z)=0$ usually is a curved surface, the space curve is often expressed by the equation set of two ternary equations

$$\begin{cases} F_1(x,y,z)=0, \\ F_2(x,y,z)=0. \end{cases}$$

For several kinds of curved surface, it's required to understand the equation's properties and judgments. The standard equation of spherical surface is

$$(x-x_0)^2+(y-y_0)^2+(z-z_0)^2=R^2,$$

and its general equation is a ternary quadric equation whose coefficients of quadric terms are same and has no cross term. The equation of the cylindrical surface whose generatrix is parallel to the coordinate axis lacks one variable. Surface of revolution's equation has the quadratic sum of two coordinates and their coefficients are same. Equation of the curved surface in center form (ellipsoidal or hyperbolic type) can be written as $Ax^2+By^2+Cz^2=1$. Equation of the curved surface in non-center form (parabolic type) can be written as $Ax^2+By^2=2z$.

It's necessary for the study of multivariate calculus and following courses to be familiar and understand equations and figures of common quadric surfaces.

2. Basic requirements

(1) Understand concepts of vector correctly and master algebraic operations of vector.

(2) Be familiar to various equations of plane and space linear, and be able to use parallel, perpendicular conditions to find equations of straight line and plane.

(3) Know equations of space curved surface and space curve, and be able to find

equations of curved surface and curve.

(4) Be able to describe spherical surface, cylindrical surface, conical surface and surface of revolution according to given equations.

(5) Be familiar to standard equations and figures of ellipsoidal surface, hyperboloid and paraboloid.

The emphasis of this chapter are as follows, concepts and operations of vector, plane and space linear equations, equations and figures of common quadric surfaces.

Quiz

1. Given that $A(1,2,1), \overrightarrow{AB}=(0,2,3)$, find:
(1) the coordinate of point B; (2) $|\overrightarrow{AB}|$; (3) $\overrightarrow{AB}°$.

2. Given that $\boldsymbol{a}=(2,-3,1), \boldsymbol{b}=(1,-2,3), \boldsymbol{c}=(2,1,2)$, find:
(1) $2\boldsymbol{a}+3\boldsymbol{b}-4\boldsymbol{c}$; (2) $\boldsymbol{a}\cdot\boldsymbol{b}$ and $\boldsymbol{a}\times\boldsymbol{b}$; (3) $(\boldsymbol{a},\boldsymbol{b},\boldsymbol{c})$.

3. Given that $A(1,1,2), B(2,2,1), C(2,1,2)$, find the area of the triangle ABC.

4. Find the plane equation which passes through the point $(0,1,4)$ and is parallel to the plane $x-3y+4z-2=0$.

5. Find the plane equation which passes through the two points $P_1(4,0,-2), P_2(5,1,7)$ and is parallel to the x-axis.

6. Find the linear equation which passes through the point $(1,1,1)$ and is parallel to the straight line $\begin{cases} x+y+3z=0, \\ x-y-z=0. \end{cases}$

7. Find the intersection of the straight line $\dfrac{x+1}{2}=\dfrac{y}{3}=\dfrac{z-3}{6}$ and the plane $10x+2y-11z+3=0$.

8. Find the names of curve surfaces or curves expressed by the following equations or simultaneous equations:

(1) $x^2+y^2-2ax=0$;

(2) $\dfrac{x^2}{25}-\dfrac{y^2}{16}+\dfrac{z^2}{25}=1$;

(3) $x^2+\dfrac{1}{4}y^2+z^2=1$;

(4) $x^2+y^2=2z$;

(5) $\begin{cases} x^2-y^2=1, \\ z=0; \end{cases}$

(6) $\begin{cases} 2x+3y+1=0, \\ x-3y+4z=0. \end{cases}$

9. Find projective curvilinear equations of curve $\begin{cases} 2x^2+3y^2+z^2=1, \\ x+y+z=1 \end{cases}$ on three coordinate planes.

Exercises

1. Use vector to show that if a quadrangle's diagonals bisect each other on the plane, then this quadrangle is a parallelogram.

2. Given that $\overrightarrow{AB}=2i-3i-k$, and the point $A(1,1,1)$, find:
(1) The coordinate of point B.　　　(2) The unit vector on $\overrightarrow{AB}$.
3. Given that $a=(1,2,-2)$ and $b=(3,4,0)$, find:
(1) $a \cdot b, a \times b, (a+b) \cdot (a-b)$.
(2) The unit vector which is perpendicular to a,b simultaneously.
4. Given that triangle's three vertexes are $A(1,2,3), B(2,-1,2), C(3,2,3)$, find the area of this triangle and the height on side AB.
5. Given that tetrahedron's four vertexes are $A(2,3,1), B(4,1,-2), C(6,3,7)$, $D(-5,4,8)$, find the volume of this tetrahedron and the height extended from vertex D.
6. Find the plane equation which passes through the two points $P_1(3,-2,9)$, $P_2(-6,0,4)$ and is perpendicular to plane $2x-y+4z-8=0$.
7. Assume that the plane is $x+ky-2z-9=0$.
(1) Find k such that the plane is perpendicular to plane $2x+4y+3z-3=0$.
(2) Find k such that the plane is parallel to plane $3x-7y-6z-1=0$.
8. Find the linear equations which satisfy the following conditions.
(1) Pass through the origin and be perpendicular to the x-axis and the straight line $\dfrac{x-3}{3}=\dfrac{y-6}{2}=\dfrac{2}{-1}$.
(2) Pass through the point $(3,-3,2)$ and be parallel to the planes $x-4y+10=0$ and $3x+5y-z-4=0$.
9. Find the intersections of the following straight lines and planes.
(1) Straight line $\dfrac{x-1}{2}=\dfrac{y-12}{3}=\dfrac{z-9}{3}$ and plane $x+3y-5z-2=0$.
(2) Straight line $2x-y-2=0, 3y-2z+2=0$ and plane $y+2z-2=0$.
10. Find the linear equation which passes through the point $(-1,0,4)$ and is parallel to the plane $3x-4y+z-10=0$ and is perpendicular to the straight line $\dfrac{x+1}{1}=\dfrac{y-3}{1}=\dfrac{z}{2}$.
11. Find the spherical equation whose center is the origin and passes through the point $(-2,1,2)$.
12. Find names of the following curved surfaces:
(1) $x^2+y^2+z^2-2x+4y+2z=0$;　　(2) $x^2+y^2=4$;
(3) $z=x^2-y^2$;　　(4) $3x^2-2y^2+5z^2=-1$.
13. Find the following curves' projective cylinders and projective curves on plane xOy:
(1) $\begin{cases} x^2+y^2=z, \\ z=2-x^2-y^2; \end{cases}$　　(2) $\begin{cases} z^2=x^2+y^2, \\ z^2=2y. \end{cases}$
14. Draw the solid figures surrounded by the following curve surface:
(1) $x^2+y^2=z$ and $z=4$;　　(2) $z=\sqrt{x^2+y^2}$ and $z=2-x^2-y^2$.

Chapter 11 Differentiation of multivariable function and its application

So far, we have discussed the function that has only one variable. And many problems in the natural science and engineering technology are often related to a variety of factors. To reflect in mathematics, the relationship is that a variable depends on many other variables. Therefore, we introduce the concepts and calculus of multivariable function. In this chapter, we will discuss the differentiation of multivariable function and its application.

Multivariable function differential calculus is an generalization of the one-variable function. This chapter mainly includes the following two parts. The first part is the concepts of differentiation of multivariable function and its geometric interpretation, which is made up of partial derivatives, directional derivatives, perfect differentials and gradients. The second part is the computation and application of Multivariable function differential calculus, which is made up of differentiation of compound function and implicit function.

The following is a network diagram about the applications of calculus in the economy.

The application in economy
- The application of series economy
  - compound interest
  - annual effective income
- The application of limit economy—continuous compound interest compound interest
- The application of derivative in economy
  - cost function—average minimum cost
  - demand function
  - supply function
  - equilibrium price
  - revenue function
  - profit function—maximum profit
  - marginal function
  - elastic function
    - supply elasticity
    - demand elasticity
- The application of integral in economy
  - present value of income stream
  - future value of income stream
  - consumer surplus
  - producer surplus
- The application of partial derivatives in economy—seek maximum profit
- The applications of ordinary differential equations and difference equations in economy.— convert some problems into ordinary equations to solve

Chapter 11 Differentiation of multivariable function and its application

11.1 The concept of multivariable function

11.1.1 Plane point set and *n*-dimensional space

When we discuss the function of one variable, we have to consider the variation range of variables by using the concept of neighborhood or interval. When we discuss the basic concept about the function of two variables, we need to extend the concept to obtain the concept of planar point set and interval. Therefore, firstly, we introduce the planar point set and the interval, and extend the neighborhood in line to that in plane.

1) Plane point set

Because there is a one-to-one correspondence between the binary ordered real numbers (x_0, y_0) and the point P in the plane, so (x_0, y_0) can be regarded as the coordinates of the point P, denoted by $P(x_0, y_0)$. By this way, the set of points with a certain property M in the plane is called the set of points E, denoted by

$$E = \{(x,y) \mid (x,y) \text{ with property } M\}.$$

The whole plane is often regarded as a two-dimensional space, which is denoted by $\mathbf{R}^2$, and $E \subset \mathbf{R}^2$ is used to denote the planar point set.

For example, $E = \{(x,y) \mid x^2 + y^2 < 1\}$ indicates all points (x, y) satisfies the inequality $x^2 + y^2 < 1$, that is, it is a set of all points in the unit circle with center at the origin. Now let's introduce the concept of neighborhood in plane.

Suppose the point $P_0(x_0, y_0)$, δ is a positive number, then the set of the point $P(x, y)$ such that $|P_0 P| < \delta$ is called the neighborhood of P_0 with radius δ, denoted by $U(P_0, \delta)$. We have $U(P_0, \delta) = \{P \mid |P_0 P| < \delta\}$, that is

$$U(P_0, \delta) = \{(x,y) \mid \sqrt{(x-x_0)^2 + (y-y_0)^2} < \delta\}.$$

The neighborhood without the point P_0 is denoted by $\overset{\circ}{U}(P_0, \delta)$, and

$$\overset{\circ}{U}(P_0, \delta) = \{P \mid 0 < |P_0 P| < \delta\}.$$

In geometry, the neighborhood $U(P_0, \delta)$ $(\delta > 0)$ is all points $P(x,y)$ in the circle with center $P_0(x_0, y_0)$ and radius δ, in which δ is called the radius of the neighborhood $U(P_0, \delta)$. If there is no need to emphasize the radius δ, we can also use $U(P_0)$ or $\overset{\circ}{U}(P_0)$ to denote the neighborhood.

Then, we will describe the relationship of points and point set by the neighborhood.

(1) Interior point. Suppose E is a point set in the plane, $P \in E$. If there is a neighborhood of point P such that $U(P) \subset E$, then P is called an interior point of E (Figure 11-1). Apparently, the interior point of E belongs to E.

(2) External point. Suppose E is a point set of a plane. If there is a neighborhood $U(P_1)$ such that $U(P_1) \cap E = \varnothing$, P_1 is called the external point of the point set E (Figure 11-1). Obviously, the external point does not belong to E.

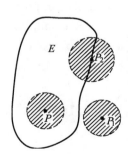

Figure 11-1

(3) Boundary point. Suppose E is a point set of a plane. For any neighborhood $U(P_2)$ if it includes the point of E and the point that does not belong to E, then the point P_2 is called the boundary point of E. (Figure 11-1)

The set of all boundary points of E is called the boundary of E, which is denoted as ∂E and the boundary points of E may belong to E or not.

For example, suppose the planar point set $E=\{(x,y)\mid 1<x^2+y^2\leqslant 2\}$, the interior points of E satisfy $1<x^2+y^2<2$, the points on the boundary $x^2+y^2=1$ don't belong to E, and the points on on the boundary $x^2+y^2=2$ belong to E.

In addition to the above three kinds of relations, there is another one, the accumulation point.

Accumulation point. Suppose E is a planar point set, P is a point in the plane, which can belong to E or not. If any neighborhood of P without center always includes the points of the point set E, P is called the accumulation point of E.

For example, in the point set $E_2=\{(x,y)\mid 0<x^2+y^2\leqslant 1\}$, the point $O(0,0)$ is the boundary point and the accumulation point, but it does not belong to E_2, while each point on the circle $x^2+y^2=1$ is the boundary point and accumulation point and they also belong to E_2 (Figure 11-2).

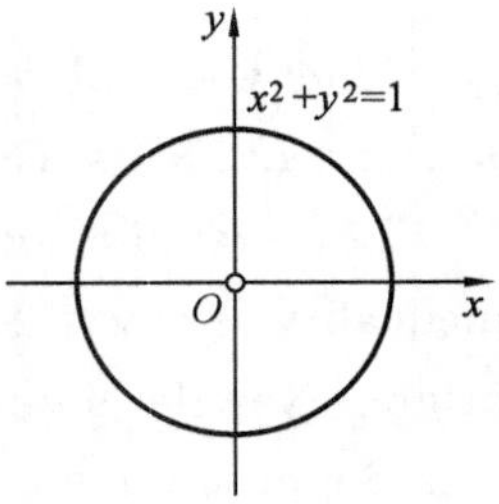

We define some planar point set according to the characteristics of the points.

(1) Open set. If any point in set E is interior point, E is called an open set.

Figure 11-2

(2) Closed set. If the complementary set E^c is an open set, E is called a closed set.

(3) Connected set. If any two points P_1 and P_2 in set E can be connected by a broken line, and all points on the line belong to E, then E is connected.

(4) The region (or open region). The connected open set is called region or open region.

(5) Closed region. An open region, along with its boundary, is called a closed region.

Obviously, if E is a region, the interior points and the boundary points are all the accumulation points of E.

For example, $\{(x,y)\mid x+y>0\}$, $\{(x,y)\mid x^2+y^2<1\}$ are both regions (Figure 11-3), while $\{(x,y)\mid x^2+y^2\leqslant 1\}$ is a closed region.

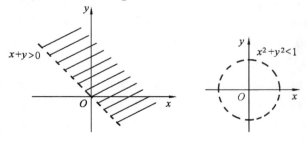

Figure 11-3

Chapter 11 Differentiation of multivariable function and its application

(6) Bounded set. If there is a positive number r in a planar point set E such thar $E \subset U(O, r)$, where O is the origin, then E is called a bounded set.

(7) Unbounded set. If a set is not a bounded set, it is an unbounded set.

(8) Bounded region. If the distance between any point $P(x, y)$ of the region D and a certain point A is always no more than a positive number M, that is, $|AP| \leqslant M$, the region D is bounded region. Otherwise, it is called an unbounded region.

For example, $\{(x,y) | x^2 + y^2 \leqslant 1\}$ is a bounded closed region and $\{(x,y) | x+y > 0\}$ is an unbounded open region.

2) n-dimensional space

We know there is the one-to-one correspondence between the point M in the number axis and a real number x, so x could represent all points in the number axis. They are called one-dimensional space, denoted by $\mathbf{R}^1$. Similarly, there is the one-to-one correspondence between the planar point M and ordered variables (x, y) in the right angle coordinate system, then (x, y) could represent all points in the plane. They are called two-dimensional space and denoted by $\mathbf{R}^2$. There is the one-to-one correspondence between the point M and ordered variables (x, y, z) in the space right angle coordinate system, (x, y, z) represents all points in the space. They are called three-dimensional space and denoted by $\mathbf{R}^3$.

In conclusion, ordered n variables $(x_1, x_2, \cdots, x_n)$ can be called n-dimensional space and denoted by $\mathbf{R}^n$. And the ordered n variables $(x_1, x_2, \cdots, x_n)$ represents a point M in n-dimensional space, which is denoted by $M(x_1, x_2, \cdots, x_n)$. The number $x_i (i=1, 2, \cdots, n)$ is called the i th coordinates.

Suppose the points $M(x_1, x_2, \cdots, x_n), N(y_1, y_2, \cdots, y_n) \in \mathbf{R}^n$ and the distance between M and N is

$$|MN| = \sqrt{(y_1 - x_1)^2 + (y_2 - x_2)^2 + \cdots + (y_n - x_n)^2}.$$

Obviously, when $n=1, 2, 3$, the above formula is the distance formula between two points in the straight line, plane or space.

Then, we can define the neighborhood in $\mathbf{R}^n$. Suppose $P \in \mathbf{R}^n$ and δ is a positive number, the point set

$$U(P_0, d) = \{P | |PP_0| < \delta, P \in \mathbf{R}^n\}$$

is called the neighborhood of P_0 with radius δ. Now, similarly, we also can define the interior point, boundary point, region, point of accumulation, etc..

11.1.2 Multivariable function

We focus on functions of two or three variables and multivariable functions in this part. The function of one variable contains one independent variable, $y = f(x)$. A multivariate function contains several independent variables, such as, the function of two variables $z = f(x, y)$, and the function of three variables $u = f(x, y, z)$, etc..

We know, if there are two variables x and y and there is only one value y corresponding to a given value x, y is the function of x, denoted by $y = f(x)$, in which x is the independent variable and y is the dependent variable.

Since there is only one independent variable in the function, it is called the function of one variable. The set of values taken by x is called the domain of the function. The set of corresponding values of y is called the range of the function.

In the analysis of economic problems, we usually use functions to describe the changing relations among economic variables. For example, in the relationship between supply and demand of commodities, we denote the price, the quantity of demand and the quantity of supply of a commodity by P, Q_D and Q_S, so the function of demand and price can be expressed as

$$Q_D = f(P), \quad Q_S = g(P).$$

However, the economic environment is very complex, because every economic variable is affected by many factors. Therefore, there is great limitation if we use the function of one variable to analyze economic problems. So we often use multivariable functions to study the economic problems. The multivariable function is a function which is determined by a number of variables in a function relationship, and always expressed in the form of $y = f(x_1, x_2, \cdots, x_n)$, which indicates that the dependent variable y depends on the value of independent variables $x_1, x_2, \cdots, x_n$.

For example, the basic hypothesis of consumption theory, every consumer has demands for a variety of commodities at the same time. "Utility" depends on the quantity of commodities consumed. That is, the utility function can be expressed as $U = f(x_1, x_2, \cdots, x_n)$, in which U represents the utility of consumers and $x_1, x_2, \cdots, x_n$ represent the consumption of these commodities. This function is called utility function. Similarly, the production function is often expressed as $y = f(L, K)$, in which y, K and L represent the level of outputs, capital and labor. It shows that the level of outputs depends on both the labor and the capital.

Example 1 Cobb-Douglas production function is $Q = A L^\alpha K^\beta$, then the graph when $A = 1, \alpha = 0.5, \beta = 0.5$ (in Figure 11-4).

In many natural phenomena and practical problems, a variable often depends on a variety of variables.

Example 2 To find the height from the roof of the half ellipsoid to the ground, we suppose the half ellipsoid with center at the origin $O(0, 0)$ and the length of the three semi-axes are a, b, c, so the height can be expressed as

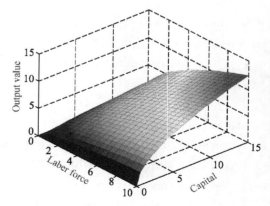

Figure 11-4

$$z = c\sqrt{1 - \frac{x^2}{a^2} - \frac{y^2}{b^2}} \quad (a > 0, b > 0, c > 0).$$

Chapter 11 Differentiation of multivariable function and its application

The variables x and y can take values freely within a certain range $\frac{x^2}{a^2}+\frac{y^2}{b^2}\leqslant 1$, so z changes with x and y. That is, for any point $P(x,y)$ in the set

$$A=\left\{(x,y)\left|\frac{x^2}{a^2}+\frac{y^2}{b^2}\leqslant 1\right.\right\},$$

there is a definite value z.

Example 3 The volume of a certain quantity ideal gas depends on the pressure p (the pressure on the unit area) and the temperature T. By the laws of Boyle, we have

$$V=R\frac{T}{p},$$

in which R is a proportionality constant. Variables p and T can take values freely within a certain range ($p>0$, $T>T_0$, in which T_0 is the liquefaction point of the gas). The variable V changes according to the variables p and T, that is, there is a definite value V corresponding to each point $P(p,T)$ in the planar point set

$$D=\{(p,T)\,|\,p>0,T>T_0\}$$

by the above formula.

The practical significance of the three examples above are different, but they have a common characteristic, the value of a variable depends on the other two variables according to certain relation.

> **Definition 1** Suppose D is a point set in the plane, if the variable z always has a definite value corresponding to each point $P(x,y)$ in set D according to certain relation, z is called the function of variables x,y (or the function of the point P), which is denoted by $z=f(x,y)$ (or $z=f(P)$).

The set D is called the domain of the function, in which x and y are called the independent variables, z is called the dependent variable, and the set $\{z\,|\,z=f(x,y),(x,y)\in D\}$ is called the range of the function which can be also denoted by $z=z(x,y),z=\varphi(x,y)$.

Similarly, we can define the function of three variables $u=f(x,y,z)$ and the function of more variables. In general, if we replace the point set D in the plane with a set in the n-dimensional space, the function of n variable(s) can be defined as $u=f(x_1,x_2,\cdots,x_n)$ or $u=f(P)$, in which the point $P(x_1,x_2,\cdots,x_n)\in D$. Obviously, when $n=1$, we get the function of one variable. The functions of two or more variables are all called multivariable functions.

The domain of the multivariable function is similar to the domain of the function of one variable, so we make the following agreement regardless of practical problems. When we discuss the multivariable function $u=f(P)$, its domain is the set in which u is a real number.

Example 4 Find the domain of the function

$$z=\ln(y-x)+\frac{\sqrt{x}}{\sqrt{1-x^2-y^2}}.$$

Solution From the definition of $\ln(y-x)$, we get $y-x>0$, from the definition of $\sqrt{x}$, we get $x\geqslant 0$, and from the definition of $\dfrac{1}{\sqrt{1-x^2-y^2}}$, we get $1-x^2-y^2>0$.

Then, we find the common solution of the inequalities set
$$\begin{cases} y-x>0, \\ x\geqslant 0, \\ 1-x^2-y^2>0. \end{cases}$$
Thus the domain of the function is
$$D=\{(x,y)\mid y>x, x\geqslant 0, x^2+y^2<1\}$$
(Figure 11-5).

Figure 11-5

We have used right angle coordinate system to sketch the graph of the function of one variable, which is a curve in the plane. We can also use space coordinate system to graph the function of two variables $z=f(x,y)$.

In the space coordinate system, the domain of the function $z=f(x,y)$ is D. There is a definite point $M(x,y,z)$ for any given point $P(x,y)\in D$. For all points in the domain D, the space point set of $M(x,y,z)$ is called the figure of the function of two variables, whose figure is usually a curved surface (Figure 11-6).

As shown in the Example 2, the figure of the function of two variables is a half ellipsoid with center at the origin and three semi-axes a, b and c (Figure 11-7).

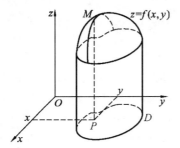
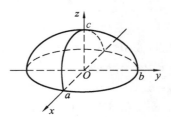

Figure 11-6　　　　　　　　Figure 11-7

In the definitions of functions above, for any point P in the set D, if there is only one value of z according to certain relation, $z=f(x,y)$ is called one-valued function, and if there are more than one values of z, $z=f(x,y)$ is called multi-valued function.

For example, the sphere surface determined by the function
$$x^2+y^2+z^2=R^2$$
is in the closed region $D=\{(x,y)\mid x^2+y^2\leqslant R^2\}$. In addition to the points on the circle $x^2+y^2=R^2$, for any point $P(x,y)\in D$, there are two real numbers $z=\sqrt{R^2-x^2-y^2}$ and $z=-\sqrt{R^2-x^2-y^2}$ corresponding to the point P. Therefore, the equation $x^2+y^2+z^2=R^2$ defines a multi-valued function, and we can separate the multi-valued function into some one-valued functions to study. As the example above, there are two one-valued functions

Chapter 11 Differentiation of multivariable function and its application

$z = \sqrt{R^2 - x^2 - y^2}$ and $z = -\sqrt{R^2 - x^2 - y^2}$. In the following sections, if we do not make a special statement, the functions discussed in this book are all one-valued functions.

11.1.3 The limit of multivariable function

Now we discuss the limit of the function $z = f(x, y)$ when the independent variable (x, y) approaches to (x_0, y_0), that is, $P(x, y) \to P_0(x_0, y_0)$.

Suppose the domain of $z = f(x, y)$ is the point set D, the point $P_0(x_0, y_0)$ is an accumulation point in D, and $P(x, y) \in D$. If the corresponding value of the function $f(x, y)$ approaches to a definite constant A as $P(x, y)$ approaches $P_0(x_0, y_0)$ in any way, A is called the double limit of the function $z = f(x, y)$ as $(x, y) \to (x_0, y_0)$, in which $P \to P_0$ means the length between the points P and P_0 approaches to zero, that is,

$$|P_0 P| = \sqrt{(x - x_0)^2 + (y - y_0)^2} \to 0.$$

Then, we will define the limit above according to the "ε-δ" definition of one variable function.

Definition 2 Suppose that the function $z = f(x, y)$ is defined in a planar point set D, the point $P_0(x_0, y_0)$ is an accumulation point of D and A is a constant. For any given positive number ε, if there exists a positive number δ, for any point $P(x, y)$ such that $0 < |P_0 P| = \sqrt{(x - x_0)^2 + (y - y_0)^2} < \delta$, we always have $|f(x) - A| < \varepsilon$. Then the constant A is called the double limit of the function $z = f(x, y)$ as $P(x, y) \to P_0(x_0, y_0)$, denoted by

$$\lim_{(x,y) \to (x_0, y_0)} f(x, y) = A \quad \text{or} \quad f(x, y) \to A (\rho = |P_0 P| \to 0).$$

Example 5 Suppose $f(x, y) = (x^2 + y^2) \cos \dfrac{1}{x^2 + y^2} (x^2 + y^2 \neq 0)$, prove that

$$\lim_{(x,y) \to (0,0)} f(x, y) = 0.$$

Proof Because

$$\left| (x^2 + y^2) \cos \dfrac{1}{x^2 + y^2} - 0 \right| = |x^2 + y^2| \cdot \left| \cos \dfrac{1}{x^2 + y^2} \right| \leqslant x^2 + y^2,$$

we know that there is an inequality

$$\left| (x^2 + y^2) \cos \dfrac{1}{x^2 + y^2} - 0 \right| < \varepsilon$$

when $\varepsilon > 0$, $\delta = \sqrt{\varepsilon}$ and $0 < \sqrt{(x-0)^2 + (y-0)^2} < \delta$, so $\lim\limits_{(x,y) \to (0,0)} f(x, y) = 0$.

We know, for the function $y = f(x)$, $\lim\limits_{x \to x_0} f(x)$ exists if and only if $f(x_0 - 0) = f(x_0 + 0)$. But from the definition of the limit of a binary function, the existence of a double limit means the function $f(x, y)$ is close to the same constant A as $P(x, y) \in D$ tends to $P_0(x_0, y_0)$ through any way. Thus, if the point $P(x, y)$ tends to $P_0(x_0, y_0)$ in a particular way, even if $f(x, y)$ is close to a certain value, we cannot conclude that the limit of function exists. But in contrary, if the function tends to different values when $P(x, y)$ tends to $P_0(x_0, y_0)$ in different ways, the function does not have limit at $P_0(x_0, y_0)$.

Example 6 $f(x,y)=\begin{cases}\dfrac{2xy}{x^2+y^2}, & x^2+y^2\neq 0,\\ 0, & x^2+y^2=0,\end{cases}$ discuss as $P(x,y)\to O(0,0)$, whether the limit of the function exists or not.

Solution As the point $P(x,y)$ tends to $O(0,0)$ along the x-axis, which means $y=0$ and $x\to 0$, we have
$$\lim_{\substack{(x,y)\to(0,0)\\y=0}}f(x,y)=\lim_{x\to 0}f(x,0)=0.$$

When the point $P(x,y)$ tends to $O(0,0)$ along the y-axis, which means $x=0$ and $y\to 0$, we have
$$\lim_{\substack{(x,y)\to(0,0)\\x=0}}f(x,y)=\lim_{y\to 0}f(0,y)=0.$$

As the point $P(x,y)$ tends to $O(0,0)$ in the above two special ways, the limits exist and are same, but when $P(x,y)$ tends to $O(0,0)$ along $y=kx$, we have
$$\lim_{\substack{(x,y)\to(0,0)\\y=kx}}\frac{2xy}{x^2+y^2}=\lim_{x\to 0}\frac{2kx^2}{x^2+k^2x^2}=\frac{2k}{1+k^2}.$$

Obviously, the limit depends on the value of k, that is, the limit is not a single one. So $f(x,y)$ does not have limit at $O(0,0)$.

We can also get that a function has partial derivative at the point $O(0,0)$ but it is not continuous at the point from the example above.

Example 7 For the function $f(x,y)=\begin{cases}\dfrac{x^2y}{x^4+y^2}, & x^2+y^2\neq 0,\\ 0, & x^2+y^2=0,\end{cases}$ as the point $P(x,y)$ tends to $O(0,0)$ alone any straight line passing through $P(x_0,y_0)$, $f(x,y)$ has the same limit. However, if $P(x,y)$ tends to $O(0,0)$ along the parabola $y=x^2$, the limit is $\dfrac{1}{2}$. So, $\lim\limits_{(x,y)\to(0,0)}f(x,y)$ does not exist.

From the two examples, we know that in the definition of limit, the real meaning of $P(x,y)\xrightarrow{\text{in any way}}P_0(x_0,y_0)$ is that its tendency routes are infinite.

If we regard n-variable function $f(x_1,x_2,\cdots,x_n)$ as the function of the points $P(x_1,x_2,\cdots,x_n)$ in n-dimension space, and then we can obtain related definition of n-variable function and it will be given in the form of point function.

Definition 3 Suppose the n-variable function $u=f(P)$ is defined on the point set D, P_0 is the accumulation point of D and A is a constant. If for any given positive number ε, there is δ, for the point P such that
$$0<|P_0P|<\delta,\ P\in D,$$
we always have $|f(P)-A|<\varepsilon$. Then A is the limit of $u=f(P)$ as $P\to P_0$, denoted by $\lim\limits_{P\to P_0}f(P)=A$.

From this definition, if $n=1$, we obtain the definition of the limit of one variable function, and if $n=2$, we obtain the definition of the limit of binary function.

In the discussion of the limit of multivariate function, it is more convenient to use point function.

Example 8 Find the limit $\lim\limits_{(x,y)\to(0,2)}\dfrac{\sin(xy)}{x}$.

Solution Because as $(x,y)\to(0,2)$, $xy\to 0$, we have $\dfrac{\sin(xy)}{xy}\to 1$.

$$\lim_{(x,y)\to(0,2)}\frac{\sin(xy)}{x}=\lim_{(x,y)\to(0,2)}\left[\frac{\sin(xy)}{xy}\cdot y\right]=\lim_{(x,y)\to(0,2)}\frac{\sin(xy)}{xy}\cdot\lim_{(x,y)\to(0,2)}y=1\times 2=2.$$

The algorithm of the limit of multivariate function is completely similar to the one variable function, such as the following four arithmetic operation.

If $\lim\limits_{(x,y)\to(x_0,y_0)}f(x,y)=A$, $\lim\limits_{(x,y)\to(x_0,y_0)}g(x,y)=B$, then

① $\lim\limits_{(x,y)\to(x_0,y_0)}[f(x,y)\pm g(x,y)]=A\pm B$;

② $\lim\limits_{(x,y)\to(x_0,y_0)}f(x,y)g(x,y)=AB$;

③ $\lim\limits_{(x,y)\to(x_0,y_0)}\dfrac{f(x,y)}{g(x,y)}=\dfrac{A}{B}(B\neq 0)$.

The algorithm of limit of compound operation is also true.

11.1.4 Continuity of multivariable function

It is easy to describe the continuity of multivariate function by its limit. Similar to the definition of continuity of one variable function, we can obtain the definition of continuity of binary function.

Definition 4 Suppose a binary function $z=f(x,y)$ is defined on a planar point set D, $P_0(x_0,y_0)$ is an accumulation point of D and $P_0\in D$. If
$$\lim_{(x,y)\to(x_0,y_0)}f(x,y)=f(x_0,y_0),$$
the binary function $f(x,y)$ is continuous at P_0.

For a set D, if $f(x,y)$ is continuous at any point on D, we call that $f(x,y)$ is continuous on D. If $f(x,y)$ is not continuous at a point $P_0(x_0,y_0)$, we call that $f(x,y)$ is discontinuous at the point $P_0(x_0,y_0)$.

From the Example 6, since the limit at $O(0,0)$ does not exist, the point $O(0,0)$ is a discontinuity point of the function. The discontinuous points of a binary function maybe also a curve, such as the function $z=\dfrac{1}{x^2-y}$, it is not defined on the parabola $y=x^2$, so each point on the parabola is discontinuous point of the function.

Similar to the property of one variable continuous function defined on a closed interval, the multivariable continuous function also has following properties on a bounded closed region.

Property 1 (Maximum and minimum theorem) The multivariable continuous function $f(P)$ on a bounded closed region D has the maximum value and the minimum value. It means there exist P_1 and P_2 on D such that $f(P_1) \leqslant f(P) \leqslant f(P_2)$ for any $P \in D$.

Property 2 (Intermediate value theorem) Suppose $f(P)$ is the multivariable continuous function on a bounded closed region D. For any two values m and M of the function on D, the function will take the value between m and M at least one time.

Specially, if μ is between the minimum m and the maximum M, there exists at least one point P_0 on D such that $f(P_0) = \mu$.

*Property 3 (Uniform continuity theorem) The multivariable continuous functions on a bounded closed region must be continuous uniformly.

It means that if $f(P)$ is continuous on a bounded closed region D, for any given positive number ε, there exists the positive number δ, and for any two points P_1 and P_2 on D such that $|P_1 P_2| < \delta$, we have $|f(P_1) - f(P_2)| < \varepsilon$.

Similar to the one variable elementary function, the multivariable elementary function is a multivariable function which can be written in one formula, and this formula is composed of the basic multivariable elementary functions after finite operation and compound steps. For example, $\sin(x+y)$ is composed of the basic elementary function $\sin u$ and the polynomial $u = x + y$.

According to the continuity of sum, difference, product, quotient of continuous functions, continuity of compound function of continuous function and continuity of basic elementary function, we get the following conclusion.

All multivariable elementary functions are continuous on their defined regions. The so-called defined regions is the open regions or the closed regions in the domain.

Generally, to find $\lim\limits_{P \to P_0} f(P)$, if $f(P)$ is an elementary function, and P is a point in the domain of $f(P)$, then $f(P)$ is continuous at the point P_0, so $\lim\limits_{P \to P_0} f(P) = f(P_0)$.

Example 9 Find $\lim\limits_{(x,y) \to (0,0)} \dfrac{xy}{\sqrt{xy+1}-1}$.

Solution
$$\lim\limits_{(x,y) \to (0,0)} \dfrac{xy}{\sqrt{xy+1}-1} = \lim\limits_{(x,y) \to (0,0)} \dfrac{xy(\sqrt{xy+1}+1)}{(xy+1)-1}$$
$$= \lim\limits_{(x,y) \to (0,0)} \dfrac{xy(\sqrt{xy+1}+1)}{xy}$$
$$= \lim\limits_{(x,y) \to (0,0)} (\sqrt{xy+1}+1) = 2.$$

Chapter 11 Differentiation of multivariable function and its application

Exercises 11-1

1. Determine the following point sets are open regions or closed regions? And draw the graphs.
   (1) $x>0, y>0, x^2+y^2<R^2$ $(R>0)$;
   (2) $y\geqslant 0, x^2+y^2\leqslant R^2$ $(R>0)$;
   (3) $x^2\leqslant y\leqslant 1$;
   (4) $\dfrac{x^2}{4}+\dfrac{y^2}{9}>1$.

2. Determine and draw the domain of the following functions:
   (1) $z=\dfrac{1}{\sqrt{x^2+y^2-1}}$;
   (2) $z=\arccos\dfrac{x^2+y^2-4}{3}$;
   (3) $z=\sqrt{x}-\sqrt{y}$;
   (4) $z=\sqrt{\sin(x^2+y^2)}$;
   (5) $z=\ln(y-x)+\dfrac{\sqrt{x}}{\sqrt{1-x^2-y^2}}$;
   (6) $z=\sqrt{R^2-x^2-y^2}+\dfrac{1}{\sqrt{x^2+y^2-r^2}}$ $(R>r>0)$.

3. Suppose $f(x,y)=\dfrac{x^2-y^2}{2xy}$, find $f(-y,x), f\left(\dfrac{1}{x},\dfrac{1}{y}\right)$ and $f[x,f(x,y)]$.

4. Suppose $z=\sqrt{y}+f(\sqrt{x}-1)$, if $y=1$, $z=x$, determine the function f and z.

5. Find the following limits:
   (1) $\lim\limits_{(x,y)\to(0,0)}\dfrac{e^{xy}\sin y}{1+x^2+y^2}$;
   (2) $\lim\limits_{(x,y)\to(0,0)}\dfrac{2-\sqrt{xy+4}}{xy}$;
   (3) $\lim\limits_{(x,y)\to(0,0)}\dfrac{\sin(xy)}{x}$;
   (4) $\lim\limits_{(x,y)\to(0,0)}\dfrac{x+y}{x^2+y^2}$;
   (5) $\lim\limits_{(x,y)\to(\infty,k)}\left(1+\dfrac{y}{x}\right)^x$;
   (6) $\lim\limits_{(x,y)\to(0,0)}\left(x\sin\dfrac{1}{y}+y\sin\dfrac{1}{x}\right)$.

6. Prove that the following limits do not exist:
   (1) $\lim\limits_{(x,y)\to(0,0)}\dfrac{x^2-y^2}{x^2+y^2}$;
   (2) $\lim\limits_{(x,y)\to(0,0)}\dfrac{x^2y}{x^4-y^2}$.

7. Discuss the continuity of the function $f(x+y)=\begin{cases}\sqrt{1-x^2-y^2}, & x^2+y^2\leqslant 1, \\ 0, & x^2+y^2>1.\end{cases}$

8. Find the discontinuous point of $z=\tan(x^2+y^2)$.

9. Discuss the continuity of $f(x,y)=\begin{cases}(x+y)\sin\dfrac{1}{x}\sin\dfrac{1}{y}, & x\neq 0, y\neq 0, \\ 0, & x=0 \text{ or } y=0\end{cases}$ at the point $(0,0)$.

10. Suppose $f(x,y)=\dfrac{x-y}{x+y}$, try to find $\lim\limits_{x\to 0}[\lim\limits_{y\to 0}f(x,y)]$, $\lim\limits_{y\to 0}[\lim\limits_{x\to 0}f(x,y)]$ (this is called quadratic limit), and determine whether the limit $\lim\limits_{(x,y)\to(0,0)}f(x,y)$ exists or not?

11.2 Differential method of multivariate function

11.2.1 Partial derivative

1) The concept of partial derivative

For one variable function, we study the change rate of function and then introduce the concept of derivative. For multivariable function, we also need to study its change rate, but the independent variable of multivariable function is more than one, the relation between dependent variable and independent variables is also more complicated. So we consider the change rate with respect to one of the independent variables first, which means we discuss the change rate of function as one independent variable changes, but the rest of independent variables fixed (regard as constants). This is the partial derivative of binary function. Let's give an example first.

The volume of a certain amount of ideal gas is V, the function of the pressure P and the temperature T is

$$V = R\frac{T}{p},$$

where R is proportionality constant, when T and P change at the same time, the change of V is more complicated. So it is usually studied in two ways as follows.

(1) Isothermal process. It means if the temperature is fixed (that is, T is constant), by considering the change rate of V with respect to the change of p, we can obtain

$$\frac{dV}{dp} = -R\frac{T}{p^2}.$$

(2) Isobaric process. It means if the pressure is fixed (that is, P is constant), by considering the change rate of V with respect to the change of T, we can obtain

$$\frac{dV}{dT} = R\frac{1}{p}.$$

As mentioned above, the technique which lets one of the independent variables change and the others be fixed is always used in studying the change rate of multivariable function. The definition of partial derivative of the binary function $z = f(x,y)$ will be given in this way.

Note If the independent variable x changes and the independent variable y is fixed, z is deemed as one variable function of x. By the definition of derivative of one variable function, we will introduce the definition of partial derivative of binary function on x.

Definition 1 Suppose the binary function $z = f(x, y)$ is defined on some neighborhood $U(P_0, \delta)$ of the point $P_0(x_0, y_0)$, when the independent variable y is fixed and the independent variable x has increment Δx at x_0, $(x_0 + \Delta x, y_0) \in U(P_0, \delta)$. Relatively, the increment of function is $\Delta_x z = f(x_0 + \Delta x, y_0) - f(x_0, y_0)$ (which is called the partial increment of the function z to x). If the limit

$$\lim_{\Delta x \to 0} \frac{\Delta_x z}{\Delta x} = \lim_{\Delta x \to 0} \frac{f(x_0 + \Delta x, y_0) - f(x_0, y_0)}{\Delta x}$$

Chapter 11 Differentiation of multivariable function and its application

exists, the limit is called the partial derivative of the function $z=f(x,y)$ with respect to x at the point $P_0(x_0,y_0)$, denoted by $\left.\dfrac{\partial z}{\partial x}\right|_{\substack{x=x_0\\y=y_0}}, \left.\dfrac{\partial f}{\partial x}\right|_{\substack{x=x_0\\y=y_0}}, \left.z_x\right|_{\substack{x=x_0\\y=y_0}}$ or $f_x(x_0,y_0)$.

It means
$$f_x(x_0,y_0)=\lim_{\Delta x\to 0}\frac{\Delta_x z}{\Delta x}=\lim_{\Delta x\to 0}\frac{f(x_0+\Delta x,y_0)-f(x_0,y_0)}{\Delta x}. \tag{1}$$

In the same way, the definition of partial derivative of $z=f(x,y)$ with respect to y at the point $P_0(x_0,y_0)$ is, if
$$\lim_{\Delta y\to 0}\frac{\Delta_y z}{\Delta y}=\lim_{\Delta y\to 0}\frac{f(x_0,y_0+\Delta y)-f(x_0,y_0)}{\Delta y}$$
exists, this limit is called the partial derivative of the function $z=f(x,y)$ with respect to y at the point $P_0(x_0,y_0)$, denoted by $\left.\dfrac{\partial z}{\partial y}\right|_{\substack{x=x_0\\y=y_0}}, \left.\dfrac{\partial f}{\partial y}\right|_{\substack{x=x_0\\y=y_0}}, \left.z_y\right|_{\substack{x=x_0\\y=y_0}}$ or $f_y(x_0,y_0)$.

If the function $z=f(x,y)$ has partial derivative $f_x(x,y)$ and $f_y(x,y)$ at each point $P(x,y)$ in a region, they are new binary function of x and y. And they are called the partial derivative function of $z=f(x,y)$, denoted by $\dfrac{\partial z}{\partial x},\dfrac{\partial z}{\partial y}$; $\dfrac{\partial f}{\partial x},\dfrac{\partial f}{\partial y}$; z_x,z_y; $f_x(x,y), f_y(x,y)$.

Note $z_x, z_y, f_x(x,y), f_y(x,y)$ can be expressed by $z'_x, z'_y, f'_x(x,y), f'_y(x,y)$.

But the partial derivative of $z=f(x,y)$ with respect to x at the point $P_0(x_0,y_0)$ is $f_x(x_0,y_0)$, and we call it as the value of the partial derivative of $f_x(x,y)$ with respect to x at the point $P_0(x_0,y_0)$, $f_y(x_0,y_0)$ is the value of the partial derivative $f_y(x,y)$ at the point $P_0(x_0,y_0)$. Same to the derivative of one variable function, in order not to be confused, partial derivative function is referred to simply as partial derivative.

From the definition of partial derivative, finding the partial derivative of the binary function $z=f(x,y)$ is actually to find the derivative of one variable function. When we find the derivative with respect to one of independent variables, the rest of independent variables keep fixed (regarded as constant).

Similar to the definition of partial derivative of binary function, the concept of partial derivative can be extended to multivariate function. Such as for the three-variable function $u=f(x,y,z)$, if the limit
$$\lim_{\Delta x\to 0}\frac{\Delta_x u}{\Delta x}=\lim_{\Delta x\to 0}\frac{f(x+\Delta x,y,z)-f(x,y,z)}{\Delta x}$$
exists, we define this limit as the partial derivative of the function $u=f(x,y,z)$ with respect to x at the point $P(x,y,z)$, denoted by $\dfrac{\partial u}{\partial x}$, $f_x(x,y,z)$, and so on.

Example 1 Suppose Cobb-Douglas production function
$$Y=AK^\alpha L^\beta.$$
Find $\dfrac{\partial Y}{\partial K}$ and $\dfrac{\partial Y}{\partial L}$.

Solution $\dfrac{\partial Y}{\partial K} = \alpha A K^{\alpha-1} L^{\beta}, \dfrac{\partial Y}{\partial L} = \beta A K^{\alpha} L^{\beta-1}$.

Next, let's discuss the economic significance of partial derivative.

Suppose there are two commodities A and B, their price is p_1 and p_2, demand is Q_1 and Q_2. The demand Q_1 and Q_2 are determined by the price p_1 and p_2, the demand functions are

$$Q_1 = Q_1(p_1, p_2),\ Q_2 = Q_2(p_1, p_2).$$

The partial derivative of Q_1 and Q_2 with respect to p_1 and p_2 are marginal demand of these two commodities.

$\dfrac{\partial Q_1}{\partial p_1}$ is the marginal demand of Q_1 for p_1, which is the change rate of the demand Q_1 as the price p_1 changes, and $\dfrac{\partial Q_1}{\partial p_2}$ is the marginal demand of Q_1 for p_2, which is the change rate of the demand Q_1 as the price p_2 changes.

We can explain $\dfrac{\partial Q_2}{\partial p_1}$ and $\dfrac{\partial Q_2}{\partial p_2}$ similarly.

For one variable function the concept of elasticity is given. For multivariate function, we can also define the concept of elasticity, and it is called partial elasticity.

If p_2 is fixed and p_1 changes, Q_1 and Q_2 changes with p_1. Partial elasticity can be defined by

$$E_{11} = \lim_{\Delta p_1 \to 0} \dfrac{\dfrac{\Delta_1 Q_1}{Q_1}}{\dfrac{\Delta p_1}{p_1}} = \dfrac{p_1}{Q_1} \dfrac{\partial Q_1}{\partial p_1},$$

$$E_{12} = \lim_{\Delta p_1 \to 0} \dfrac{\dfrac{\Delta_1 Q_2}{Q_2}}{\dfrac{\Delta p_1}{p_1}} = \dfrac{p_1}{Q_2} \dfrac{\partial Q_2}{\partial p_1},$$

where $\Delta_1 Q_i = Q_i(p_1 + \Delta p_1, p_2) - Q_i(p_1, p_2)$ $(i=1, 2)$. Similarly, if p_1 is fixed and p_2 changes, we define

$$E_{21} = \lim_{\Delta p_2 \to 0} \dfrac{\dfrac{\Delta_2 Q_1}{Q_1}}{\dfrac{\Delta p_2}{p_2}} = \dfrac{p_2}{Q_1} \dfrac{\partial Q_1}{\partial p_2},$$

$$E_{22} = \lim_{\Delta p_2 \to 0} \dfrac{\dfrac{\Delta_2 Q_2}{Q_2}}{\dfrac{\Delta p_2}{p_2}} = \dfrac{p_2}{Q_2} \dfrac{\partial Q_2}{\partial p_2},$$

where $\Delta_2 Q_i = Q_i(p_1, p_2 + \Delta p_2) - Q_i(p_1, p_2)$ $(i=1, 2)$. E_{11} (E_{22}) is partial elasticity of direct price of the demand Q_1 (Q_2) with respect to the price p_1 (p_2). E_{12} (E_{21}) is partial elasticity of cross price of the demand Q_1 (Q_2) with respect to the relevant price p_2 (p_1).

The prices of commodity A and B are at a certain level, if p_2 is fixed and p_1 increases 1%, E_{11} is the percent of change (increase or decrease) of Q_1. It reflects the sensitivity of

Chapter 11 Differentiation of multivariable function and its application

change of Q_1 when p_2 is fixed and p_1 changes. E_{22} has similar explanation.

The price of commodity A and B are at a certain level, if p_1 is fixed and p_2 increases 1%, E_{12} is the percent of change (increase or decrease) of Q_1. It reflects the sensitivity of change of Q_1 when p_1 is fixed and p_2 changes. E_{21} has similar explanation.

The cross elasticity of demand for price can be used to analyze the correlation of two commodities.

If $E_{12} < 0$, that is, the partial elasticity of cross price of A to B is negative, it means that when the price of A is fixed and the price of B raises, the demand of A will decrease relatively. The relationship between A and B is mutually complementary.

If $E_{12} > 0$, that is, the partial elasticity of cross price of A to B is positive, it means that when the price of A is fixed and the price of B raises, the demand of A will increase relatively. The relationship between A and B is competing with (substitution for) each other.

Example 2 Find the partial elasticity of direct price and cross price of the demand function

$$Q_1 = 1\,000 p_1^{-\frac{1}{2}} p_2^{\frac{1}{5}}$$

at the point $(p_1, p_2) = (4, 32)$. Then determine A and B are compete with each other or complement with each other, and give the reason.

Solution When $p_1 = 4$, $p_2 = 32$ and $Q_1 = 1\,000$, we have

$$\frac{\partial Q_1}{\partial p_1} = -500 p_1^{-\frac{3}{2}} p_2^{\frac{1}{5}}, \quad \frac{\partial Q_1}{\partial p_2} = 200 p_1^{-\frac{1}{2}} p_2^{-\frac{4}{5}},$$

$$E_{11}\bigg|_{\substack{p_1=4 \\ p_2=32}} = \left(\frac{p_1}{Q_1}\frac{\partial Q_1}{\partial p_1}\right)_{\substack{p_1=4 \\ p_2=32}} = -\frac{1}{2},$$

$$E_{12}\bigg|_{\substack{p_1=4 \\ p_2=32}} = \left(\frac{p_2}{Q_1}\frac{\partial Q_1}{\partial p_2}\right)_{\substack{p_1=4 \\ p_2=32}} = \frac{1}{5}.$$

Because $E_{12}\bigg|_{\substack{p_1=4 \\ p_2=32}} = \frac{1}{5} > 0$, A and B are compete with each other.

Example 3 We know that the demand Q of some commodity is a function of its price p_1, its relevant commodity's price p_2 and the income of customer x,

$$Q_1 = \frac{1}{200} p_1^{-\frac{3}{8}} p_2^{-\frac{2}{5}} x^{\frac{5}{2}}.$$

Find E_{11}, E_{12} and partial elasticity of income E_{1x}.

Solution Because the demand function Q_1 is the form of product, take logarithm on either side of the function first, we have

$$\ln Q_1 = -\ln 200 - \frac{3}{8}\ln p_1 - \frac{2}{5}\ln p_2 + \frac{5}{2}\ln x.$$

Then, differentiate either side

$$\frac{1}{Q_1}\frac{\partial Q_1}{\partial p_1} = -\frac{3}{8}\frac{1}{p_1}.$$

So $E_{11} = \dfrac{p_1}{Q_1}\dfrac{\partial Q_1}{\partial p_1} = -\dfrac{3}{8}$. Similarly, we have

$$E_{12} = -\frac{2}{5}, \quad E_{1x} = \frac{5}{2}.$$

Example 4 The sales of some digital camera is Q_A, in addition to its own price P_A, it is also related to the price of color inkjet printer P_B, the relationship is

$$Q_A = 120 + \frac{250}{P_A} - 10P_B - P_B^2.$$

If $P_A = 50, P_B = 5$, find:

(1) elasticity of Q_A to P_A;

(2) cross elasticity of Q_A to P_B.

Solution (1) The elasticity of Q_A to P_A is

$$\frac{EQ_A}{EP_A} = \frac{\partial Q_A}{\partial P_A} \frac{P_A}{Q_A}$$

$$= -\frac{250}{P_A^2} \cdot \frac{P_A}{120 + \frac{250}{P_A} - 10P_B - P_B^2}$$

$$= -\frac{250}{120 P_A + 250 - P_A(10P_B + P_B^2)}.$$

If $P_A = 50$ and $P_B = 5$, we have

$$\frac{EQ_A}{EP_A} = -\frac{250}{120 \times 50 + 250 - 50 \times (50 + 25)} = -\frac{1}{10}.$$

(2) The cross elasticity of Q_A to P_B is

$$\frac{EQ_A}{EP_B} = \frac{\partial Q_A}{\partial P_B} \frac{P_B}{Q_A}$$

$$= -(10 + 2P_B) \frac{P_B}{120 + \frac{250}{P_A} - 10P_B - P_B^2}.$$

If $P_A = 50$ and $P_B = 5$, we have

$$\frac{EQ_A}{EP_B} = (-20) \times \frac{5}{120 + 5 - 50 - 25} = -2.$$

2) Elastic analysis

(1) The concept of elastic analysis.

Definition 2 The ratio of the relative change of the function $y = f(x)$, which is $\frac{\Delta y}{y_0} = \frac{f(x_0 + \Delta x) - f(x_0)}{y_0}$, to the relative change of independent variable $\frac{\Delta x}{x_0}$ is $\frac{\frac{\Delta y}{y_0}}{\frac{\Delta x}{x_0}}$, this ration is called relative change rate or elasticity of $f(x)$ between $x = x_0$ and $x = x_0 + \Delta x$.

If $f'(x_0)$ exists, the limit

$$\lim_{\Delta x \to 0} \frac{\frac{\Delta y}{y_0}}{\frac{\Delta x}{x_0}} = \lim_{\Delta x \to 0} \frac{x_0}{y_0} \cdot \frac{\Delta y}{\Delta x} = f'(x_0) \frac{x_0}{y_0}$$

Chapter 11 Differentiation of multivariable function and its application

is called rate of relative change or relative derivative or elasticity of $f(x)$ at the point x_0, denoted by $\left.\dfrac{Ey}{Ex}\right|_{x=x_0}$ or $\dfrac{E}{Ex}f(x_0)$. That is,

$$\left.\frac{Ey}{Ex}\right|_{x=x_0}=\frac{E}{Ex}f(x_0)=f'(x_0)\frac{x_0}{y_0}.$$

If $f'(x)$ exists, the limit

$$\frac{Ey}{Ex}=\frac{E}{Ex}f(x)=\lim_{\Delta x\to 0}\frac{\frac{\Delta y}{y}}{\frac{\Delta x}{x}}=\lim_{\Delta x\to 0}\frac{x}{y}\cdot\frac{\Delta y}{\Delta x}=f'(x)\frac{x}{y}$$

(it is a function of x) is called elasticity function of $f(x)$.

Because $\lim\limits_{\Delta x\to 0}\dfrac{\frac{\Delta y}{y_0}}{\frac{\Delta x}{x_0}}=\dfrac{E}{Ex}f(x_0)$, if $|\Delta x|$ is small enough, we have $\dfrac{\frac{\Delta y}{y_0}}{\frac{\Delta x}{x_0}}\approx\dfrac{E}{Ex}f(x_0)$, so $\dfrac{\Delta y}{y_0}\approx\dfrac{\Delta x}{x_0}\cdot\dfrac{E}{Ex}f(x_0)$. If $\dfrac{\Delta x}{x_0}=1\%$, we have $\dfrac{\Delta y}{y_0}\approx\dfrac{E}{Ex}f(x_0)\%$.

The economic significance of elasticity is that if $f'(x_0)$ exists, $\dfrac{E}{Ex}f(x_0)$ means that $f(x)$ changes $\dfrac{E}{Ex}f(x_0)\%$ approximately as x changes 1% at the point x_0 (the word approximately is always omitted).

Therefore, the elasticity of $f(x)$ at the point x is $\dfrac{E}{Ex}f(x)$, and it reflects the range of change of $f(x)$ caused by the change of x, which is also the intensity or sensitivity of $f(x)$ to the change of x.

Example 5 Suppose $y=a^x(a>0,a\neq 1)$, find $\dfrac{Ey}{Ex},\left.\dfrac{Ey}{Ex}\right|_{x=1}$.

Solution Because $\dfrac{Ey}{Ex}=y'\cdot\dfrac{x}{y}=a^x\cdot\ln a\cdot\dfrac{x}{a^x}=x\ln a$, we have

$$\left.\frac{Ey}{Ex}\right|_{x=1}=\ln a.$$

Example 6 Suppose $y=x^a$, find $\dfrac{Ey}{Ex}$.

Solution $\dfrac{Ey}{Ex}=y'\cdot\dfrac{x}{y}=ax^{a-1}\dfrac{x}{x^a}=a.$

(2) Demand elasticity.

Demand elasticity reflects the intensity of the change of demand as the price changes. Since the demand function $q=f(p)$ is a decreasing function, we have $f'(p)\leqslant 0$, then $f'(p_0)\dfrac{p_0}{Q_0}$ is negative. Generally, economists denote the demand elasticity by positive number. Therefore, we always use the opposite number of the rate of relative change of demand function to define the demand elasticity.

Definition 3 Suppose the demand function of some commodity is $q=f(p)$, then
$$\bar{\eta}(p_0, p_0+\Delta p) = -\frac{\Delta q}{\Delta p} \cdot \frac{p_0}{Q_0}$$
is called the demand elasticity of the commodity between $p=p_0$ and $p=p_0+\Delta p$. If $f'(p_0)$ exists,
$$\eta\big|_{p=p_0} = \eta(p_0) = -f'(p_0) \cdot \frac{p_0}{f(p_0)}$$
is called the demand elasticity of the commodity at $p=p_0$.

Example 7 The demand function of some commodity is
$$q=f(p)=\frac{1\,200}{p}.$$

① Find the demand elasticity at any point from $p=30$ to $p=20$ and 50.

② Find the demand elasticity at the point $p=30$, and explain its economic significance.

Solution ① If $p=30$, we have $q=40$, so
$$\bar{\eta}(30,20) = -\frac{\frac{1\,200}{20} - \frac{1\,200}{30}}{20-30} \cdot \frac{30}{40} = 1.5,$$
$$\bar{\eta}(30,50) = -\frac{\frac{1\,200}{50} - \frac{1\,200}{30}}{50-30} \cdot \frac{30}{40} = 0.6.$$

$\bar{\eta}(30,20)$ means that the demand increases 1.5% (from 40) as the price p decreases 1% (from 30).

$\bar{\eta}(30,50)$ means that the demand decreases 0.6% (from 40) as the price p increases 1% (from 30).

② Because
$$\eta(p) = -f'(p)\frac{p}{f(p)} = -\frac{-1\,200}{p^2} \cdot \frac{p}{\frac{1\,200}{p}} = 1,$$

$\eta(30)=1$. It means that the demand decreases 1% as the price increases 1% from $p=30$, and the demand increases 1% as the price decreases 1% from $p=30$.

(3) Supply elasticity.

The definition of supply elasticity is similar to the demand elasticity.

Definition 4 Suppose the supply function of some commodity is $q=\varphi(p)$, then
$$\bar{\varepsilon}(p_0, p_0+\Delta p) = \frac{\Delta q}{\Delta p} \cdot \frac{p_0}{Q_0}$$
is called the supply elasticity of the commodity between $p=p_0$ and $p=p_0+\Delta p$. If $\varphi'(p_0)$ exists,

Chapter 11 Differentiation of multivariable function and its application

$$\varepsilon\big|_{p=p_0}=\varepsilon(p_0)=\varphi'(p_0)\cdot\frac{p_0}{\varphi(p_0)}$$

is the elasticity at the point $p=p_0$.

Example 8 Suppose $q=e^{2p}$, find $\varepsilon(2)$, and explain its economic significance.

Solution Because $(e^{2p})'=2e^{2p}$, we have

$$\varepsilon(p)=\varphi'(p)\cdot\frac{p}{\varphi(p)}=2e^{2p}\cdot\frac{p}{e^{2p}}=2p,$$

and $\varepsilon(2)=4$. It means that at $p=2$ the supply increases 4% as the price increases 1% and the supply decreases 4% as the price decreases 1%.

Example 9 Suppose the demand function of some commodity is $q=q(p)$, the revenue function is $R=pq$, where p is the price of the commodity, q is the demand (output) and $q(p)$ is a monotony decrease function. If the price is p_0, the relative output is q_0, then the marginal revenue is

$$\frac{\partial R}{\partial q}\bigg|_{q=Q_0}=a>0,$$

the marginal effect of revenue on price is

$$\frac{\partial R}{\partial p}\bigg|_{p=p_0}=c>0,$$

the elasticity of the demand q to the price p is $\eta_p=b>1$, find p_0 and q_0.

Solution Because the revenue is $R=pq$, we have

$$\frac{\partial R}{\partial q}=p+q\frac{\partial p}{\partial q}=p+\left(-\frac{1}{\frac{\partial q}{\partial p}\cdot\frac{p}{q}}\right)(-p)$$

$$=p\left(1-\frac{1}{\eta_p}\right).$$

So we get $\dfrac{\partial R}{\partial q}\bigg|_{q=Q_0}=p_0\left(1-\dfrac{1}{b}\right)=a$, and then, $p_0=\dfrac{ab}{b-1}$. Because

$$\frac{\partial R}{\partial p}=q+p\cdot\frac{\partial q}{\partial p}=q-\left(-\frac{\partial q}{\partial p}\cdot\frac{p}{q}\right)q=q(1-\eta_p),$$

and $\dfrac{\partial R}{\partial p}\bigg|_{p=p_0}=Q_0(1-\eta_p)=c$, we can obtain $Q_0=\dfrac{c}{1-b}$.

Example 10 Suppose the demand of some commodity Q is a monotony decrease function of the price p, $Q=Q(p)$, and the demand elasticity is $\eta=\dfrac{2p^2}{192-p^2}>0$.

① Suppose R is the total revenue function, prove $\dfrac{\partial R}{\partial p}=Q(1-\eta)$.

② Find the elasticity of total revenue to price at $p=6$, and explain its economic significance.

Solution ① Since $R(p)=pQ(p)$, differentiate either side of the equation, we get

$$\frac{\partial R}{\partial p}=Q+p\frac{\partial Q}{\partial p}=Q\left(1+\frac{p}{Q}\frac{\partial Q}{\partial p}\right)=Q(1-\eta).$$

② Since $\dfrac{ER}{Ep}=\dfrac{p}{R}\dfrac{\partial R}{\partial p}=\dfrac{p}{pQ}Q(1-\eta)=1-\eta=1-\dfrac{2p^2}{192-p^2}=\dfrac{192-3p^2}{192-p^2}$, we have

$$\left.\frac{ER}{Ep}\right|_{p=6}=\frac{192-3\times 6^2}{192-6^2}=\frac{7}{13}\approx 0.54.$$

The economic significance is that the total revenue increases 0.54% if the price increases 1%.

3) Geometric significance of partial derivative

For one variable function, we know that the derivative of $y=f(x)$ is $\dfrac{\mathrm{d}y}{\mathrm{d}x}$, which is the slope of the tangent of the curve $y=f(x)$ at the point (x,y). The partial derivative of $z=f(x,y)$ at the point (x_0,y_0) has similar geometric significance.

Suppose the point $M_0(x_0,y_0,f(x_0,y_0))$ is a point on the curve surface $z=f(x,y)$, the plane $y=y_0$ cut this curve surface through the point M_0 and we can draw a curve whose equation is $z=f(x,y_0)$. So the partial derivative $f_x(x_0,y_0)$ is the slope of the tangent $M_0 T_x$ (to the x-axis) of the curve $z=f(x,y_0)$ on $y=y_0$. In the same way, the geometric significance of partial derivative $f_y(x_0,y_0)$ is the slope of the tangent $M_0 T_y$ (to the y-axis) of the curve $z=f(x_0,y)$ on $x=x_0$ (Figure 11-8).

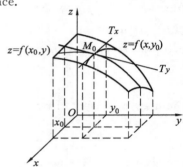

Figure 11-8

As we know, for one variable function, if a function is derivable at some point, the function must be continuous at this point. But for multivariable function, even if each partial derivative at the point $P(x_0,y_0)$ exists, there is no guarantee that the function is continuous at this point. That is because the existence of partial derivative only can guarantee the point P tends to the point P_0 along the direction parallel to the axis, $f(P)$ tends to $f(P_0)$, but cannot guarantee $f(P)$ always tends to $f(P_0)$ as the point P tends to P_0 in any way.

Example 11 Suppose $f(x,y)=\begin{cases}\dfrac{2xy}{x^2+y^2}, & x^2+y^2\neq 0,\\ 0, & x^2+y^2=0,\end{cases}$ find the partial derivative $f_x(0,0)$ and $f_y(0,0)$.

Solution We know that this function is discontinuous at the point $(0,0)$, but the partial derivatives are

$$f_x(0,0)=\lim_{\Delta x\to 0}\frac{f(0+\Delta x,0)-f(0,0)}{\Delta x}=\lim_{\Delta x\to 0}\frac{0}{\Delta x}=0,$$

$$f_y(0,0)=\lim_{\Delta y\to 0}\frac{f(0,0+\Delta y)-f(0,0)}{\Delta y}=\lim_{\Delta y\to 0}\frac{0}{\Delta y}=0.$$

So for multivariable function, the existence of the partial derivative of the function at a point cannot derives that the function is continuous at this point.

4) High-order partial derivative

Suppose the function $z=f(x,y)$ has partial derivatives $f_x(x,y)$ and $f_y(x,y)$ on D. Generally, $f_x(x,y)$ and $f_y(x,y)$ are still binary functions of independent variables x and y. If partial derivatives of these two functions exist, we call them the second partial derivative of $z=f(x,y)$. According to the order of derivatives, the second partial derivative of the function has following notations.

$$\frac{\partial}{\partial x}\left(\frac{\partial z}{\partial x}\right) \triangleq \frac{\partial^2 z}{\partial x^2} \triangleq f_{xx}(x,y), \quad \frac{\partial}{\partial y}\left(\frac{\partial z}{\partial x}\right) \triangleq \frac{\partial^2 z}{\partial x \partial y} \triangleq f_{xy}(x,y),$$

$$\frac{\partial}{\partial x}\left(\frac{\partial z}{\partial y}\right) \triangleq \frac{\partial^2 z}{\partial y \partial x} \triangleq f_{yx}(x,y), \quad \frac{\partial}{\partial y}\left(\frac{\partial z}{\partial y}\right) \triangleq \frac{\partial^2 z}{\partial y^2} \triangleq f_{yy}(x,y).$$

$f_{xy}(x,y)$ and $f_{yx}(x,y)$ are called the second mixed partial derivative of $f(x,y)$.

Example 12 Suppose $z = x^3 y + 3x^2 y^3 - xy + 2$, find $\dfrac{\partial^2 z}{\partial x^2}, \dfrac{\partial^2 z}{\partial x \partial y}, \dfrac{\partial^2 z}{\partial y \partial x}, \dfrac{\partial^2 z}{\partial y^2}$ and $\dfrac{\partial^3 z}{\partial x^3}$.

Solution $\dfrac{\partial z}{\partial x} = 3x^2 y + 6xy^3 - y$, $\dfrac{\partial z}{\partial y} = x^3 + 9x^2 y^2 - x$. Then

$$\frac{\partial^2 z}{\partial x^2} = 6xy + 6y^3, \quad \frac{\partial^2 z}{\partial y^2} = 18x^2 y,$$

$$\frac{\partial^2 z}{\partial x \partial y} = 3x^2 + 18xy^2 - 1, \quad \frac{\partial^2 z}{\partial y \partial x} = 3x^2 + 18xy^2 - 1,$$

$$\frac{\partial^3 z}{\partial x^3} = 6y.$$

From Example 12, although the orders of the partial derivatives $\dfrac{\partial^2 z}{\partial x \partial y}$ and $\dfrac{\partial^2 z}{\partial y \partial x}$ are different, they are equal. But it is not always correct.

Example 13 Suppose $f(x,y) = \begin{cases} xy\dfrac{x^2 - y^2}{x^2 + y^2}, & x^2 + y^2 \neq 0, \\ 0, & x^2 + y^2 = 0, \end{cases}$ find $f_{xy}(0,0)$ and $f_{yx}(0,0)$.

Solution If $x^2 + y^2 \neq 0$, we obtain

$$f_x(x,y) = \frac{y(x^4 + 4x^2 y^2 - y^4)}{(x^2 + y^2)^2},$$

$$f_y(x,y) = \frac{x(x^4 - 4x^2 y^2 - y^4)}{(x^2 + y^2)^2}.$$

If $x^2 + y^2 = 0$, we have

$$f_x(0,0) = \lim_{\Delta x \to 0} \frac{f(0+\Delta x, 0) - f(0,0)}{\Delta x} = \lim_{\Delta x \to 0} \frac{0}{\Delta x} = 0.$$

$$f_y(0,0) = \lim_{\Delta y \to 0} \frac{f(0+\Delta y, 0) - f(0,0)}{\Delta y} = 0.$$

So,

$$f_x(x,y) = \begin{cases} \dfrac{y(x^4 + 4x^2 y^2 - y^4)}{(x^2 + y^2)^2}, & x^2 + y^2 \neq 0, \\ 0, & x^2 + y^2 = 0, \end{cases}$$

$$f_y(x,y) = \begin{cases} \dfrac{x(x^4 - 4x^2 y^2 - y^4)}{(x^2 + y^2)^2}, & x^2 + y^2 \neq 0, \\ 0, & x^2 + y^2 = 0, \end{cases}$$

Then,
$$f_{xy}(0,0) = \lim_{\Delta y \to 0} \frac{f_x(0, 0+\Delta y) - f_x(0,0)}{\Delta y} = \lim_{\Delta y \to 0} \frac{\Delta y \left[\frac{-(\Delta y)^4}{(\Delta y)^4} \right]}{\Delta y} = -1,$$

$$f_{yx}(0,0) = \lim_{\Delta x \to 0} \frac{f_y(0+\Delta x, 0) - f_y(0,0)}{\Delta x} = \lim_{\Delta x \to 0} \frac{\Delta x \left[\frac{(\Delta x)^4}{(\Delta x)^4} \right]}{\Delta x} = 1.$$

Here $f_{xy}(0,0) \neq f_{yx}(0,0)$.

When will two second mixed partial derivatives be equal? We give the following conclusion without proof.

Theorem 1 If two second mixed partial derivatives of the function $z = f(x,y)$, $\frac{\partial^2 z}{\partial x \partial y}$ and $\frac{\partial^2 z}{\partial y \partial x}$, are continuous on D, these two second mixed partial derivatives must be equal on this region.

The Theorem 1 also means that it has nothing to do with the order of derivation if the second mixed partial derivative is continuous. This conclusion can also be generalized to the higher order partial derivative. For example, if the third-order mixed partial derivative of $z = f(x,y)$ is continuous, then
$$f_{xxy}(x,y) = f_{xyx}(x,y) = f_{yxx}(x,y), \quad f_{xyy}(x,y) = f_{yxy}(x,y) = f_{yyx}(x,y).$$

Similarly, the high order partial derivative of multivariable function can be defined, and it also has nothing to do with the order of derivation as the partial derivative is continuous.

Example 14 Prove that the function $u = \frac{1}{r}$ satisfies the equations
$$\frac{\partial^2 u}{\partial x^2} + \frac{\partial^2 u}{\partial y^2} + \frac{\partial^2 u}{\partial z^2} = 0,$$
and
$$r = \sqrt{x^2 + y^2 + z^2}.$$

Proof By $u = \frac{1}{r}$, we have $\frac{\partial u}{\partial x} = -\frac{1}{r^2} \cdot \frac{\partial r}{\partial x} = -\frac{1}{r^2} \cdot \frac{x}{r} = -\frac{x}{r^3}$ and then
$$\frac{\partial^2 u}{\partial x^2} = -\frac{1}{r^3} + \frac{3x}{r^4} \cdot \frac{\partial r}{\partial x} = -\frac{1}{r^3} + \frac{3x^2}{r^5}.$$

Because of the symmetry of the function, we also have
$$\frac{\partial^2 u}{\partial y^2} = -\frac{1}{r^3} + \frac{3y^2}{r^5},$$
$$\frac{\partial^2 u}{\partial z^2} = -\frac{1}{r^3} + \frac{3z^2}{r^5}.$$

Thus,
$$\frac{\partial^2 u}{\partial x^2} + \frac{\partial^2 u}{\partial y^2} + \frac{\partial^2 u}{\partial z^2} = -\frac{3}{r^3} + \frac{3(x^2+y^2+z^2)}{r^5} = -\frac{3}{r^3} + \frac{3r^2}{r^5} = 0.$$

The equation in Example 14 is the Laplace Equation which is one of important equations studied in Mathematical Physics.

11.2.2 Perfect differential and its applications

1) The concept of perfect differential

From the differential calculus of one variable function, the differential of $y=f(x)$ is $dy=f'(x)dx$ which is the linear principal part of the increment Δy of the function $f(x)$. It describes the approximation of the change of the function $y=f(x)$ as the independent variable has a change Δx. Replace the increment Δy by the differential dy, as $\Delta x \to 0$, the high order infinitesimal is omitted.

There is similar definition for multivariable function, then we will discuss the definition of binary function.

Suppose the binary function $z=f(x,y)$ is defined in the neighborhood of the point $P(x,y)$, when independent variable has increment Δx and Δy at the point $P(x,y)$, the increment of the function

$$\Delta z = f(x+\Delta x, y+\Delta y) - f(x,y)$$

is called the total increment of the increment Δx, Δy of $z=f(x,y)$ at the point $P(x,y)$.

Generally, it is complicated to find the total increment Δz. Thus, we hope to obtain a linear function of the increment Δx and Δy, which is similar to one variable function. Let's discuss the following example.

Example 15 Suppose there is a rectangular sheet metal, the length is x and the width is y. The sheet metal is expanded by heat, the length increases Δx and the width increases Δy. Find the increment of the area of sheet metal.

Solution Denote the area of sheet metal by A, then $A=xy$.

Because the length and width increase Δx and Δy respectively, the increment of area is

$$\Delta A = (x+\Delta x)(y+\Delta y) - xy$$
$$= y\Delta x + x\Delta y + \Delta x \Delta y.$$

So, the total increment ΔA of the binary function $A=xy$ consists of two parts (Figure 11-9). Let $\rho = \sqrt{(\Delta x)^2 + (\Delta y)^2}$, then $\rho \to 0$ is equal to $(\Delta x, \Delta y) \to (0,0)$. The first part $y\Delta x + x\Delta y$ is a linear function of Δx and Δy, the second part $\Delta x \Delta y$ is the area of a rectangle at the upper right corner. As $\rho \to 0$, since $\Delta x \Delta y$ is a high order infinitesimal of $\rho = \sqrt{(\Delta x)^2 + (\Delta y)^2}$, we get

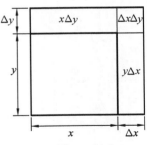

Figure 11-9

$$\lim_{\rho \to 0} \frac{\Delta x \Delta y}{\rho} = \lim_{\rho \to 0} \frac{\Delta x \Delta y}{\sqrt{(\Delta x)^2 + (\Delta y)^2}} = 0.$$

Thus, the first part $y\Delta x + x\Delta y$ is the linear main body of the total increment ΔA, which is the approximation of ΔA. As $\rho \to 0$, the infinitesimal $\Delta x \Delta y$ has the higher order than $\rho = \sqrt{(\Delta x)^2 + (\Delta y)^2}$, so it could be omitted.

If the length x and the width y of the sheet metal are constants, $y=a$ and $x=b$, the increment of area ΔA is the function of Δx and Δy, that is, $\Delta A = a\Delta x + b\Delta y + o(\rho)$, where a and b are independent with Δx and Δy. Since $A = xy$, we have $A_x = y$ and $A_y = x$, that is, $\Delta A = A_x(b,a)\Delta x + A_y(b,a)\Delta y + o(\rho)$.

Does this conclusion have universal significance? Now we introduce the definition of perfect differential of binary function.

> **Definition 5** Suppose the total increment $\Delta z = f(x+\Delta x, y+\Delta y) - f(x,y)$ of $z = f(x,y)$ at the point $P(x,y)$ can be denoted as
> $$\Delta z = A\Delta x + B\Delta y + o(\rho), \qquad (2)$$
> where A and B are independent with Δx and Δy, but related to x and y, and $\rho = \sqrt{(\Delta x)^2 + (\Delta y)^2}$. Then the function $z = f(x,y)$ is differentiable at the point $P(x,y)$ and $A\Delta x + B\Delta y$ is called the perfect differential of $z = f(x,y)$ at $P(x,y)$, denoted by dz or $df(x,y)$, that is,
> $$dz = A\Delta x + B\Delta y.$$

If the function $z = f(x,y)$ is differentiable at each point on D, we call the function $z = f(x,y)$ is differentiable on D.

We have known that even if each partial derivative at some point exists, there is no guarantee that the function is continuous at this point. But if the function $z = f(x,y)$ is differentiable at the point $P_0(x_0, y_0)$, the function $z = f(x,y)$ must be continuous at this point. This conclusion will be given by the following discussion.

Next we will discuss the necessary condition and the sufficient condition of the differentiability of the function $z = f(x,y)$.

> **Theorem 2 (Necessary condition)** If the function $z = f(x,y)$ is differentiable at the point (x_0, y_0),
> (1) $f(x,y)$ is continuous at the point $P_0(x_0, y_0)$;
> (2) the partial derivatives of $f(x,y)$ at the point $P_0(x_0, y_0)$ all exist, and
> $$A = f_x(x_0, y_0), B = f_y(x_0, y_0),$$
> which means the perfect differential of $z = f(x,y)$ at the point $P_0(x_0, y_0)$ can be denoted by
> $$dz = f_x(x_0, y_0)\Delta x + f_y(x_0, y_0)\Delta y.$$

Proof (1) Since $z = f(x,y)$ is differentiable, we have
$$\Delta z = f(x_0+\Delta x, y_0+\Delta y) - f(x_0, y_0) = A\Delta x + B\Delta y + o(\rho),$$
so
$$\lim_{\rho \to 0} \Delta z = \lim_{\rho \to 0} [f(x_0+\Delta x, y_0+\Delta y) - f(x_0, y_0)] = 0,$$
which means
$$\lim_{\rho \to 0} f(x_0+\Delta x, y_0+\Delta y) = f(x_0, y_0).$$

Thus, $f(x,y)$ is continuous at the point $P_0(x_0, y_0)$.

(2) Let $\Delta y = 0$, then $\rho = |\Delta x|$, so $\rho \to 0$ is same to $\Delta x \to 0$. Then we have
$$f(x_0+\Delta x, y_0) - f(x_0, y_0) = A\Delta x + o(|\Delta x|).$$

Divide the two sides by Δx, and let $\Delta x \to 0$, we obtain

Chapter 11 Differentiation of multivariable function and its application

$$\lim_{\Delta x \to 0} \frac{f(x_0+\Delta x, y_0)-f(x_0,y_0)}{\Delta x}=A.$$

In the same way, we have

$$\lim_{\Delta y \to 0} \frac{f(x_0, y_0+\Delta y)-f(x_0,y_0)}{\Delta y}=B.$$

So the partial derivatives of $f(x,y)$ at the point $P_0(x_0,y_0)$ exist and $f_x(x_0,y_0)=A$, $f_y(x_0,y_0)=B$.

If the function $z=f(x,y)$ is differentiable at each point $P(x,y)$ on the region D and its complete differential is

$$dz=f_x(x,y)\Delta x+f_y(x,y)\Delta y \quad \text{or} \quad dz=\frac{\partial z}{\partial x}\Delta x+\frac{\partial z}{\partial y}\Delta y.$$

The necessary condition of the differentiability of the function at one point is given in the above theorem. It is different from one variable function that these conditions are not sufficient to guarantee the differentiability of the function. If each partial derivative of the function exists, although we can write out $\frac{\partial z}{\partial x}\Delta x+\frac{\partial z}{\partial y}\Delta y$, the difference between it and Δz is not necessarily the higher order infinitesimal of ρ. Thus, the sum above may be not the perfect differential of the function, in other words, the existence of each partial differential is necessary but not sufficient condition for the existence of perfect differential.

Example 16 Suppose $f(x,y)=\begin{cases} \dfrac{xy}{\sqrt{x^2+y^2}}, & x^2+y^2 \neq 0, \\ 0, & x^2+y^2=0, \end{cases}$ find $f_x(0,0), f_y(0,0)$ and discuss the differentiability of $z=f(x,y)$ at the point $O(0,0)$.

Solution It is easy to get that the limit of $f(x,y)$ at the point $(0,0)$ exists and $f(x,y)$ is continuous at $(0,0)$. Then, we have

$$f_x(0,0)=\lim_{\Delta x \to 0}\frac{f(0+\Delta x,0)-f(0,0)}{\Delta x}=\lim_{\Delta x \to 0}\frac{0}{\Delta x}=0,$$

$$f_y(0,0)=\lim_{\Delta y \to 0}\frac{f(0,0+\Delta y)-f(0,0)}{\Delta y}=\lim_{\Delta y \to 0}\frac{0}{\Delta y}=0,$$

that is, the partial derivatives of $f(x,y)$ at the point $(0,0)$ exist. Since

$$\Delta z-[f_x(0,0)\Delta x+f_y(0,0)\Delta y]=[f(0+\Delta x,0+\Delta y)-f(0,0)]-0$$

$$=f(\Delta x,\Delta y)=\frac{\Delta x \Delta y}{\sqrt{(\Delta x)^2+(\Delta y)^2}}.$$

as $(\Delta x, \Delta y)$ tends to $O(0,0)$ along the line $y=x$, we have

$$\frac{\frac{\Delta x \Delta y}{\sqrt{(\Delta x)^2+(\Delta y)^2}}}{\rho}=\frac{\Delta x \Delta y}{(\Delta x)^2+(\Delta y)^2}\xrightarrow{\Delta y=\Delta x}\frac{\Delta x \cdot \Delta x}{(\Delta x)^2+(\Delta x)^2}=\frac{1}{2}.$$

Obviously, as $\rho \to 0$, the equality above will not tend to 0. This indicates that as $\rho \to 0$, the difference

$$\Delta z-[f_x(0,0)\Delta x+f_y(0,0)\Delta y]$$

is not infinitesimal whose order is higher than ρ. So the function $z=f(x,y)$ is not differentiable at the point $O(0,0)$.

From Theorem 2 and Example 9, the existence of partial derivative is the necessary condition of the differentiability, but not sufficient condition. The following theorem is the sufficient condition of the differentiability.

Theorem 3 (Sufficient condition) If the partial derivatives of the function $z = f(x,y)$, $\frac{\partial z}{\partial x} = f_x(x,y)$, $\frac{\partial z}{\partial y} = f_y(x,y)$ are continuous at the point $P(x,y)$, then $z = f(x,y)$ is differentiable at this point.

Proof To prove the differentiability of $z = f(x,y)$ at $P(x,y)$, we should prove $\Delta z = f_x(x,y)\Delta x + f_y(x,y)\Delta y + o(\rho)$ first.

Since the partial derivative $f_x(x,y)$ and $f_y(x,y)$ are continuous at the point $P(x,y)$, suppose the point $P'(x+\Delta x, y+\Delta y)$ is a point on the neighborhood of $P(x,y)$, the total increment of the function is

$$\Delta z = f(x+\Delta x, y+\Delta y) - f(x,y)$$
$$= [f(x+\Delta x, y+\Delta y) - f(x, y+\Delta y)] + [f(x, y+\Delta y) - f(x,y)].$$

Because $y + \Delta y$ keeps fixed, the term in the first bracket can be deemed as the increment of the one variable function $f(x, y+\Delta y)$ of x. Because $f_x(x,y)$ exists on the neighborhood of the point $P(x,y)$, which means the derivative of the function $f(x, y+\Delta y)$ with respect to x exists on the interval $[x, x+\Delta x]$ or $[x+\Delta x, x]$. From the conclusion that a derivable function must be continuous, the function $f(x, y+\Delta y)$ of x satisfies Lagrange's Mean Value Theorem between the interval $[x, x+\Delta x]$ or $[x+\Delta x, x]$, so we get

$$f(x+\Delta x, y+\Delta y) - f(x, y+\Delta y) = f_x(x+\theta_1 \Delta x, y+\Delta y)\Delta x \ (0<\theta_1<1).$$

Because x keeps fixed, the term in the second bracket can be deemed as the increment of the one variable function $f(x,y)$ of y. In the same way, the function $f(x,y)$ of y satisfies Lagrange's Mean Value Theorem on the interval $[y, y+\Delta y]$ or $[y+\Delta y, y]$. By the Lagrange's Mean Value Theorem, we obtain that

$$f(x, y+\Delta y) - f(x, y) = f_y(x, y+\theta_2 \Delta y)\Delta y \ (0<\theta_2<1).$$

From the question, the partial derivatives $f_x(x,y)$ and $f_y(x,y)$ are continuous at the point $P(x,y)$, so

$$\lim_{\rho \to 0} f_x(x+\theta_1 \Delta x, y+\Delta y) = f_x(x,y),$$

$$\lim_{\rho \to 0} f_y(x, y+\theta_2 \Delta y) = f_y(x,y).$$

So we have

$$f_x(x+\theta_1 \Delta x, y+\Delta y) = f_x(x,y) + \alpha, \text{ where } \lim_{\rho \to 0} \alpha = 0,$$
$$f_y(x, y+\theta_2 \Delta y) = f_y(x,y) + \beta, \text{ where } \lim_{\rho \to 0} \beta = 0.$$

The expression of total increment is

$$\Delta z = f_x(x,y)\Delta x + f_y(x,y)\Delta y + \alpha \Delta x + \beta \Delta y. \tag{3}$$

Chapter 11 Differentiation of multivariable function and its application

Because $\left|\dfrac{\Delta x}{\rho}\right| = \left|\dfrac{\Delta x}{\sqrt{(\Delta x)^2+(\Delta y)^2}}\right| \leqslant 1$, $\lim\limits_{\rho \to 0} \alpha = 0$, and the product of bounded function and infinitesimal is still infinitesimal, we have

$$\lim_{\rho \to 0} \frac{\alpha \cdot \Delta x}{\rho} = 0, \quad \lim_{\rho \to 0} \frac{\beta \cdot \Delta y}{\rho} = 0.$$

So $\lim\limits_{\rho \to 0} \dfrac{\alpha \Delta x + \beta \Delta y}{\rho} = 0$, which means

$$\Delta z = f_x(x,y)\Delta x + f_y(x,y)\Delta y + o(\rho).$$

The converse of Theorem 3 is not true, which means the differentiability of $z = f(x,y)$ at the point $P(x,y)$ cannot guarantee that its partial derivatives $f_x(x,y)$ and $f_y(x,y)$ are continuous at this point.

Example 17 Suppose $f(x,y) = \begin{cases} (x^2+y^2)\sin \dfrac{1}{\sqrt{x^2+y^2}}, & x^2+y^2 \neq 0, \\ 0, & x^2+y^2 = 0. \end{cases}$

(1) Find the partial derivative $f_x(x,y), f_y(x,y)$.
(2) Whether $f_x(x,y), f_y(x,y)$ are continuous at $O(0,0)$ or not.
(3) Whether the function $f(x,y)$ is differentiable at the point $O(0,0)$ or not.

Solution (1) When $(x,y) \neq (0,0)$, we have

$$f_x(x,y) = 2x\sin\frac{1}{\sqrt{x^2+y^2}} - \frac{x}{\sqrt{x^2+y^2}}\cos\frac{1}{\sqrt{x^2+y^2}},$$

$$f_y(x,y) = 2y\sin\frac{1}{\sqrt{x^2+y^2}} - \frac{y}{\sqrt{x^2+y^2}}\cos\frac{1}{\sqrt{x^2+y^2}}.$$

But

$$f_x(0,0) = \lim_{\Delta x \to 0} \frac{f(0+\Delta x, 0) - f(0,0)}{\Delta x} = \lim_{\Delta x \to 0} \frac{(\Delta x)^2 \sin\frac{1}{|\Delta x|}}{\Delta x}$$

$$= \lim_{\Delta x \to 0} \Delta x \cdot \sin\frac{1}{|\Delta x|} = 0.$$

In the same way, $f_y(0,0) = 0.$

So

$$f_x(x,y) = \begin{cases} 2x\sin\dfrac{1}{\sqrt{x^2+y^2}} - \dfrac{x}{\sqrt{x^2+y^2}}\cos\dfrac{1}{\sqrt{x^2+y^2}}, & x^2+y^2 \neq 0, \\ 0, & x^2+y^2 = 0, \end{cases}$$

$$f_y(x,y) = \begin{cases} 2y\sin\dfrac{1}{\sqrt{x^2+y^2}} - \dfrac{y}{\sqrt{x^2+y^2}}\cos\dfrac{1}{\sqrt{x^2+y^2}}, & x^2+y^2 \neq 0, \\ 0, & x^2+y^2 = 0. \end{cases}$$

(2) As $P(x,y)$ tends to $O(0,0)$ along the x-axis, since

$$\lim_{\substack{(x,y) \to (0,0) \\ y=0}} f_x(x,y) = \lim_{x \to 0}\left(2x\sin\frac{1}{x} - \cos\frac{1}{x}\right)$$

does not exist, $f_x(x,y)$ is not continuous at the point $O(0,0)$.

In the same way, $f_y(x,y)$ is not continuous at the point $O(0,0)$.

(3) Because
$$\lim_{\rho \to 0} \frac{\Delta z - f_x(0,0)\Delta x - f_y(0,0)\Delta y}{\rho}$$
$$= \lim_{\rho \to 0} \frac{\Delta z}{\rho} = \lim_{\rho \to 0} \frac{[(\Delta x)^2 + (\Delta y)^2]\sin\frac{1}{\sqrt{(\Delta x)^2 + (\Delta y)^2}}}{\sqrt{(\Delta x)^2 + (\Delta y)^2}}$$
$$= \lim_{\rho \to 0} \rho \sin\frac{1}{\rho} = 0,$$

$f(x,y)$ is differentiable at $O(0,0)$.

The above is the definition of perfect differential of binary function and the necessary and sufficient condition of differentiability, it can also be generalized to multivariable function.

Same to one variable function, the increment of independent variable is equal to the differential, that is, $\Delta x = dx, \Delta y = dy$, so the perfect differential of $z = f(x,y)$ can be denoted by
$$dz = \frac{\partial z}{\partial x}dx + \frac{\partial z}{\partial y}dy.$$

Generally, the equation above is called the binary function meeting superposition principle, which means the perfect differential of binary function is equal to the sum of its two partial differentials. The superposition principle is also applicable for multivariable function. For example, if the third-variable function $u = f(x,y,z)$ is differentiable at the point $P(x,y,z)$, then the perfect differential is
$$du = \frac{\partial u}{\partial x}dx + \frac{\partial u}{\partial y}dy + \frac{\partial u}{\partial z}dz.$$

Example 18 Find the perfect differential of $z = x^2 y + y^2$ at $(1,2)$.

Solution Because $\frac{\partial z}{\partial x} = 2xy, \frac{\partial z}{\partial y} = x^2 + 2y$, we have
$$\left.\frac{\partial z}{\partial x}\right|_{\substack{x=1\\y=2}} = 4, \left.\frac{\partial z}{\partial y}\right|_{\substack{x=1\\y=2}} = 5.$$

Then the perfect differential is
$$\left.dz\right|_{\substack{x=1\\y=2}} = 4dx + 5dy.$$

Example 19 Find the perfect differential of
$$u = e^{xyz} + xy + z^2.$$

Solution It is easy to get $\frac{\partial u}{\partial x} = yze^{xyz} + y, \frac{\partial u}{\partial y} = xze^{xyz} + x, \frac{\partial u}{\partial z} = xye^{xyz} + 2z$, so
$$du = (yze^{xyz} + y)dx + (xze^{xyz} + x)dy + (xye^{xyz} + 2z)dz.$$

2) The application of perfect differential in approximation calculation

(1) Find approximation of the function.

Suppose $z = f(x,y)$ is a differentiable function, the total increment at the point $P_0(x_0, y_0)$ is
$$\Delta z = f(x_0 + \Delta x, y_0 + \Delta y) - f(x_0, y_0)$$

Chapter 11 Differentiation of multivariable function and its application

$$= f_x(x_0, y_0)\Delta x + f_y(x_0, y_0)\Delta y + o(\rho).$$

If $|\Delta x|$ and $|\Delta y|$ are both small, there is approximation formula

$$\Delta z \approx dz = f_x(x_0, y_0)\Delta x + f_y(x_0, y_0)\Delta y,$$

which means $f(x_0 + \Delta x, y_0 + \Delta y) \approx f(x_0, y_0) + f_x(x_0, y_0)\Delta x + f_y(x_0, y_0)\Delta y.$ (4)

We can calculate the approximation of the function by the Formula (4).

Example 20 Find the approximation of $(1.04)^{2.02}$.

Solution Regard $(1.04)^{2.02}$ as the value of the function $f(x,y) = x^y$ when $x = 1.04$, $y = 2.02$.

Let $x_0 = 1$, $y_0 = 2$, $\Delta x = 0.04$, $\Delta y = 0.02$ and because

$$f_x(x,y) = yx^{y-1}, f_x(1,2) = 2,$$
$$f_y(x,y) = x^y \ln x, f_y(1,2) = 0,$$

from the Formula (4), we obtain

$$(1.04)^{2.02} \approx f(1,2) + f_x(1,2) \times 0.04 + f_y(1,2) \times 0.02$$
$$= 1 + 2 \times 0.04 + 0 \times 0.02 = 1.08.$$

(2) Error estimation.

For binary function $z = f(x,y)$, if the absolute error of independent variable x, y is δ_x, δ_y which means $|\Delta x| \leqslant \delta_x$, $|\Delta y| \leqslant \delta_y$, so the error of z can be calculated by $z = f(x,y)$ is

$$|\Delta z| \approx |dz| = \left|\frac{\partial z}{\partial x}\Delta x + \frac{\partial z}{\partial y}\Delta y\right| \leqslant \left|\frac{\partial z}{\partial x}\right||\Delta x| + \left|\frac{\partial z}{\partial y}\right||\Delta y|$$
$$\leqslant \left|\frac{\partial z}{\partial x}\right|\delta_x + \left|\frac{\partial z}{\partial y}\right|\delta_y.$$

We call

$$\delta_z = \left|\frac{\partial z}{\partial x}\right|\delta_x + \left|\frac{\partial z}{\partial y}\right|\delta_y$$

as the absolute error limit of variable z and call

$$\frac{\delta_z}{|z|} = \left|\frac{1}{z}\frac{\partial z}{\partial x}\right|\delta_x + \left|\frac{1}{z}\frac{\partial z}{\partial y}\right|\delta_y$$

as relative error limit of variable z.

The following is the specific application of differential in economic phenomena.

Example 21 The relation between the cost C and the quantity x, y of the product A and B is

$$C = x^2 - 0.5xy + y^2.$$

The output of A increases from 100 to 105, and the output of B increases from 50 to 52. How much does the cost increase?

Solution Because

$$\Delta C \approx dC = C_x \Delta x + C_y \Delta y$$
$$= (2x - 0.5y)\Delta x + (2y - 0.5x)\Delta y,$$

and $x = 100$, $\Delta x = 5$, $y = 50$, $\Delta y = 2$, we get

$$\Delta C = (2 \times 100 - 0.5 \times 50) \times 5 + (2 \times 50 - 0.5 \times 100) \times 2 = 975.$$

So the cost increases 975.

11.2.3 Differential method of multivariable compound function

1) The Partial derivative of multivariate compound function

For the compound function of one variable, $y=f[\varphi(x)]$, if $y=f(u)$ is derivable at u and $u=\varphi(x)$ is derivable at x, then, the differential rule of one variable compound function is

$$\frac{dy}{dx}=\frac{dy}{du}\cdot\frac{du}{dx}.$$

We can use the following linkage diagram to express the process of derivation,

$$y(\text{function})\to u(\text{intermediate variable})\to x(\text{independent variable}).$$

We will generalize the differential rule to the multivariable compound function.

Suppose the function $z=f(u,v)$ is a compound function $z=f[\varphi(x,y),\psi(x,y)]$ of the variables x and y through intermediate variable $u=\varphi(x,y)$ and $v=\psi(x,y)$.

Now we discuss the derivation of the compound function $z=f[\varphi(t),\psi(t)]$ with respect to t.

Theorem 4 Suppose $u=\varphi(t)$ and $v=\psi(t)$ are derivable at the point t, the function $z=f(u,v)$ has continuous partial derivative at corresponding point (u,v), then the compound function $z=f[\varphi(t),\psi(t)]$ is derivable at the point t, and its derivative can be found by the following formula,

$$\frac{dz}{dt}=\frac{\partial z}{\partial u}\cdot\frac{du}{dt}+\frac{\partial z}{\partial v}\cdot\frac{dv}{dt}. \tag{5}$$

Proof Suppose the increment of t is Δt, and relative increment of $u=\varphi(t), v=\psi(t)$ are $\Delta u, \Delta v$, the function $z=f(u,v)$ get the increment Δz relatively. Since $z=f(u,v)$ has continuous partial derivative at the point (u,v), by Formula (3), we obtain

$$\Delta z=\frac{\partial z}{\partial u}\Delta u+\frac{\partial z}{\partial v}\Delta v+\alpha\Delta u+\beta\Delta v,$$

where $\alpha\to 0, \beta\to 0$ as $\Delta u\to 0, \Delta v\to 0$.

Divide the above equation by Δt on both sides, we obtain

$$\frac{\Delta z}{\Delta t}=\frac{\partial z}{\partial u}\frac{\Delta u}{\Delta t}+\frac{\partial z}{\partial v}\frac{\Delta v}{\Delta t}+\alpha\frac{\Delta u}{\Delta t}+\beta\frac{\Delta v}{\Delta t}.$$

Since $\Delta u\to 0, \Delta v\to 0, \frac{\Delta u}{\Delta t}\to\frac{du}{dt}, \frac{\Delta v}{\Delta t}\to\frac{dv}{dt}$ as $\Delta t\to 0$, we have

$$\lim_{\Delta t\to 0}\frac{\Delta z}{\Delta t}=\frac{\partial z}{\partial u}\frac{du}{dt}+\frac{\partial z}{\partial v}\frac{dv}{dt}.$$

That is, the compound function $z=f[\varphi(t),\psi(t)]$ is derivable at the point t, and the derivative can be found by the Formula (5).

Theorem 4 tells us that the function z relies on independent variable t through two intermediate variables u and v.

Chapter 11 Differentiation of multivariable function and its application

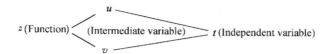

By Formula (5), the method of drawing "linkage diagram" of function structure will help us to master the differential rule of compound function. For example, the compound function $z=f[\varphi(t),\psi(t)]$ is composed by $z=f(u,v)$, $u=\varphi(t)$ and $v=\psi(t)$, then we can use the diagram to express intermediate variable, independent variable and the number of them in Formula (5). Finding the derivative of z with respect to t is same as that there are two path to t, through u or v.

Because there are only one independent variable in Formula (5), the derivative is called the total derivative. In the same way, the Theorem 4 can be generalized to the compound function with two or more intermediate variables.

Corollary The compound function is composed by $z=f(u,v,w)$, $u=\varphi(t)$, $v=\psi(t)$ and $w=\omega(t)$, suppose the functions $u=\varphi(t), v=\psi(t), w=\omega(t)$ are all derivable at the point t, $z=f(u,v,w)$ has continuous partial derivatives at corresponding point (u,v,w), the compound function is derivable at the point t, and the linkage diagram is

$$z \leftarrow \begin{matrix} u \\ v \\ w \end{matrix} \leftarrow t.$$

So the derivation formula of z with respect to t is

$$\frac{dz}{dt}=\frac{\partial z}{\partial u}\frac{du}{dt}+\frac{\partial z}{\partial v}\frac{dv}{dt}+\frac{\partial z}{\partial w}\frac{dw}{dt}. \tag{6}$$

The above theorem can be generalized to the derivation of the compound function with multivariable intermediate functions, the following theorem is actually the basic theorem of derivation of multivariable compound function.

Theorem 5 Suppose the function $u=\varphi(x,y), v=\psi(x,y)$ have partial derivatives at the point (x,y), $z=f(u,v)$ is differentiable at corresponding point (u,v), the partial derivatives $\frac{\partial z}{\partial x}$ and $\frac{\partial z}{\partial y}$ of the compound function $z=f[\varphi(x,y),\psi(x,y)]$ at the point (x,y) exist, and there are following formulas

$$\frac{\partial z}{\partial x}=\frac{\partial z}{\partial u}\frac{\partial u}{\partial x}+\frac{\partial z}{\partial v}\frac{\partial v}{\partial x},$$
$$\frac{\partial z}{\partial y}=\frac{\partial z}{\partial u}\frac{\partial u}{\partial y}+\frac{\partial z}{\partial v}\frac{\partial v}{\partial y}. \tag{7}$$

The linkage diagram of the variables in Theorem 5 is as follows,

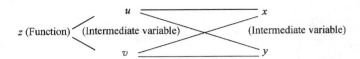

The linkage diagram shows that there are two intermediate variables, two independent variables and the relationship of these variables, then the equality (7) expresses the derivation of z with respect to x (or y).

Proof Give the increment Δx and let y keep fixed, then the partial increments of $u = \varphi(x, y)$, $v = \psi(x, y)$ are

$$\Delta_x u = \varphi(x + \Delta x, y) - \varphi(x, y),$$
$$\Delta_x v = \psi(x + \Delta x, y) - \psi(x, y).$$

Because the partial derivatives of the function $u = \varphi(x, y), v = \psi(x, y)$ exist, from the fact that the derivable function must be continuous, we know that $u = \varphi(x, y), v = \psi(x, y)$ are continuous functions of x. So we have $\Delta_x u \to 0$, $\Delta_x v \to 0$ as $\Delta x \to 0$. Because the function $z = f(u, v)$ is differentiable at the corresponding point (u, v), the total increment of $z = f(u, v)$ at the point (u, v) is

$$\Delta z = f(u + \Delta u, v + \Delta v) - f(u, v)$$
$$= \frac{\partial z}{\partial u} \Delta u + \frac{\partial z}{\partial v} \Delta v + o(\rho),$$

where $o(\rho)$ is higher-order infinitesimal of $\rho = \sqrt{(\Delta u)^2 + (\Delta v)^2} \to 0$.

The partial increment of the compound function $z = f[\varphi(x, y), \psi(x, y)]$ with respect to x at the point (x, y) is

$$\Delta_x z = f[\varphi(x + \Delta x, y), \psi(x + \Delta x, y)] - f[\varphi(x, y), \psi(x, y)]$$
$$= f(u + \Delta_x u, v + \Delta_x v) - f(u, v)$$
$$= \frac{\partial z}{\partial u} \Delta_x u + \frac{\partial z}{\partial v} \Delta_x v + o(\rho),$$

and then we obtain that

$$\frac{\Delta_x z}{\Delta x} = \frac{\partial z}{\partial u} \frac{\Delta_x u}{\Delta x} + \frac{\partial z}{\partial v} \frac{\Delta_x v}{\Delta x} + \frac{o(\rho)}{\Delta x}.$$

Because

$$\frac{o(\rho)}{\Delta x} = \frac{o(\rho)}{\rho} \frac{|\Delta x|}{\Delta x} \sqrt{\left(\frac{\Delta_x u}{\Delta x}\right)^2 + \left(\frac{\Delta_x v}{\Delta x}\right)^2},$$

and $\Delta_x u \to 0$, $\Delta_x v \to 0$ as $\Delta x \to 0$, that is,

$$\rho = \sqrt{(\Delta_x u)^2 + (\Delta_x v)^2} \to 0,$$

then

$$\frac{o(\rho)}{\sqrt{(\Delta_x u)^2 + (\Delta_x v)^2}} \to 0.$$

And as $\Delta x \to 0$,

$$\sqrt{\left(\frac{\Delta_x u}{\Delta x}\right)^2 + \left(\frac{\Delta_x v}{\Delta x}\right)^2} \to \sqrt{\left(\frac{\partial u}{\partial x}\right)^2 + \left(\frac{\partial v}{\partial x}\right)^2},$$

Chapter 11 Differentiation of multivariable function and its application

so $\dfrac{|\Delta x|}{\Delta x}\sqrt{\left(\dfrac{\Delta_x u}{\Delta x}\right)^2+\left(\dfrac{\Delta_x v}{\Delta x}\right)^2}$ is bounded.

So
$$\lim_{\Delta x\to 0}\dfrac{\Delta_x z}{\Delta x}=\dfrac{\partial z}{\partial u}\cdot\dfrac{\partial u}{\partial x}+\dfrac{\partial z}{\partial v}\cdot\dfrac{\partial v}{\partial x},$$

Then
$$\dfrac{\partial z}{\partial x}=\dfrac{\partial z}{\partial u}\cdot\dfrac{\partial u}{\partial x}+\dfrac{\partial z}{\partial v}\cdot\dfrac{\partial v}{\partial x}.$$

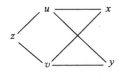

In the same way, we obtain
$$\dfrac{\partial z}{\partial y}=\dfrac{\partial z}{\partial u}\cdot\dfrac{\partial u}{\partial y}+\dfrac{\partial z}{\partial v}\cdot\dfrac{\partial v}{\partial y}.$$

From the linkage diagram, we can obtain two laws: one is that the number of partial derivative is same to the number of independent variables, the other is that the number of addition terms in the formula is same to the number of the intermediate variables.

This theorem can be generalized from the following aspects.

(1) Suppose $u=\varphi(x,y)$, $v=\psi(x,y)$ and $w=(x,y)$ all have partial derivatives with respect to x and y at the point (x,y), the function $z=f(u,v,w)$ has continuous partial derivatives at the corresponding point (u,v,w), then the partial derivatives of the compound function
$$z=f[\varphi(x,y),\psi(x,y),w(x,y)]$$
at the point (x,y) exist, and they can be found by the following formula
$$\dfrac{\partial z}{\partial x}=\dfrac{\partial z}{\partial u}\dfrac{\partial u}{\partial x}+\dfrac{\partial z}{\partial v}\dfrac{\partial v}{\partial x}+\dfrac{\partial z}{\partial w}\dfrac{\partial w}{\partial x},$$
$$\dfrac{\partial z}{\partial y}=\dfrac{\partial z}{\partial u}\dfrac{\partial u}{\partial y}+\dfrac{\partial z}{\partial v}\dfrac{\partial v}{\partial y}+\dfrac{\partial z}{\partial w}\dfrac{\partial w}{\partial y}.$$

The linkage diagram is shown on the right side.

(2) Suppose $u=\varphi(x,y,t)$ and $v=\psi(x,y,t)$ have partial derivatives at the point (x,y,t), and $z=f(u,v)$ has continuous partial derivatives at the corresponding point (u,v). Then the partial derivatives of $z=f[\varphi(x,y,t),\psi(x,y,t)]$ with respect to x,y,t all exist and
$$\dfrac{\partial z}{\partial x}=\dfrac{\partial z}{\partial u}\dfrac{\partial u}{\partial x}+\dfrac{\partial z}{\partial v}\dfrac{\partial v}{\partial x},$$
$$\dfrac{\partial z}{\partial y}=\dfrac{\partial z}{\partial u}\dfrac{\partial u}{\partial y}+\dfrac{\partial z}{\partial v}\dfrac{\partial v}{\partial y},$$
$$\dfrac{\partial z}{\partial t}=\dfrac{\partial z}{\partial u}\dfrac{\partial u}{\partial t}+\dfrac{\partial z}{\partial v}\dfrac{\partial v}{\partial t}.$$

The linkage diagram is as follows

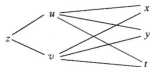

Especially, if $z=f(u,x,y)$ has continuous partial derivatives, and $u=\varphi(x,y)$ has partial derivatives, the compound function $z=f[\varphi(x,y),x,y]$ can be regarded as a special case in which $v=x$, $w=y$. Thus,

$$\frac{\partial v}{\partial x}=1, \frac{\partial w}{\partial x}=0,$$

$$\frac{\partial v}{\partial y}=0, \frac{\partial w}{\partial y}=1.$$

So the compound function has partial derivatives with respect to the independent variables x and y, then we obtain

$$\frac{\partial z}{\partial x}=\frac{\partial f}{\partial u}\frac{\partial u}{\partial x}+\frac{\partial f}{\partial x},$$

$$\frac{\partial z}{\partial y}=\frac{\partial f}{\partial u}\frac{\partial u}{\partial y}+\frac{\partial f}{\partial y}.$$

Note In the formula above, $\frac{\partial z}{\partial x}$ and $\frac{\partial f}{\partial x}$ are different. $\frac{\partial z}{\partial x}$ is the partial derivative of the compound function with respect to x while y is fixed, $\frac{\partial f}{\partial x}$ is the partial derivative of $f(u,x,y)$ with respect to x while u and y are both fixed. $\frac{\partial z}{\partial y}$ and $\frac{\partial f}{\partial y}$ have similar difference.

Example 22 Suppose $z=e^{x-2y}$, $x=\sin t$, $y=t^3$, find $\frac{\mathrm{d}z}{\mathrm{d}t}$.

Solution Let $u=x-2y$, the total derivative of the function z is

$$\frac{\mathrm{d}z}{\mathrm{d}t}=\frac{\mathrm{d}z}{\mathrm{d}u}\frac{\partial u}{\partial x}\frac{\mathrm{d}x}{\mathrm{d}t}+\frac{\mathrm{d}z}{\mathrm{d}u}\frac{\partial u}{\partial y}\frac{\mathrm{d}y}{\mathrm{d}t}$$

$$=e^u \cdot 1 \cdot \cos t+e^u \cdot (-2) \cdot 3t^2 = e^{\sin t-2t^3}(\cos t-6t^2).$$

Another solution of this question is to eliminate intermediate variable. That is, plug intermediate variable into the function z, we have

$$z=e^{\sin t-2t^3}.$$

Then the question is the derivation of one variable function, we get the derivative

$$z_t=e^{\sin t-2t^3}(\cos t-6t^2).$$

This method is often used to find the total derivative of multivariable function.

Example 23 We know that the production cost of x products is

$$C=25\ 000+200x+\frac{1}{40}x^2\ (\text{Yuan}).$$

(1) How many products should be produced to make the average cost minimum.

(2) If the unit price of this product is 500 Yuan, how many products should be produced to make the profit maximum.

Solution (1) Suppose the average cost is y, then

$$y=\frac{25\ 000}{x}+200+\frac{x}{40}.$$

Let $y'=-\frac{25\ 000}{x^2}+\frac{1}{40}=0$, we get $x_1=1\ 000$, $x_2=-1\ 000$ (it is negative, omitted). And

Chapter 11 Differentiation of multivariable function and its application

$$y''=\frac{50\,000}{x^3},\ y''|_{x=1\,000}=5\times10^{-5}>0,$$

so y reaches the unique minima as $x=1\,000$.

Thus, in order to reach the average cost minimum, 1 000 products should be produced.

(2) The profit function is

$$L(x)=500x-(25\,000+200x+\frac{x^2}{40})=300x-\frac{x^2}{40}-25\,000,$$

and $L'(x)=300-\frac{x}{20}$. Let $L'(x)=0$, we have $x=6\,000$. Because

$$L''(x)=-\frac{1}{20},\ L''(6\,000)<0.$$

L obtains unique maxima at $x=6\,000$, which is also the maximum.

Thus, in order to make the profit maximum, 6 000 products should be produced.

Example 24 The price of product P satisfies the formula $P=7-0.2x$, x is the sales (the unit is kg), the cost function (the unit is hundred Yuan) is $C=3x+1$.

(1) If we sale 1kg goods, the government will tax t (the unit is hundred Yuan). Find the sales when the seller gets the maximum profit.

(2) Find t such that the total tax revenue of government is maximum.

Solution (1) It is easy to get the total tax revenue $T=tx$, and the income of total sales is

$$R=px=(7-0.2x)x,$$

the profit function is

$$L=R-C-T=-0.2x^2+(4-t)x-1.$$

Since $\frac{\partial L}{\partial x}=-0.4x+4-t$, let $\frac{\partial L}{\partial x}=0$, we obtain

$$x=\frac{5}{2}(4-t).$$

And because $\frac{\partial^2 L}{\partial x^2}<0$, we get that $x=\frac{5}{2}(4-t)$ is the sales when the profit is maximum.

(2) Plug $x=\frac{5}{2}(4-t)$ into $T=tx$, we obtain

$$T=10t-\frac{5}{2}t^2,\ \frac{\partial T}{\partial t}=10-5t.$$

Let $\frac{\partial T}{\partial t}=0$, we obtain $t=2$.

$\frac{\partial^2 T}{\partial t^2}=-5<0$, so T has the unique maxima and the maximum at $t=2$. Meanwhile, the tax revenue of government is the maximum.

Example 25 The purchase price of some good is a (Yuan/piece), when the sales price is b (Yuan/piece), the sales is c pieces (a,b,c are all positive constants, and $b\geqslant\frac{4}{3}a$). The market research shows that when the sales price decreases 10%, the sales will increase 40%. Now we decide to reduce the price at one time. Find the sales price such that the maximum profit will be obtained? And find the maximum profit.

Solution Suppose p is the price after reducing, x is the increased sales, $L(x)$ is the total profit, then we have $\dfrac{x}{b-p}=\dfrac{0.4c}{0.1b}$, so

$$p=b-\dfrac{b}{4c}x,$$

$$L(x)=\left(b-\dfrac{b}{4c}x-a\right)(c+x).$$

Then, we obtain

$$L'(x)=-\dfrac{b}{2c}x+\dfrac{3}{4}b-a.$$

Let $L'(x)=0$, we can obtain the unique stationary point

$$x_0=\dfrac{(3b-4a)c}{2b}.$$

From the actual meaning of the question or

$$L''(x_0)=-\dfrac{b}{2c}<0,$$

we know that x_0 is the maxima point and also maximum point.

So, at the price

$$p=b-\left(\dfrac{3}{8}b-\dfrac{1}{2}a\right)=\dfrac{5}{8}b+\dfrac{1}{2}a(\text{Yuan}),$$

we obtain the maximum profit

$$L(x_0)=\dfrac{c}{16b}(5b-4a)^2(\text{Yuan}).$$

Example 26 Suppose $z=e^u\sin v$ and $u=xy, v=x^2+y^2$, find $\dfrac{\partial z}{\partial x}$ and $\dfrac{\partial z}{\partial y}$.

Solution From the differential rule of compound function, we obtain

$$\dfrac{\partial z}{\partial x}=\dfrac{\partial z}{\partial u}\dfrac{\partial u}{\partial x}+\dfrac{\partial z}{\partial v}\dfrac{\partial v}{\partial x}$$

$$=ye^u\sin v+2xe^u\cos v$$

$$=e^{xy}[y\sin(x^2+y^2)+2x\cos(x^2+y^2)],$$

$$\dfrac{\partial z}{\partial y}=xe^u\sin v+2ye^u\cos v$$

$$=e^{xy}[x\sin(x^2+y^2)+2y\cos(x^2+y^2)].$$

In this question we can also find the partial derivatives by eliminating intermediate variables, which is often used to test the correctness of the result. But it is not applicable to the abstract functions.

Example 27 Suppose $u=f(x,y,z)=e^{x^2+y^2+z^2}$ and $z=x^2\sin y$, find $\dfrac{\partial u}{\partial x}$ and $\dfrac{\partial u}{\partial y}$.

Solution $\dfrac{\partial u}{\partial x}=\dfrac{\partial f}{\partial x}+\dfrac{\partial f}{\partial z}\dfrac{\partial z}{\partial x}$

$$= 2xe^{x^2+y^2+z^2} + 2ze^{x^2+y^2+z^2} \cdot 2x\sin y$$
$$= 2x(1+2x^2\sin^2 y)e^{x^2+y^2+x^4\sin^2 y},$$
$$\frac{\partial u}{\partial y} = \frac{\partial f}{\partial y} + \frac{\partial f}{\partial z} \cdot \frac{\partial z}{\partial y} = 2ye^{x^2+y^2+z^2} + 2ze^{x^2+y^2+z^2} x^2 \cos y$$
$$= 2(y+x^4\sin y\cos y)e^{x^2+y^2+x^4\sin^2 y}.$$

Example 28 Suppose $z = f(u,v)$ and $u = x^2 - y^2$, $v = e^{xy}$, f has the partial derivative, find $\frac{\partial z}{\partial x}, \frac{\partial z}{\partial y}$.

Solution $\frac{\partial z}{\partial x} = \frac{\partial f}{\partial u}\frac{\partial u}{\partial x} + \frac{\partial f}{\partial v}\frac{\partial v}{\partial x} = \frac{\partial f}{\partial u}2x + \frac{\partial f}{\partial v}ye^{xy} = 2xf_u + ye^{xy}f_v,$

$$\frac{\partial z}{\partial y} = -2y\frac{\partial f}{\partial u} + xe^{xy}\frac{\partial f}{\partial v} = -2yf_u + xe^{xy}f_v.$$

2) The high order partial derivative of multivariate compound function

The definition of high order partial derivative has been given, there are some examples to illustrate the method of finding the high order partial derivative of multivariate compound function.

Example 29 Suppose $w = f(x+y+z, xyz)$, where f has the second continuous partial derivative, find $\frac{\partial w}{\partial x}$ and $\frac{\partial^2 w}{\partial x \partial z}$.

Solution Let $u = x+y+z$, $v = xyz$, then $w = f(u,v)$.

In short, the following notations are introduced

$$f'_1 = \frac{\partial f(u,v)}{\partial u}, \quad f''_{12} = \frac{\partial^2 f(u,v)}{\partial u \partial v},$$

where the subscript 1 indicates the partial derivative of the first variable v, and the subscript 2 indicates the partial derivative of the second variable v. It is similar to $f''_{12}, f''_{11}, f''_{22}$ and so on.

Because the given function is compounded by $w = f(u,v)$ and $u = x+y+z$, $v = xyz$, according to the derivation rule of compound function, we have

$$\frac{\partial w}{\partial x} = \frac{\partial f}{\partial u}\frac{\partial u}{\partial x} + \frac{\partial f}{\partial v}\frac{\partial v}{\partial x} = f'_1 + yzf'_2,$$

$$\frac{\partial^2 w}{\partial x \partial z} = \frac{\partial}{\partial z}(f'_1 + yzf'_2) = \frac{\partial f'_1}{\partial z} + yf'_2 + yz\frac{\partial f'_2}{\partial z}.$$

To find $\frac{\partial f'_1}{\partial z}$ and $\frac{\partial f'_2}{\partial z}$, we should notice that f'_1 and f'_2 are still functions of u,v and u,v are still functions of x,y, which means f'_1 and f'_2 have the same intermediate variables and independent variables as the function f, then

$$\frac{\partial f'_1}{\partial z} = \frac{\partial f'_1}{\partial u}\frac{\partial u}{\partial z} + \frac{\partial f'_1}{\partial v}\frac{\partial v}{\partial z} = f''_{11} + xyf''_{12},$$

$$\frac{\partial f'_2}{\partial z} = \frac{\partial f'_2}{\partial u}\frac{\partial u}{\partial z} + \frac{\partial f'_2}{\partial v}\frac{\partial v}{\partial z} = f''_{21} + xyf''_{22}.$$

So,

$$\frac{\partial^2 w}{\partial x \partial z} = f''_{11} + xyf''_{12} + yf'_2 + yzf''_{21} + xy^2 zf''_{22}$$

$$= f''_{11} + y(x+z)f''_{12} + xy^2 zf''_{22} + yf'_2.$$

Example 30 Suppose $z = xf(2x + 3y, xy)$ and f has the second order continuous partial derivative, find $\dfrac{\partial^2 z}{\partial x \partial y}$.

Solution Let $u = 2x + 3y$, $v = xy$, so

$$\frac{\partial z}{\partial x} = f(u,v) + x\left(f'_1 \frac{\partial u}{\partial x} + f'_2 \frac{\partial v}{\partial x}\right)$$

$$= f(u,v) + 2xf'_1 + xyf'_2,$$

$$\frac{\partial^2 z}{\partial x \partial y} = \frac{\partial}{\partial y}[f(u,v) + 2xf'_1 + xyf'_2] = \frac{\partial}{\partial y}f(u,v) + 2x\frac{\partial}{\partial y}f'_1 + x\frac{\partial}{\partial y}(yf'_2)$$

$$= f'_1 \frac{\partial u}{\partial y} + f'_2 \frac{\partial v}{\partial y} + 2x\left(f''_{11} \frac{\partial u}{\partial y} + f''_{12} \frac{\partial v}{\partial y}\right) + x\left[f'_2 + y\left(f''_{21} \frac{\partial u}{\partial y} + f''_{22} \frac{\partial v}{\partial y}\right)\right]$$

$$= 3f'_1 + xf'_2 + 2x(3f''_{11} + xf''_{12}) + x[f'_2 + y(3f''_{21} + xf''_{22})]$$

$$= 3f'_1 + 2xf'_2 + 6xf''_{11} + (2x^2 + 3xy)f''_{12} + x^2 yf''_{22}.$$

Example 31 Suppose $z = f(x,y)$ has the second order continuous partial derivatives, prove the following equality by polar coordinate transformation,

$$\frac{\partial^2 z}{\partial x^2} + \frac{\partial^2 z}{\partial y^2} = \frac{\partial^2 z}{\partial r^2} + \frac{1}{r}\frac{\partial z}{\partial r} + \frac{1}{r^2}\frac{\partial^2 z}{\partial \theta^2}.$$

Proof Because z is the function of x, y, and on polar coordinate, $x = r\cos\theta$, $y = r\sin\theta$, z is a compound function of r, from differential rule of compound function, we obtain that

$$\frac{\partial z}{\partial r} = \frac{\partial z}{\partial x}\frac{\partial x}{\partial r} + \frac{\partial z}{\partial y}\frac{\partial y}{\partial r} = \frac{\partial z}{\partial x}\cos\theta + \frac{\partial z}{\partial y}\sin\theta,$$

$$\frac{\partial z}{\partial \theta} = \frac{\partial z}{\partial x}\frac{\partial x}{\partial \theta} + \frac{\partial z}{\partial y}\frac{\partial y}{\partial \theta} = \frac{\partial z}{\partial x}(-r\sin\theta) + \frac{\partial z}{\partial y}(r\cos\theta).$$

And then find the second order partial derivative,

$$\frac{\partial^2 z}{\partial r^2} = \frac{\partial}{\partial r}\left(\frac{\partial z}{\partial r}\right) = \frac{\partial}{\partial r}\left(\frac{\partial z}{\partial x}\cos\theta + \frac{\partial z}{\partial y}\sin\theta\right)$$

$$= \left(\frac{\partial^2 z}{\partial x^2}\cos\theta + \frac{\partial^2 z}{\partial x \partial y}\sin\theta\right)\cos\theta + \left(\frac{\partial^2 z}{\partial y \partial x}\cos\theta + \frac{\partial^2 z}{\partial y^2}\sin\theta\right)\sin\theta$$

$$= \frac{\partial^2 z}{\partial x^2}\cos^2\theta + 2\frac{\partial^2 z}{\partial x \partial y}\sin\theta\cos\theta + \frac{\partial^2 z}{\partial y^2}\sin^2\theta,$$

$$\frac{\partial^2 z}{\partial \theta^2} = \frac{\partial}{\partial \theta}\left(\frac{\partial z}{\partial \theta}\right) = \frac{\partial}{\partial \theta}\left[\frac{\partial z}{\partial x}(-r\sin\theta) + \frac{\partial z}{\partial y}(r\cos\theta)\right]$$

$$= \left[\frac{\partial^2 z}{\partial x^2}(-r\sin\theta) + \frac{\partial^2 z}{\partial x \partial y}(r\cos\theta)\right](-r\sin\theta) - r\frac{\partial z}{\partial x}\cos\theta +$$

$$\left[\frac{\partial^2 z}{\partial y \partial x}(-r\sin\theta) + \frac{\partial^2 z}{\partial y^2}(r\cos\theta)\right](r\cos\theta) - r\frac{\partial z}{\partial y}\sin\theta$$

$$= \frac{\partial^2 z}{\partial x^2}r^2\sin^2\theta - 2\frac{\partial^2 z}{\partial x \partial y}r^2\sin\theta\cos\theta + \frac{\partial^2 z}{\partial y^2}r^2\cos^2\theta - r\frac{\partial z}{\partial x}\cos\theta - r\frac{\partial z}{\partial y}\sin\theta.$$

Thus,

$$\frac{\partial^2 z}{\partial r^2}+\frac{1}{r}\frac{\partial z}{\partial r}+\frac{1}{r^2}\frac{\partial^2 z}{\partial \theta^2}=\frac{\partial^2 z}{\partial x^2}(\cos^2\theta+\sin^2\theta)+\frac{\partial^2 z}{\partial y^2}(\sin^2\theta+\cos^2\theta)$$
$$=\frac{\partial^2 z}{\partial x^2}+\frac{\partial^2 z}{\partial y^2}.$$

3) The invariance of perfect differential forms

For one variable function, we know that in the formula $dy = y_u du$, whether u is independent variable or intermediate variable, dy can always be expressed by $y_u du$. We call this the invariance of total differential forms. This character is also true for multivariable function.

Suppose the function $z = f(u,v)$ has continuous partial derivatives, so the perfect differential of the compound function $z = f[\varphi(x,y), \psi(x,y)]$ is

$$dz = \frac{\partial z}{\partial x}dx + \frac{\partial z}{\partial y}dy,$$

where $\frac{\partial z}{\partial x}$ and $\frac{\partial z}{\partial y}$ can be found by Formula (7). Then plug $\frac{\partial z}{\partial x}$ and $\frac{\partial z}{\partial y}$ into the above formula, we obtain that

$$dz = \left(\frac{\partial z}{\partial u}\frac{\partial u}{\partial x}+\frac{\partial z}{\partial v}\frac{\partial v}{\partial x}\right)dx + \left(\frac{\partial z}{\partial u}\frac{\partial u}{\partial y}+\frac{\partial z}{\partial v}\frac{\partial v}{\partial y}\right)dy$$
$$= \frac{\partial z}{\partial u}\left(\frac{\partial u}{\partial x}dx+\frac{\partial u}{\partial y}dy\right)+\frac{\partial z}{\partial v}\left(\frac{\partial v}{\partial x}dx+\frac{\partial v}{\partial y}dy\right)$$
$$= \frac{\partial z}{\partial u}du + \frac{\partial z}{\partial v}dv.$$

Thus, whether z is the function of independent variables u,v or of the intermediate variables u,v, the form of its perfect differential is the same. This character is also the invariance of total differential forms.

Example 32 Suppose $z = f(x^2 - y^2, e^{xy})$, and f has the first continuous partial derivative, find $dz, \frac{\partial z}{\partial x}, \frac{\partial z}{\partial y}$ by the invariance of total differential forms.

Solution
$$dz = df(x^2 - y^2, e^{xy})$$
$$= f'_1 d(x^2 - y^2) + f'_2 \cdot de^{xy}$$
$$= f'_1(2xdx - 2ydy) + f'_2 e^{xy} d(xy)$$
$$= 2xf'_1 dx - 2yf'_1 dy + f'_2 e^{xy}(ydx + xdy)$$
$$= (2xf'_1 + ye^{xy}f'_2)dx + (-2yf'_1 + xe^{xy}f'_2)dy,$$

so
$$\frac{\partial z}{\partial x} = 2xf'_1 + ye^{xy}f'_2, \quad \frac{\partial z}{\partial y} = -2yf'_1 + xe^{xy}f'_2.$$

11.2.4 Derivative of implicit function

1) The differential method of implicit function determined by an equation

The concept of implicit function has been given in Chapter 2, derivative and differential, and the so-called implicit function is a function $f(x)$ determined by an equation

$$F(x,y) = 0, \tag{8}$$

and this function satisfies $F[x, f(x)] \equiv 0$ on some interval.

But we should notice that not all equation $F(x,y)=0$ can determine y as a function of x. Only under certain conditions, $F(x,y)=0$ can determine y as a function of x, this is the existence theorem of implicit function which will be introduced below.

> **Theorem 6 (The existence theorem of implicit function 1)** Suppose the function $F(x,y)$ has continuous partial derivatives on the neighborhood of the point (x_0, y_0), and $F(x_0, y_0)=0$, $F_y(x_0, y_0) \neq 0$, then
> (1) the equation $F(x,y)=0$ always determine a single-valued continuous implicit function $y=f(x)$ on the neighborhood of the point (x_0, y_0), and $y_0 = f(x_0)$ at $x=x_0$.
> (2) $f(x)$ has continuous derivative and
> $$\frac{dy}{dx} = -\frac{F_x(x,y)}{F_y(x,y)}. \tag{9}$$

The Formula (9) is the derivative formula of implicit function.

We will not prove this theorem, and just explain the Formula (9) as follows.

Plug the function $y=f(x)$ determined by the Equation (8) into the Equation (8), we get the equality
$$F[x, f(x)] \equiv 0.$$

The left side of the identity above can be regarded as a compound function of x, to find the perfect derivative of this function. Because both sides of the identity are still identical after the derivation, we can obtain that
$$\frac{\partial F}{\partial x} + \frac{\partial F}{\partial y}\frac{dy}{dx} = 0.$$

Because F_y is continuous, and $F_y(x_0, y_0) \neq 0$, there is a neighborhood of (x_0, y_0) on which $F_y \neq 0$, then we obtain that
$$\frac{dy}{dx} = -\frac{F_x(x,y)}{F_y(x,y)}.$$

Notice that $F_x(x,y), F_y(x,y)$ in the Formula (9) are the partial derivatives of the binary function $F(x,y)$.

If the second order partial derivatives of $F(x,y)$ are all continuous, we regard two sides of the Formula (9) as the compound function of x, and then find the derivative, we obtain that
$$\frac{d^2 y}{dx^2} = \frac{\partial}{\partial x}\left(-\frac{F_x}{F_y}\right) + \frac{\partial}{\partial y}\left(-\frac{F_x}{F_y}\right)\frac{dy}{dx}$$
$$= -\frac{F_{xx}F_y - F_{yx}F_x}{F_y^2} - \frac{F_{xy}F_y - F_{yy}F_x}{F_y^2}\left(-\frac{F_x}{F_y}\right)$$
$$= -\frac{F_{xx}F_y^2 - 2F_{xy}F_xF_y + F_{yy}F_x^2}{F_y^3}.$$

Example 33 Prove that the equation $x^2 + y^2 - 1 = 0$ can determine a single-value and continuous function on the neighborhood of the point $(0,1)$. Find the first and the second order derivative of this function at $x=0$.

Solution Suppose $F(x,y)=x^2+y^2-1$, we have
$$F_x=2x,\ F_y=2y,\ F(0,1)=0,\ F_y(0,1)=2\neq 0,$$
which satisfies the Theorem 6. So the equation $x^2+y^2-1=0$ can determine a single-valued continuous derivative, when $x=0$, the value $y=1$.

Let's find the first and the second order derivative of this function as follows
$$\frac{dy}{dx}=-\frac{F_x}{F_y}=-\frac{x}{y},\ \left.\frac{dy}{dx}\right|_{x=0}=0;$$

$$\frac{d^2y}{dx^2}=-\frac{y-xy'}{y^2}=-\frac{y-x\left(-\frac{x}{y}\right)}{y^2}=-\frac{y^2+x^2}{y^3}=-\frac{1}{y^3},$$

$$\left.\frac{d^2y}{dx^2}\right|_{x=0}=-1.$$

Same to Theorem 6, the three-variable function $F(x,y,z)=0$ can determine a binary implicit function $z=f(x,y)$ under some conditions, and there is a partial derivative formula of binary function similar to the Formula (9).

Theorem 7 (Existence theorem of implicit function 2) Suppose the function $F(x,y,z)$ has continuous partial derivatives on the neighborhood of the point (x_0,y_0,z_0), and $F(x_0,y_0,z_0)=0, F_z(x_0,y_0,z_0)\neq 0$, then

(1) on the neighborhood of the point (x_0,y_0), the equation $F(x,y,z)=0$ can determine a single-valued continuous implicit function $z=f(x,y)$ and $z_0=f(x_0,y_0)$.

(2) $f(x,y)$ has continuous partial derivatives and
$$\frac{\partial z}{\partial x}=-\frac{F_x(x,y,z)}{F_z(x,y,z)},\ \frac{\partial z}{\partial y}=-\frac{F_y(x,y,z)}{F_z(x,y,z)}. \tag{10}$$

Formula (10) is the derivation formula of binary implicit function. Similar to the Theorem 6, we will not prove it but just explain the Formula (10) as follows.

If the function $F(x,y,z)$ satisfies the condition of Theorem 7, then a implicit function $z=f(x,y)$ is defined by the equation $F(x,y,z)=0$. By plugging $z=f(x,y)$ into $F(x,y,z)=0$, we obtain the identity
$$F[x,y,f(x,y)]\equiv 0.$$
By derivation rule of compound function, find partial derivative with respect to x and y respectively on both sides, we obtain
$$F_x+F_z\frac{\partial z}{\partial x}=0,\ F_y+F_z\frac{\partial z}{\partial y}=0.$$

Because the partial derivative $F_z(x,y,z)$ is continuous and $F_z(x_0,y_0,z_0)\neq 0$, there is a neighborhood of the point (x_0,y_0,z_0), $F_z(x,y,z)\neq 0$ on this neighborhood, then
$$\frac{\partial z}{\partial x}=-\frac{F_x(x,y,z)}{F_z(x,y,z)},\ \frac{\partial z}{\partial y}=-\frac{F_y(x,y,z)}{F_z(x,y,z)}.$$

Example 34 Suppose $2x^2+y^2+z^2-2z=0$ determines an implicit function $z=f(x,y)$, find $\dfrac{\partial^2 z}{\partial x^2}$.

Solution Suppose $F(x,y,z)=2x^2+y^2+z^2-2z$, then
$$F_x=4x,\ F_z=2z-2,$$
so
$$\frac{\partial z}{\partial x}=-\frac{F_x}{F_z}=-\frac{4x}{2z-2}=\frac{2x}{1-z}.$$

Find the partial derivative with respect to x again, we obtain
$$\frac{\partial^2 z}{\partial x^2}=\frac{2(1-z)+2x\cdot\dfrac{\partial z}{\partial x}}{(1-z)^2}=\frac{2(1-z)+2x\cdot\dfrac{2x}{1-z}}{(1-z)^2}$$
$$=\frac{2(1-z)^2+4x^2}{(1-z)^3}.$$

Example 35 Suppose z is the implicit function of x, y, and it is determined by $F\left(\dfrac{z}{x},\dfrac{z}{y}\right)=0$, where $F\left(\dfrac{z}{x},\dfrac{z}{y}\right)$ is differentiable function, find $\dfrac{\partial z}{\partial x},\dfrac{\partial z}{\partial y}$.

Solution Suppose $u=\dfrac{z}{x}, v=\dfrac{z}{y}$, then
$$F_x=F'_1\cdot u_x+F'_2\cdot v_x=-\frac{z}{x^2}F'_1,$$
$$F_y=F'_1\cdot u_y+F'_2\cdot v_y=-\frac{z}{y^2}F'_2,$$
$$F_z=F'_1\cdot u_z+F'_2\cdot v_z=\frac{1}{x}F'_1+\frac{1}{y}F'_2.$$

By the Formula (10) of finding partial derivative of binary implicit function, we obtain
$$\frac{\partial z}{\partial x}=-\frac{F_x}{F_z}=-\frac{-\dfrac{z}{x^2}F'_1}{\dfrac{1}{x}F'_1+\dfrac{1}{y}F'_2}=\frac{yzF'_1}{x(yF'_1+xF'_2)},$$

$$\frac{\partial z}{\partial y}=-\frac{F_y}{F_z}=-\frac{-\dfrac{z}{y^2}F'_2}{\dfrac{1}{x}F'_1+\dfrac{1}{y}F'_2}=\frac{xzF'_2}{y(yF'_1+xF'_2)}.$$

2) Differential method of implicit function determined by equation set

Now we will study a more generalized situation. Suppose
$$\begin{cases} F(x,y,u,v)=0, \\ G(x,y,u,v)=0. \end{cases} \tag{11}$$

There are four variables in the two equations, so two variables can vary independently (they are called free unknowns in algebra). The Equation (11) can determine two binary functions under the below conditions.

Chapter 11 Differentiation of multivariable function and its application

Theorem 8 (Existence theorem of implicit function 3) Suppose the function $F(x, y, u, v)$ and $G(x, y, u, v)$ have continuous partial derivatives with respect to each variable on some neighborhood of (x_0, y_0, u_0, v_0). Because $F(x_0, y_0, u_0, v_0) = 0$, $G(x_0, y_0, u_0, v_0) = 0$ and the determinant of the derivatives

$$J = \frac{\partial(F,G)}{\partial(u,v)} = \begin{vmatrix} \frac{\partial F}{\partial u} & \frac{\partial F}{\partial v} \\ \frac{\partial G}{\partial u} & \frac{\partial G}{\partial v} \end{vmatrix}$$

(it is Jacobi Determinant) is not zero at the point (x_0, y_0, u_0, v_0), then

(1) from the equation set

$$\begin{cases} F(x,y,u,v) = 0, \\ G(x,y,u,v) = 0 \end{cases}$$

the only single-valued continuous function $u = u(x, y)$, $v = v(x, y)$ are determined on some neighborhood of (x_0, y_0) which satisfies $u_0 = u(x_0, y_0), v_0 = v(x_0, y_0)$.

(2) the function $u(x, y), v(x, y)$ have continuous partial derivatives, and

$$\frac{\partial u}{\partial x} = -\frac{1}{J}\frac{\partial(F,G)}{\partial(x,v)} = -\frac{1}{J}\begin{vmatrix} F_x & F_v \\ G_x & G_v \end{vmatrix},$$

$$\frac{\partial v}{\partial x} = -\frac{1}{J}\frac{\partial(F,G)}{\partial(u,x)} = -\frac{1}{J}\begin{vmatrix} F_u & F_x \\ G_u & G_x \end{vmatrix},$$

$$\frac{\partial u}{\partial y} = -\frac{1}{J}\frac{\partial(F,G)}{\partial(y,v)} = -\frac{1}{J}\begin{vmatrix} F_y & F_v \\ G_y & G_v \end{vmatrix},$$

$$\frac{\partial v}{\partial y} = -\frac{1}{J}\frac{\partial(F,G)}{\partial(u,y)} = -\frac{1}{J}\begin{vmatrix} F_u & F_y \\ G_u & G_y \end{vmatrix}.$$

(12)

Similar to the previous theorems, we do not prove it, but just explain the Formula (12) as follows.

Because

$$\begin{cases} F[x,y,u(x,y),v(x,y)] \equiv 0, \\ G[x,y,u(x,y),v(x,y)] \equiv 0, \end{cases}$$

by derivation rule of compound function, find partial derivative with respect to x on either side of the identity, then we can obtain the linear system of equations

$$\begin{cases} F_x + F_u \dfrac{\partial u}{\partial x} + F_v \dfrac{\partial v}{\partial x} = 0, \\ G_x + G_u \dfrac{\partial u}{\partial x} + G_v \dfrac{\partial v}{\partial x} = 0. \end{cases}$$

From the Theorem 8, we know that the coefficient determinant on some neighborhood of (x_0, y_0, u_0, v_0) is

$$J = \begin{vmatrix} F_u & F_v \\ G_u & G_v \end{vmatrix} \neq 0,$$

then we obtain

$$\frac{\partial u}{\partial x} = -\frac{1}{J}\frac{\partial(F,G)}{\partial(x,v)}, \quad \frac{\partial v}{\partial x} = -\frac{1}{J}\frac{\partial(F,G)}{\partial(u,x)}.$$

In the same way, we get

$$\frac{\partial u}{\partial y} = -\frac{1}{J}\frac{\partial(F,G)}{\partial(y,v)}, \quad \frac{\partial v}{\partial y} = -\frac{1}{J}\frac{\partial(F,G)}{\partial(u,y)}.$$

Example 36 Suppose $\begin{cases} xu - yv = 0, \\ yu + xv = 1, \end{cases}$ find $\frac{\partial u}{\partial x}, \frac{\partial u}{\partial y}, \frac{\partial v}{\partial x}$ and $\frac{\partial v}{\partial y}$.

Solution Since u, v are functions of x, y, find the partial derivative with respect to x on either side of the given equation set, we obtain

$$\begin{cases} x\dfrac{\partial u}{\partial x} - y\dfrac{\partial v}{\partial x} = -u, \\ y\dfrac{\partial u}{\partial x} + x\dfrac{\partial v}{\partial x} = -v. \end{cases}$$

The coefficient determinant of linear non-homogeneous equations of $\frac{\partial u}{\partial x}, \frac{\partial v}{\partial x}$ is

$$\begin{vmatrix} x & -y \\ y & x \end{vmatrix}.$$

If $J = \begin{vmatrix} x & -y \\ y & x \end{vmatrix} = x^2 + y^2 \neq 0$, then

$$\frac{\partial u}{\partial x} = \frac{\begin{vmatrix} -u & -y \\ -v & x \end{vmatrix}}{J} = -\frac{xu + yv}{x^2 + y^2},$$

$$\frac{\partial v}{\partial x} = \frac{\begin{vmatrix} x & -u \\ y & -v \end{vmatrix}}{J} = \frac{yu - xv}{x^2 + y^2}.$$

In the same way, find the partial derivative with respect to y on either side of the equation set. If $J = x^2 + y^2 \neq 0$, we obtain

$$\frac{\partial u}{\partial y} = \frac{xv - yu}{x^2 + y^2}, \quad \frac{\partial v}{\partial y} = -\frac{xu + yv}{x^2 + y^2}.$$

Example 37 Suppose the function $x = x(u,v), y = y(u,v)$ are continuous and have continuous partial derivatives on some neighborhood of (u,v), and $\frac{\partial(x,y)}{\partial(u,v)} \neq 0$.

(1) Prove the equation set

$$\begin{cases} x = x(u,v), \\ y = y(u,v), \end{cases}$$

determines a couple of single-valued continuous inverse function $u = u(x,y), v = v(x,y)$, which have continuous partial derivatives on some neighborhood of the point (x,y,u,v).

(2) Find the partial derivative of the inverse function $u = u(x,y), v = v(x,y)$ with respect to x, y.

Solution (1) Rewrite the equation set into

Chapter 11 Differentiation of multivariable function and its application

Suppose
$$\begin{cases} F(x,y,u,v)=x-x(u,v)=0, \\ G(x,y,u,v)=y-y(u,v)=0. \end{cases}$$
$$J=\frac{\partial(F,G)}{\partial(u,v)}=\frac{\partial(x,y)}{\partial(u,v)}\neq 0,$$

from Theorem 8, we obtain the conclusion that we want.

(2) Plug the inverse function $u=u(x,y), v=v(x,y)$ into the primitive equation set, we get
$$\begin{cases} x\equiv x[u(x,y),v(x,y)], \\ y\equiv y[u(x,y),v(x,y)]. \end{cases}$$

By the rule of finding partial derivative of compound function, find the derivatives with respect to x on either side of the identity, we obtain
$$\begin{cases} 1=\dfrac{\partial x}{\partial u}\dfrac{\partial u}{\partial x}+\dfrac{\partial x}{\partial v}\dfrac{\partial v}{\partial x}, \\ 0=\dfrac{\partial y}{\partial u}\dfrac{\partial u}{\partial x}+\dfrac{\partial y}{\partial v}\dfrac{\partial v}{\partial x}. \end{cases}$$

Because
$$J=\begin{vmatrix} \dfrac{\partial x}{\partial u} & \dfrac{\partial x}{\partial v} \\ \dfrac{\partial y}{\partial u} & \dfrac{\partial y}{\partial v} \end{vmatrix}=\frac{\partial(x,y)}{\partial(u,v)}\neq 0,$$

we obtain
$$\frac{\partial u}{\partial x}=\frac{1}{J}\frac{\partial y}{\partial v},\quad \frac{\partial v}{\partial x}=-\frac{1}{J}\frac{\partial y}{\partial u}.$$

In the same way, we have
$$\frac{\partial u}{\partial y}=-\frac{1}{J}\frac{\partial x}{\partial v},\quad \frac{\partial v}{\partial y}=\frac{1}{J}\frac{\partial x}{\partial u}.$$

Exercises 11-2

1. Find partial derivative of the following functions:

(1) $z=xy+\dfrac{x}{y}$;

(2) $z=\dfrac{x}{\sqrt{x^2+y^2}}$;

(3) $z=\arctan(x-y^2)$;

(4) $z=x\sin(x+y)$;

(5) $z=\tan\dfrac{x^2}{y}$;

(6) $z=(1+xy)^y$;

(7) $u=(xy)^z$;

(8) $u=\left(\dfrac{x}{y}\right)^z$;

(9) $u=e^{x(x^2+y^2+z^2)}$;

(10) $u=\arctan(x-y)^z$.

2. Suppose $f(x,y,z)=\ln(xy+z)$, find $f_x(1,2,0), f_y(1,2,0)$ and $f_z(1,2,0)$.

3. Suppose $z=xy+xe^{\frac{y}{x}}$, prove $x\dfrac{\partial z}{\partial x}+y\dfrac{\partial z}{\partial y}=xy+z$.

4. Suppose $f(x,y)=\sqrt{x^2+y^4}$, find $f_x(x,1)$.

5. Find the angle between the tangent of the curve $\begin{cases} z=\sqrt{1+x^2+y^2}, \\ x=1 \end{cases}$ at $(1,1,\sqrt{3})$ and

the positive direction of the y-axis.

6. Find the second partial derivative $\dfrac{\partial^2 z}{\partial x^2}, \dfrac{\partial^2 z}{\partial y^2}$ and $\dfrac{\partial^2 z}{\partial x \partial y}$.

(1) $z=\sqrt{2xy+y^2}$;

(2) $z=\arctan\dfrac{x+y}{1-xy}$;

(3) $z=y^x$;

(4) $z=\sin^2(ax+by)$.

7. Suppose $u=x^\alpha y^\beta z^\gamma$, find $\dfrac{\partial^3 u}{\partial x \partial y \partial z}$.

8. If $u=z\arctan\dfrac{x}{y}$, prove $\dfrac{\partial^2 u}{\partial x^2}+\dfrac{\partial^2 u}{\partial y^2}+\dfrac{\partial^2 u}{\partial z^2}=0$.

9. Find perfect differential of the following functions:

(1) $z=x^2 y^3$;

(2) $z=\dfrac{x^2-y^2}{x^2+y^2}$;

(3) $z=yx^y$;

(4) $z=\sin^2 x+\cos^2 y$;

(5) $z=\arctan\dfrac{y}{x}+\arctan\dfrac{x}{y}$;

(6) $u=\left(xy+\dfrac{x}{y}\right)^z$.

10. If $f(x,y,z)=\dfrac{z}{\sqrt{x^2+y^2}}$, find $df(3,4,5)$.

11. Find total increment and perfect differential of $z=\dfrac{xy}{x^2+y^2}$ if $x=2, y=1, \Delta x=0.01, \Delta y=0.03$.

12. Find the perfect differential of $u=z\sqrt{\dfrac{x}{y}}$ at the point $M_0(1,1,1)$.

13. By perfect differential, find the approximation of the following value:

(1) $1.002\times(2.003)^2\times(3.004)^3$;

(2) $\sqrt{(1.02)^3+(1.97)^3}$.

14. Suppose $f(x,y)=\begin{cases}(x^2+y^2)\sin\dfrac{1}{\sqrt{x^2+y^2}}, & x^2+y^2\neq 0, \\ 0, & x^2+y^2=0.\end{cases}$

(1) Find the partial derivative $f_x(x,y), f_y(x,y)$.

(2) Whether $f_x(x,y), f_y(x,y)$ are continuous at the point $(0,0)$ or not?

(3) Whether the function $f(x,y)$ is differentiable at the point $(0,0)$?

15. We measure the side of triangle $a=200$ m, and its maximum error is 2 m, the side $b=300$ m, its maximum error is 5 m. The angle between a and b is $\angle C=60°$, and the maximum error is $1°$. Find the maximum error of the third side.

16. There are 100 metal cylinders with radius $R=5$ cm, height $H=20$ cm. Now, the surface of the cylinder is to be plated with nickel of the thickness 0.05 cm, how many nickels is needed?

17. Prove that the absolute error of the sum of two functions is equal to the sum of their respective absolute error.

18. Prove that the relative error of the product of two functions is equal to the sum of the relative error of each factor.

Chapter 11 Differentiation of multivariable function and its application

19. Suppose the function $z=\dfrac{y}{x}, x=e^t, y=1-e^{2t}$, find $\dfrac{dz}{dt}$.

20. Suppose $z=u^2v-uv^2, u=x\cos y, v=x\sin y$, find $\dfrac{\partial z}{\partial x}, \dfrac{\partial z}{\partial y}$.

21. Suppose $z=\arctan(xy), y=e^x$, find $\dfrac{dz}{dx}$.

22. (1) Suppose $z=f(u)$ and $\dfrac{y}{x}$, prove $x\dfrac{\partial z}{\partial x}+y\dfrac{\partial z}{\partial y}=0$.

    (2) Suppose $u=\dfrac{1}{x}f\left(\dfrac{y}{x}\right)$, prove $x\dfrac{\partial u}{\partial x}+y\dfrac{\partial u}{\partial y}+u=0$.

23. Find the first partial derivative of the following function (f has the first continuous partial derivative):

    (1) $u=f(x^2+y^2+z^2)$;

    (2) $u=f\left(\dfrac{x}{y},\dfrac{y}{z}\right)$;

    (3) $u=f(x^2+y^2, xy, xyz)$.

24. Find the second partial derivative of the following function (f has the second continuous partial derivative):

    (1) $z=f\left(x,\dfrac{x}{y}\right)$;

    (2) $z=f(xy^2, x^2y)$.

25. Suppose $z=xy+xF(u), u=\dfrac{y}{x}$, F is a differentiable function, prove $x\dfrac{\partial z}{\partial x}+y\dfrac{\partial z}{\partial y}=z+xy$.

26. Suppose f has the second continuous partial derivative, φ is a differentiable function, if $z=f[x+\varphi(y)]$, prove $\dfrac{\partial z}{\partial x}\dfrac{\partial^2 z}{\partial x\partial y}=\dfrac{\partial z}{\partial y}\dfrac{\partial^2 z}{\partial x^2}$.

27. Suppose φ, ψ have the second continuous partial derivatives, prove that the function $z=x\varphi\left(\dfrac{y}{x}\right)+\psi\left(\dfrac{y}{z}\right)$ satisfies the equation $x^2\dfrac{\partial^2 z}{\partial x^2}+2xy\dfrac{\partial^2 z}{\partial x\partial y}+y^2\dfrac{\partial^2 z}{\partial y^2}=0$.

28. The equation $(x^2+y^2)^3-3(x^2+y^2)+1=0$ determines y is the function of x, find $\dfrac{dy}{dx}$ and $\dfrac{d^2y}{dx^2}$.

29. Suppose $\dfrac{x}{z}+\dfrac{y}{z}=\ln\dfrac{z}{x}$, find $\dfrac{\partial z}{\partial x}, \dfrac{\partial z}{\partial y}$.

30. Suppose $F(u,v)$ has continuous partial derivative, prove that the function $z=f(x,y)$ determined by $F(cx+az, cy-bz)=0$ satisfies the equation $a\dfrac{\partial z}{\partial x}+b\dfrac{\partial z}{\partial y}=c$.

31. Suppose the function $F(u,v)$ is a differentiable function, and the function $F\left(x+\dfrac{z}{y}, y+\dfrac{z}{x}\right)=0$ determines that z is the function of x,y. Prove $x\dfrac{\partial z}{\partial x}+y\dfrac{\partial z}{\partial y}=z-xy$.

32. Suppose $\begin{cases} z=x^2+y^3, \\ x^2+5y^2+6z^2=5, \end{cases}$ find $\dfrac{dy}{dx}, \dfrac{dz}{dx}$.

33. Suppose $\begin{cases} x=e^u\cos v, \\ y=e^u\sin v, \\ z=uv, \end{cases}$ find $\dfrac{\partial z}{\partial x}, \dfrac{\partial z}{\partial y}$.

34. Suppose $z=f(x,y)+g(u,v)$, $u=x^3$, $v=x^y$, and f,g have the first continuous partial derivative, find $\dfrac{\partial z}{\partial x}, \dfrac{\partial z}{\partial y}$.

35. Suppose $u=f(z)$, the equation $z=x+y\varphi(z)$ determines $z=z(x,y)$ and f, φ are all differentiable, prove $\dfrac{\partial u}{\partial y}=\varphi(z)\dfrac{\partial u}{\partial x}$.

36. Company X and Y are competitors in the machine tool industry, the supply function of their major production are $P_X=1\,000-5Q_X$ and $P_Y=1\,600-4Q_Y$. The sales of X and Y are 100 units and 250 units.

(1) What is the current price elasticity of X and Y?

(2) If the price of X is reduced, Q_Y increases to 300 units and the sales of X, Q_X reduces to 75 unites. What is the cross price elasticity of the X's products?

37. Make a product from raw material A and B, the amount of A, B used is a and b respectively, the output of products is Q, which means the production function is $Q=Q(a,b)$. We know that the unit price of raw material is P_a and P_b. Prove that for the given output $Q=Q_0$, the input combination $(\bar{a},\bar{b})$, which makes the cost minimum, satisfies $\dfrac{\dfrac{\partial Q}{\partial a}}{\dfrac{\partial Q}{\partial b}}=\dfrac{P_a}{P_b}$.

11.3 Direction derivative and gradient

11.3.1 Direction derivative

For the function $z=f(x,y)$, the partial directive $f_x(x,y)$ and $f_y(x,y)$ are the change of rate of $z=f(x,y)$ at $P(x,y)$ along the x-axis and the y-axis respectively. They only describe the change of $z=f(x,y)$ in special directions.

But in practice, we have to find the change of rate of $z=f(x,y)$ along a specified direction sometimes, which is the direction derivative introduced now.

Definition 1 Suppose the function $z=f(x,y)$ is defined on some neighborhood of $P(x,y)$, draw a ray l from the point P, the angle between the positive direction of the x-axis and the ray l is α. Suppose another point $P'(x+\Delta x, y+\Delta y)$ is on the ray l (Figure 11-10), the distance between P and P' is $\rho=\sqrt{(\Delta x)^2+(\Delta y)^2}$.

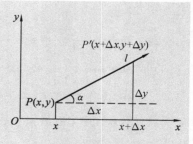

Figure 11-10

For the ratio of the total increment
$$\Delta z=f(x+\Delta x, y+\Delta y)-f(x,y)$$
with respect to ρ, as P' tends to P along the ray l, if the limit
$$\lim_{\rho\to 0}\dfrac{\Delta z}{\rho}=\lim_{\rho\to 0}\dfrac{f(x+\Delta x, y+\Delta y)-f(x,y)}{\rho}$$

exist, it is called the direction derivative of $z=f(x,y)$ at the point $P(x,y)$ along l, and denote it by $\frac{\partial z}{\partial l}$, that is,

$$\frac{\partial z}{\partial l}=\lim_{\rho\to 0}\frac{f(x+\Delta x,y+\Delta y)-f(x,y)}{\rho}. \tag{1}$$

From the definition of the direction derivative, if the partial derivatives $f_x(x,y)$ and $f_y(x,y)$ exist, they are the direction derivatives of $z=f(x,y)$ at $P(x,y)$ along the positive direction of the x-axis $e_1=(1,0)$ and the positive direction of the y-axis $e_2=(0,1)$. Similarly, the direction derivatives along the negative direction of the x-axis $e_1=(-1,0)$ and the negative direction of y-axis $e_2=(0,-1)$ also exist, and they are $-f_x(x,y)$ and $-f_y(x,y)$.

For the existence and calculation method of direction derivative along an arbitrary direction, there are following theorems.

Theorem If the function $z=f(x,y)$ is differentiable at the point $P(x,y)$, the direction derivatives of $f(x,y)$ at this point along arbitrary direction l exist, and

$$\frac{\partial z}{\partial l}=\frac{\partial z}{\partial x}\cos\alpha+\frac{\partial z}{\partial y}\cos\beta, \tag{2}$$

where $\cos\alpha,\cos\beta$ are direction cosine of l.

Proof Since $z=f(x,y)$ is differentiable at the point $P(x,y)$, the total increment at $P(x,y)$ can be expressed by

$$\Delta z=f(x+\Delta x,y+\Delta y)-f(x,y)=\frac{\partial z}{\partial x}\Delta x+\frac{\partial z}{\partial y}\Delta y+o(\rho),$$

where $\rho=\sqrt{(\Delta x)^2+(\Delta y)^2}$. Divide either side of the equality above by ρ, we have

$$\frac{\Delta z}{\rho}=\frac{f(x+\Delta x,y+\Delta y)-f(x,y)}{\rho}=\frac{\partial z}{\partial x}\frac{\Delta x}{\rho}+\frac{\partial z}{\partial y}\frac{\Delta y}{\rho}+\frac{o(\rho)}{\rho}$$

$$=\frac{\partial z}{\partial x}\cos\alpha+\frac{\partial z}{\partial y}\cos\beta+\frac{o(\rho)}{\rho},$$

then

$$\lim_{\rho\to 0}\frac{f(x+\Delta x,y+\Delta y)-f(x,y)}{\rho}=\frac{\partial z}{\partial x}\cos\alpha+\frac{\partial z}{\partial y}\cos\beta.$$

Example 1 Find the direction derivative of $z=xy+\sin(x+2y)$ along the direction from $O(0,0)$ to $P(1,2)$.

Solution The direction is along the vector $\overrightarrow{OP}=(1,2)$, and the unit vector on $\overrightarrow{OP}$ is $e_l=\left(\frac{1}{\sqrt{5}},\frac{2}{\sqrt{5}}\right)$. Since

$$\frac{\partial z}{\partial x}=y+\cos(x+2y),\frac{\partial z}{\partial y}=x+2\cos(x+2y),$$

we have $\frac{\partial z}{\partial x}=1, \frac{\partial z}{\partial y}=2$, at $(0,0)$.

So, the direction derivative is
$$\frac{\partial z}{\partial l}=\frac{\partial z}{\partial x}\cos\alpha+\frac{\partial z}{\partial y}\cos\beta=1\frac{1}{\sqrt{5}}+2\frac{2}{\sqrt{5}}=\sqrt{5}.$$

Similarly, if the three-variable function $u=f(x,y,z)$ is differentiable at $P(x,y,z)$, then the direction derivative of $u=f(x,y,z)$ along arbitrary direction at this point exists and
$$\frac{\partial u}{\partial l}=\frac{\partial u}{\partial x}\cos\alpha+\frac{\partial u}{\partial y}\cos\beta+\frac{\partial u}{\partial z}\cos\gamma,$$
where $\cos\alpha$, $\cos\beta$, $\cos\gamma$ are the cosine of the direction l.

Example 2 A point charge q is at the origin $O(0,0,0)$, the electric potential of any point $P(x,y,z)$ (x,y,z are not zero at the same time) in the electric field produced by q is $u=\frac{kq}{r}$, where k is a constant and r is the distance between the origin and the point P. Find the change of rate of the electric potential at the point P along the direction $l=(\cos\alpha,\cos\beta,\cos\gamma)$.

Solution Because $u=\frac{kq}{r}, r=\sqrt{x^2+y^2+z^2}$, then
$$\frac{\partial u}{\partial x}=-\frac{kq}{r^2}\cdot\frac{\partial r}{\partial x}=-\frac{kq}{r^2}\cdot\frac{x}{\sqrt{x^2+y^2+z^2}}=-\frac{kqx}{r^3}.$$

In the same way, we obtain
$$\frac{\partial u}{\partial y}=-\frac{kqy}{r^3}, \frac{\partial u}{\partial z}=-\frac{kqz}{r^3}.$$

So
$$\frac{\partial z}{\partial l}=-\frac{kq}{r^3}(x\cos\alpha+y\cos\beta+z\cos\gamma).$$

11.3.2 Gradient

Generally, the direction derivatives of a binary function at a given point along the different directions are always different. The maximum direction derivative is needed to be found in many practical problems. Now we introduce the concept of gradient.

> **Definition 2** Suppose the partial derivatives of $z=f(x,y)$ at the point (x,y) exist, the vector $\frac{\partial z}{\partial x}i+\frac{\partial z}{\partial y}j$ is called the gradient of $z=f(x,y)$ at the point (x,y), which is denoted by **grad** $f(x,y)$, that is,
> $$\mathbf{grad}\ f(x,y)=\frac{\partial z}{\partial x}i+\frac{\partial z}{\partial y}j. \tag{3}$$

From the definition of quantitative product of vectors, the Formula (2) can be written as

Chapter 11 Differentiation of multivariable function and its application

$$\frac{\partial z}{\partial l} = \left(\frac{\partial z}{\partial x}, \frac{\partial z}{\partial y}\right) \cdot (\cos \alpha, \cos \beta) = \mathbf{grad}\ f(x,y) \cdot \mathbf{e}_l,$$

where $\mathbf{e}_l = (\cos \alpha, \cos \beta)$. This formula above shows that the direction derivative of $f(x,y)$ at the point (x,y) along the direction l is equal to the quantitative product of the gradient at this point and the unit vector $\mathbf{e}_l$ along the direction l. Thus, if the function $z = f(x,y)$ is differentiable at the point (x,y), then $\frac{\partial z}{\partial l}$ is the projection of the gradient along the ray l.

When the direction l is same to the direction of gradient, then $\left|\frac{\partial z}{\partial l}\right| = |\mathbf{grad}\ f(x,y)|$, so $\frac{\partial z}{\partial l}$ has maximum. We can conclude that the direction derivative reaches its maximum value if the direction of it is along the direction of gradient. In brief, the rate of growth of a differentiable function reaches its maximum value along the direction of gradient and the maximum growth rate is equal to the length of gradient. So, we get the following conclusions.

The gradient of a function at some point is a vector such that its direction is same to the direction in which the direction derivative reaches its maximum value, its length is equal to the maximum growth rate, that is,

$$|\mathbf{grad}\ f(x,y)| = \sqrt{\left(\frac{\partial f}{\partial x}\right)^2 + \left(\frac{\partial f}{\partial y}\right)^2}.$$

If $\frac{\partial f}{\partial x}$ is not zero, the tangent of the angle from the x-axis to the gradient is

$$\tan \theta = \frac{\frac{\partial f}{\partial y}}{\frac{\partial f}{\partial x}}.$$

From the definition of gradient, it is easy to prove the following properties of gradient.

(1) $\mathbf{grad}\ (u+v) = \mathbf{grad}\ u + \mathbf{grad}\ v$,

(2) $\mathbf{grad}\ (uv) = u\ \mathbf{grad}\ v + v\ \mathbf{grad}\ u$,

(3) $\mathbf{grad}\ (f(u)) = f'(u)\mathbf{grad}\ u$, where $f(u)$ is a differentiable function.

Example 3 Suppose $z = f(x,y) = xe^y$.

(1) Find the rate of change of $z = f(x,y) = xe^y$ at the point $P(2,0)$ along the direction from P to $Q\left(\frac{1}{2}, 2\right)$.

(2) What is the direction such that the function f reach the maximum growth rate at the point $P(2,0)$? And what is the maximum growth rate?

Solution (1) Suppose $\mathbf{e}_l$ is the unit vector on $\overrightarrow{PQ}$, since $\overrightarrow{PQ} = \left(-\frac{3}{2}, 2\right)$, we have

$$\mathbf{e}_l = \left(-\frac{3}{5}, \frac{4}{5}\right).$$

And $\mathbf{grad}\ f(x,y) = (e^y, xe^y)$, so

$$\left.\frac{\partial f}{\partial l}\right|_{(2,0)} = \mathbf{grad}\ f(2,0) \cdot \mathbf{e}_l = (1,2) \cdot \left(-\frac{3}{5}, \frac{4}{5}\right) = 1.$$

(2) $f(x,y)$ has the maximum growth rate at the point $P(2,0)$ along the direction **grad** $f(2,0)=(1,2)$, and the maximum growth rate is

$$|\text{\textbf{grad}}\ f(2,0)|=\sqrt{5}.$$

To understand the gradient more clear, we will observe the direction of **grad** f geometrically.

Generally, the binary function $z=f(x,y)$ is a curve surface in geometry. This curve surface is cut by the plane $Z=C$ (C is constant) and the intersection curve L is

$$\begin{cases} z=f(x,y), \\ z=C. \end{cases}$$

Suppose the plane curve L^* is the projection of the curve L on the xOy plane, then the equation of L^* on the xOy plane is $f(x,y)=C$ (Figure 11-11). For any point on L^*, the value of $z=f(x,y)$ is the same constant C. So, the plane curve L^* is called the contour line (isoline) of the function $z=f(x,y)$.

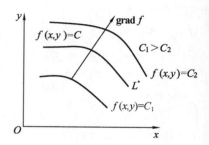

Figure 11-11

If f_x, f_y are not zero at the same time, then the unit normal vector at an arbitrary point $P_0(x_0,y_0)$ on the isoline $f(x,y)=C$ is

$$n=\frac{1}{\sqrt{f_x^2(x_0,y_0)+f_y^2(x_0,y_0)}}(f_x(x_0,y_0),f_y(x_0,y_0)).$$

This means that the direction of **grad** $f(x_0,y_0)$ is the same as the direction of a normal vector at this point on isoline, and the direction derivative $\frac{\partial f}{\partial n}$ along this direction is equal to $|\text{\textbf{grad}}\ f(x_0,y_0)|$, so

$$\text{\textbf{grad}}\ f(x_0,y_0)=\frac{\partial f}{\partial n}n.$$

This formula indicates the relationship of the gradient at some point and the isoline, the direction derivative at this point. That is to say, the direction of the gradient at some point is the same as the direction of the normal vector at this point on isoline and the direction is the isoline of large value to the isoline of small value, and the length of gradient is the direction derivative along the direction of the normal vector.

The concept of gradient discussed above can be generalized to three-variable function. Suppose the function $f(x,y,z)$ has the first continuous partial derivatives in a space region G, for each point $P_0(x_0,y_0,z_0)\in G$, a vector can be determined

$$f_x(x_0,y_0,z_0)\boldsymbol{i}+f_y(x_0,y_0,z_0)\boldsymbol{j}+f_z(x_0,y_0,z_0)\boldsymbol{k},$$

which is called the gradient of $f(x,y,z)$ at $P_0(x_0,y_0,z_0)$, and denote it by **grad** $f(x_0,y_0,z_0)$, that is,

$$\text{\textbf{grad}}\ f(x_0,y_0,z_0)=f_x(x_0,y_0,z_0)\boldsymbol{i}+f_y(x_0,y_0,z_0)\boldsymbol{j}+f_z(x_0,y_0,z_0)\boldsymbol{k}.$$

Similar to the binary function, the gradient is a vector such that the direction derivative reaches the maximum value and its length is the maximum value of the direction derivative.

If we define the curve surface $f(x,y,z)=C$ as the iso-surface of the function $f(x,y,z)$, the direction of the gradient of $f(x,y,z)$ at $P_0(x_0,y_0,z_0)$ is same to the normal line of the iso-surface $f(x,y,z)=C$ at the point P_0, which is from the small-valued iso-surface to the large-valued iso-surface, and the length of gradient is the direction derivative along the direction of this normal.

Example 4 Suppose the function $u=xy^2z$, find the direction such that the direction derivative of $u=xy^2z$ at $P_0(1,-1,2)$ along it will be maximum? What is the maximum value of the direction derivative?

Solution The direction derivative of the function u at P_0 is maximum if and only if along the direction of the gradient

$$\mathbf{grad}\ u = \left(\frac{\partial u}{\partial x},\frac{\partial u}{\partial y},\frac{\partial u}{\partial z}\right)\bigg|_{(1,-1,2)} = (y^2z, 2xyz, xy^2)\bigg|_{(1,-1,2)} = (2,-4,1).$$

The maximum value is

$$|\mathbf{grad}\ u(1,-1,2)| = \sqrt{2^2+(-4)^2+1^2} = \sqrt{21}.$$

Now, we will introduce the concept of the scalar field and the vector field.

For any point M in a space region G, there is a fixed number $f(M)$, then we call that a scalar field (such as temperature field, density field) is defined in the space region G. That is, a scalar field can be determined by a scalar function $f(M)$. If the point M corresponds to a vector $\mathbf{F}(M)$, then we call that a vector field (such as force field, speed field) is defined in space region G. That is, a vector field can be determined by a vector function $\mathbf{F}(M)$, and

$$\mathbf{F}(M) = P(M)\mathbf{i} + Q(M)\mathbf{j} + R(M)\mathbf{k},$$

where $P(M), Q(M), R(M)$ are scalar functions of the point M.

From the concept of field, we can say the vector function $\mathbf{grad}\ f(M)$ determines a vector filed, which is called the gradient field generated by the scalar field $f(M)$. The function $f(M)$ is generally called the potential of this vector field, and this vector field is also called the potential field. It is necessary to notice that an arbitrary vector field is not necessarily a potential field, because it is not necessary to be a gradient field of some scalar function.

Exercises 11-3

1. Find the direction derivative of the function $z=x^2+y^2$ at the point $(1, 2)$ along the direction from the point $(1, 2)$ to the point $(2, 2+\sqrt{3})$.

2. Find the direction derivative of the function $u=xyz$ at the point $(5, 1, 2)$ along the direction from the point $(5, 1, 2)$ to the point $(9, 4, 14)$.

3. Find the direction derivative of the function $z=x^2-xy+y^2$ at the point $(1, 1)$ along the direction l such that the angle between l and the positive direction of the x-axis is α. Find the direction such that the direction derivative has (1) maximum; (2) minimum;

(3) 0.

4. Find the direction derivative of $z=\ln(x+y)$ at the point $(1, 2)$ on the parabola $y^2=4x$ along the direction of the tangent at this point.

5. Find the maximum value and the minimum value of the direction derivative of $u=xy^2z^3$ at the point $(1, 1, 1)$.

6. Find the direction derivative of the function $u=x+y+z$ at the point (x_0, y_0, z_0) on spherical surface $x^2+y^2+z^2=R^2$ along the direction of the normal vector of the spherical surface.

7. Suppose $f(x,y,z)=x^2+2y^2+3z^2+xy+3x-2y-6z$, find **grad** $f(1,1,1)$ and **grad** $f(2,2,2)$.

8. Find the value and the direction of the gradient of $u=\dfrac{1}{\gamma}(\gamma=\sqrt{x^2+y^2+z^2})$ at the point (x_0, y_0, z_0).

9. Suppose u, v are functions of x, y, z, and u, v have continuous partial derivatives, prove that

(1) **grad**$(\alpha u+\beta v)=\alpha$ **grad** $u+\beta$ **grad** v (α, β are constants);

(2) **grad**$(uv)=u$ **grad** $v+v$ **grad** u;

(3) **grad** $F(u)=F'(u)$**grad** u.

11.4 Geometric applications of the differential of multivariable function

11.4.1 Tangent line and normal plane of space curve

1) The definition of tangent and normal plane of space curve

Suppose $M_0(x_0, y_0, z_0)$ and $M(x, y, z)$ are two points on the space curve Γ, the straight line $\overline{M_0M}$ is called the secant line of Γ.

The definition of tangent:

As the point $M(x,y,z)$ tends to M_0 along the curve Γ, the limit position $\overline{M_0M}$ of these secant $\overline{M_0T}$ is called the tangent of the curve Γ at $M_0(x_0, y_0, z_0)$.

The definition of normal plane:

The plane which passes through $M_0(x_0, y_0, z_0)$ and is perpendicular to the tangent is called the normal plane of the curve Γ at $M_0(x_0, y_0, z_0)$.

2) The method of finding tangent curve and normal plane

(1) Suppose the parametric equation of space curve Γ is

$$\Gamma: \begin{cases} x=x(t), \\ y=y(t), \\ z=z(t), \end{cases} \quad (1)$$

where $x(t), y(t), z(t)$ are derivable, and their derivatives $x'(t), y'(t), z'(t)$ are not zero at the same time (Figure 11-12).

Suppose at $t=t_0$, the corresponding point on the curve Γ is

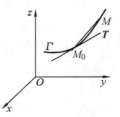

Figure 11-12

Chapter 11 Differentiation of multivariable function and its application

$M_0(x_0, y_0, z_0)$. And now find the tangent of the curve at M_0, where $x_0 = x(t_0), y_0 = y(t_0), z_0 = z(t_0)$. Suppose the parameter t has an increment Δt at the point t_0, then the corresponding point of $t = t_0 + \Delta t$ on the curve Γ is $M(x_0 + \Delta x, y_0 + \Delta y, z_0 + \Delta z)$. From space analytic geometry, the equation of secant $\overline{M_0 M}$ is

$$\frac{x-x_0}{\Delta x} = \frac{y-y_0}{\Delta y} = \frac{z-z_0}{\Delta z}.$$

By dividing the denominator of each quotient above by Δt, we obtain

$$\frac{x-x_0}{\frac{\Delta x}{\Delta t}} = \frac{y-y_0}{\frac{\Delta y}{\Delta t}} = \frac{z-z_0}{\frac{\Delta z}{\Delta t}}.$$

Then, let the point M tend to the point M_0 along the curve Γ, the limit position of the secant $M_0 M$ is the tangent of the curve Γ at the point M_0. Thus, let $M \to M_0$ (now $\Delta t \to 0$), take the limit of the above equality, we obtain the equation of the tangent of the curve Γ at the point M_0

$$\frac{x-x_0}{x'(t_0)} = \frac{y-y_0}{y'(t_0)} = \frac{z-z_0}{z'(t_0)}, \qquad (2)$$

and the direction vector of the tangent $\boldsymbol{T} = (x'(t_0), y'(t_0), z'(t_0))$ is called the tangent vector of the curve at point M_0.

The normal plane of T at point M_0:

$$x'(t_0)(x-x_0) + y'(t_0)(y-y_0) + z'(t_0)(z-z_0) = 0. \qquad (3)$$

Example 1 Find the equation of the tangent line and the normal plane of the curve $x = t - \sin t, y = 1 - \cos t, z = 4\sin \frac{t}{2}$ at $M_0\left(\frac{\pi}{2} - 1, 1, 2\sqrt{2}\right)$.

Solution If the parameter $t_0 = \frac{\pi}{2}$, the corresponding point is $M_0\left(\frac{\pi}{2} - 1, 1, 2\sqrt{2}\right)$ on the curve, at M_0 we have

$$x'\left(\frac{\pi}{2}\right) = (1-\cos t)\Big|_{t=\frac{\pi}{2}} = 1,$$

$$y'\left(\frac{\pi}{2}\right) = \sin t\Big|_{t=\frac{\pi}{2}} = 1,$$

$$z'\left(\frac{\pi}{2}\right) = 2\cos\frac{t}{2}\Big|_{t=\frac{\pi}{2}} = \sqrt{2}.$$

Thus, the tangent vector at M_0 is $\boldsymbol{T} = (1, 1, \sqrt{2})$.

So, the tangent equation at M_0 is

$$\frac{x-\left(\frac{\pi}{2}-1\right)}{1} = \frac{y-1}{1} = \frac{z-2\sqrt{2}}{\sqrt{2}}.$$

The equation of the normal plane at M_0 is

$$\left(x-\frac{\pi}{2}+1\right) + (y-1) + \sqrt{2}(z-2\sqrt{2}) = 0,$$

or

$$x+y+\sqrt{2}z-\frac{\pi}{2}-4=0.$$

Then, let's discuss the tangent equation and the normal plane equation if the space curve is the intersection of two surfaces.

(2) Suppose the space curve Γ is the intersection of two cylinders, whose equation is
$$\begin{cases} y=f(x), \\ z=g(x). \end{cases}$$

Now take x as the parameter, then the equation of the curve Γ can be expressed by the form of parametric equations,
$$\begin{cases} x=x, \\ y=f(x), \\ z=g(x). \end{cases}$$

Suppose $f(x), g(x)$ are derivable at $x=x_0$, from the above discussion, we know the tangent vector of the curve Γ at $M_0(x_0, y_0, z_0)$ is $\boldsymbol{T}=(1, f'(x_0), g'(x_0))$. Thus, the tangent equation of the curve Γ at the point M_0 is

$$\frac{x-x_0}{1}=\frac{y-y_0}{f'(x_0)}=\frac{z-z_0}{g'(x_0)}. \qquad (4)$$

The normal plane equation of the curve Γ at the point M_0 is
$$(x-x_0)+f'(x_0)(y-y_0)+g'(x_0)(z-z_0)=0. \qquad (5)$$

(3) Suppose the space curve Γ is the intersection of two curve surfaces, whose equation is
$$\begin{cases} F(x,y,z)=0, \\ G(x,y,z)=0. \end{cases} \qquad (6)$$

$M_0(x_0, y_0, z_0)$ is a point on the curve Γ. Suppose F, G have continuous partial derivatives with respect to each variable, and
$$\left.\frac{\partial(F,G)}{\partial(y,z)}\right|_{(x_0,y_0,z_0)} \neq 0.$$

The equation set (6) determines a set of functions $y=y(x), z=z(x)$ in the neighborhood of the point $M_0(x_0, y_0, z_0)$. To find the equation of the tangent line and the normal plane of the curve Γ at the point M_0, we just need to find $y'(x_0)$ and $z'(x_0)$, and then plug them into the Formula (4) and (5). For the identity
$$\begin{cases} F[x, y(x), z(x)] \equiv 0, \\ G[x, y(x), z(x)] \equiv 0, \end{cases}$$
find the perfect derivative with respect to x on either side, we obtain
$$\begin{cases} \dfrac{\partial F}{\partial x}+\dfrac{\partial F}{\partial y}\dfrac{dy}{dx}+\dfrac{\partial F}{\partial z}\cdot\dfrac{dz}{dx}=0, \\ \dfrac{\partial G}{\partial x}+\dfrac{\partial G}{\partial y}\dfrac{dy}{dx}+\dfrac{\partial G}{\partial z}\cdot\dfrac{dz}{dx}=0. \end{cases}$$

From the hypothesis, within some neighborhood of the point M_0,
$$J=\frac{\partial(F,G)}{\partial(y,z)}\neq 0.$$

Chapter 11 Differentiation of multivariable function and its application

We obtain
$$\frac{dy}{dx} = \frac{\begin{vmatrix} F_z & F_x \\ G_z & G_x \end{vmatrix}}{\begin{vmatrix} F_y & F_z \\ G_y & G_z \end{vmatrix}}, \quad \frac{dz}{dx} = \frac{\begin{vmatrix} F_x & F_y \\ G_x & G_y \end{vmatrix}}{\begin{vmatrix} F_y & F_z \\ G_y & G_z \end{vmatrix}}.$$

Thus, $T = (1, y'(x_0), z'(x_0))$ is a tangent vector of the curve Γ at M_0, and

$$y'(x_0) = \frac{\begin{vmatrix} F_z & F_x \\ G_z & G_x \end{vmatrix}_0}{\begin{vmatrix} F_y & F_z \\ G_y & G_z \end{vmatrix}_0}, \quad z'(x_0) = \frac{\begin{vmatrix} F_x & F_y \\ G_x & G_y \end{vmatrix}_0}{\begin{vmatrix} F_y & F_z \\ G_y & G_z \end{vmatrix}_0}.$$

The determinant with subscript 0 in quotient above is the value of the determinant at the point $M_0(x_0, y_0, z_0)$. In terms of multiplying the above tangent vector T by $\begin{vmatrix} F_y & F_z \\ G_y & G_z \end{vmatrix}_0$, we have

$$T_1 = \left(\begin{vmatrix} F_y & F_z \\ G_y & G_z \end{vmatrix}_0, \begin{vmatrix} F_z & F_x \\ G_z & G_x \end{vmatrix}_0, \begin{vmatrix} F_x & F_y \\ G_x & G_y \end{vmatrix}_0 \right),$$

which is also a general representation of a tangent vector. Then the tangent equation at M_0 is

$$\frac{x - x_0}{\begin{vmatrix} F_y & F_z \\ G_y & G_z \end{vmatrix}_0} = \frac{y - y_0}{\begin{vmatrix} F_z & F_x \\ G_z & G_x \end{vmatrix}_0} = \frac{z - z_0}{\begin{vmatrix} F_x & F_y \\ G_x & G_y \end{vmatrix}_0}.$$

The normal plane equation of the curve Γ at $M_0(x_0, y_0, z_0)$ is

$$\begin{vmatrix} F_y & F_z \\ G_y & G_z \end{vmatrix}_0 (x - x_0) + \begin{vmatrix} F_z & F_x \\ G_z & G_x \end{vmatrix}_0 (y - y_0) + \begin{vmatrix} F_x & F_y \\ G_x & G_y \end{vmatrix}_0 (z - z_0) = 0.$$

Especially, if $\dfrac{\partial(F, G)}{\partial(y, z)}\bigg|_0 = 0$, and $\dfrac{\partial(F, G)}{\partial(z, x)}\bigg|_0$ or $\dfrac{\partial(F, G)}{\partial(x, y)}\bigg|_0$ is not zero, we can get the same result.

Example 2 Find the equation of the tangent line and the normal plane of $x^2 + y^2 + z^2 = 6$, $x + y + z = 0$ at the point $(1, -2, 1)$.

Solution Find the derivative with respect to x on both sides of the equation, we obtain

$$\begin{cases} y \dfrac{dy}{dx} + z \dfrac{dz}{dx} = -x, \\ \dfrac{dy}{dx} + \dfrac{dz}{dx} = -1. \end{cases}$$

So

$$\frac{dy}{dx} = \frac{\begin{vmatrix} -x & z \\ -1 & 1 \end{vmatrix}}{\begin{vmatrix} y & z \\ 1 & 1 \end{vmatrix}} = \frac{z - x}{y - z}, \quad \frac{dz}{dx} = \frac{\begin{vmatrix} y & -x \\ 1 & -1 \end{vmatrix}}{\begin{vmatrix} y & z \\ 1 & 1 \end{vmatrix}} = \frac{x - y}{y - z},$$

$$\left.\frac{dy}{dx}\right|_{(1,-2,1)}=0, \quad \left.\frac{dz}{dx}\right|_{(1,-2,1)}=-1.$$

Thus, $T=(1,0,-1)$ and the tangent equation is

$$\frac{x-1}{1}=\frac{y+2}{0}=\frac{z-1}{-1}.$$

The normal plane equation is

$$(x-1)+0(y+2)-(z-1)=0 \quad \text{or} \quad x-z=0.$$

11.4.2 Tangent plane and normal line of curve surface

1) Definition of tangent plane of curve surface

The curve surface Σ and the point $M_0(x_0,y_0,z_0)$ on the surface are given.

The definition of tangent plane:

At $M_0(x_0,y_0,z_0)$, suppose tangent lines of an arbitrary curve on Σ passing through $M_0(x_0,y_0,z_0)$ are on the same plane, which is called the tangent plane of the curve surface Σ at $M_0(x_0,y_0,z_0)$.

The definition of normal:

The straight line passing through $M_0(x_0,y_0,z_0)$ and perpendicular to the tangent plane is called the normal line of the curve Σ at M_0.

2) Find tangent plane and normal line of the curve surface

Firstly, we discuss the curve surface equation $F(x,y,z)=0$ is given by an implicit function, and then regard the curve surface equation $z=f(x,y)$ as special situation to obtain relative result.

Suppose the equation of the curve surface Σ is $F(x,y,z)=0$ and $M_0(x_0,y_0,z_0)$ is a point on Σ, the partial derivatives of $F(x,y,z)$ at M_0 is continuous and not zero at the same time.

On the curve surface Σ, draw a curve Γ (Figure 11-13) passing through M_0, suppose its parameter equation is

$$\begin{cases} x=x(t), \\ y=y(t), \\ z=z(t), \end{cases} \quad (7)$$

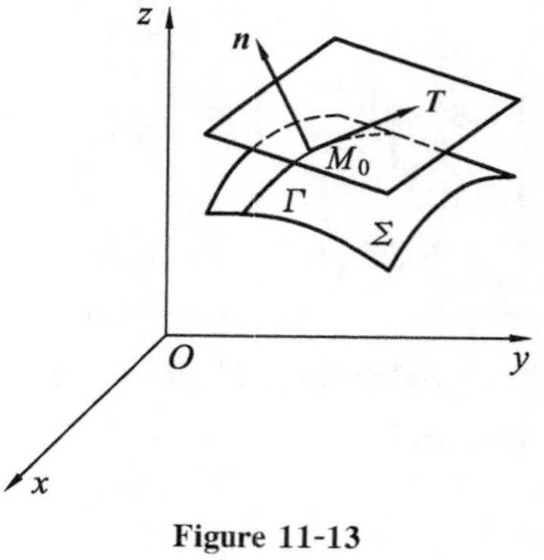

and $t=t_0$, which is corresponding to the point $M_0(x(t_0), y(t_0),z(t_0))$ on Γ (which means $x_0=x(t_0)$, $y_0=y(t_0)$, $z_0=z(t_0)$), $x'(t_0),y'(t_0),z'(t_0)$ are not zero at the same time, so the tangent vector of the curve Γ at M_0 is

$$T=(x'(t_0),y'(t_0),z'(t_0)).$$

Figure 11-13

On the other hand, the curve Γ is on the curve surface Σ, so coordinate of all points on Γ satisfies the equation of the curve surface Σ. There is an identity

$$F[x(t),y(t),z(t)]=0.$$

Find the perfect derivative, we obtain

$$\left.\frac{dF}{dt}\right|_{t=t_0}=F_x(x_0,y_0,z_0)x'(t_0)+F_y(x_0,y_0,z_0)y'(t_0)+F_z(x_0,y_0,z_0)z'(t_0)=0. \quad (8)$$

Chapter 11 Differentiation of multivariable function and its application

Introduce the vector
$$n=(F_x(x_0,y_0,z_0),F_y(x_0,y_0,z_0),F_z(x_0,y_0,z_0)),$$
then the Formula (8) indicates that, at M_0, the tangent vector $T=(x'(t_0),y'(t_0),z'(t_0))$ is perpendicular to the vector n. Because the curve (7) is an arbitrary curve passing through M_0 on the curve surface, their tangent line are perpendicular to the same vector n, so tangent lines at M_0 of all curves through M_0 on the curve surface are on the same plane (Figure 11-13). Thus, the tangent plane equation of the curve surface through M_0 is
$$F_x(x_0,y_0,z_0)(x-x_0)+F_y(x_0,y_0,z_0)(y-y_0)+F_z(x_0,y_0,z_0)(z-z_0)=0. \tag{9}$$
The normal line equation is
$$\frac{x-x_0}{F_x(x_0,y_0,z_0)}=\frac{y-y_0}{F_y(x_0,y_0,z_0)}=\frac{z-z_0}{F_z(x_0,y_0,z_0)}. \tag{10}$$
And the vector
$$n=(F_x(x_0,y_0,z_0),F_y(x_0,y_0,z_0),F_z(x_0,y_0,z_0))$$
is called a normal vector of the curve surface at M_0.

If the equation of the curve surface Σ is given in the form of explicit function
$$z=f(x,y), \tag{11}$$
let
$$F(x,y,z)=f(x,y)-z=0,$$
then
$$F_x(x,y,z)=f_x(x,y), F_y(x,y,z)=f_y(x,y), F_z(x,y,z)=-1.$$
Thus, if the partial derivatives $f_x(x,y), f_y(x,y)$ of the function $z=f(x,y)$ is continuous at the point (x_0,y_0), the normal vector of the tangent plane is
$$n=(f_x(x_0,y_0),f_y(x_0,y_0),-1).$$
Thus, the normal line equation of the curve surface Σ at M_0 is
$$\frac{x-x_0}{f_x(x_0,y_0)}=\frac{y-y_0}{f_y(x_0,y_0)}=\frac{z-z_0}{-1}. \tag{12}$$
By the tangent plane equation of the curve surface Σ at $M_0(x_0,y_0,z_0)$
$$z-z_0=f_x(x_0,y_0)(x-x_0)+f_y(x_0,y_0)(y-y_0),$$
we can explain the geometric meaning of the perfect differential of $z=f(x,y)$ at (x_0,y_0) clearly. In fact, the right side of tangent plane equation is the perfect differential of $z=f(x,y)$ at (x_0,y_0), and the left side is the increment on the vertical coordinate of the points on tangent plane. Thus, the perfect differential of the function $z=f(x,y)$ at (x_0,y_0) indicates the increment on the vertical coordinate of the points on tangent plane of $\Sigma: z=f(x,y)$, at the point $M_0(x_0,y_0,z_0)$.

Example 3 Find the tangent plane equation and normal line equation of the paraboloid $z=1-x^2-y^2$.

Solution Suppose $z=f(x,y)=1-x^2-y^2$, then
$$f_x(x,y)=-2x, f_y(x,y)=-2y,$$
$$f_x(1,1)=-2, f_y(1,1)=-2.$$
The normal vector of the paraboloid at $M_0(1,1,-1)$ is
$$n=(-2,-2,-1).$$
So, the tangent plane is

$$-2(x-1)-2(y-1)-(z+1)=0 \quad \text{or} \quad 2x+2y+z-3=0,$$

the normal line equation is

$$\frac{x-1}{2}=\frac{y-1}{2}=\frac{z+1}{1}.$$

Example 4 Prove that the sum of the intercepts on three coordinate axes of the tangent plane of the curve surface $\sqrt{x}+\sqrt{y}+\sqrt{z}=\sqrt{a}\,(a>0)$ at an arbitrary point (x_0,y_0,z_0) is constant $(x_0>0,y_0>0,z_0>0)$.

Proof Let $F(x,y,z)=\sqrt{x}+\sqrt{y}+\sqrt{z}-\sqrt{a}$, then

$$F_x(x_0,y_0,z_0)=\frac{1}{2\sqrt{x_0}},$$

$$F_y(x_0,y_0,z_0)=\frac{1}{2\sqrt{y_0}},$$

$$F_z(x_0,y_0,z_0)=\frac{1}{2\sqrt{z_0}}.$$

The normal vector at arbitrary point (x_0,y_0,z_0) on the curve surface is

$$\boldsymbol{n}=\left(\frac{1}{2\sqrt{x_0}},\frac{1}{2\sqrt{y_0}},\frac{1}{2\sqrt{z_0}}\right).$$

The tangent equation at this point is

$$\frac{1}{2\sqrt{x_0}}(x-x_0)+\frac{1}{2\sqrt{y_0}}(y-y_0)+\frac{1}{2\sqrt{z_0}}(z-z_0)=0,$$

or

$$\frac{x}{\sqrt{x_0}}+\frac{y}{\sqrt{y_0}}+\frac{z}{\sqrt{z_0}}=\sqrt{a}.$$

Write the above tangent equation into intercept formula, we have

$$\frac{1}{\sqrt{ax_0}}x+\frac{1}{\sqrt{ay_0}}y+\frac{1}{\sqrt{az_0}}z=1.$$

Then the intercepts of tangent plane on three coordinate axes are

$$\sqrt{ax_0},\sqrt{ay_0},\sqrt{az_0}.$$

So, the sum of these intercepts is

$$\sqrt{ax_0}+\sqrt{ay_0}+\sqrt{az_0}=\sqrt{a}(\sqrt{x_0}+\sqrt{y_0}+\sqrt{z_0})=\sqrt{a}\cdot\sqrt{a}=a.$$

Exercises 11-4

1. Find the equation of the tangent line and the normal plane of the curve $x=\frac{t}{1+t}, y=\frac{1+t}{t}$ at $t=1$.

2. Find the equation of the tangent line and the normal plane of the curve $x=a\cos\alpha\cos t$, $y=a\sin\alpha\cos t, z=a\sin t$ at $t=t_0$.

3. Find a point on $x=t, y=t^2, z=t^3$ at which the tangent line is parallel to the plane $x+2y+z=4$.

4. Find the equation of the tangent line and the normal plane of the curve
$\begin{cases} x^2+y^2+z^2=6, \\ x+y+z=0 \end{cases}$ at $M_0(1,-2,1)$.

5. Find a point on $z=xy$ at which the normal line is perpendicular to the plane $x+3y+z+9=0$ and find the normal equation.

6. Find the cosine of the angle between the tangent plane of the rotation ellipsoid $3x^2+y^2+z^2=16$ at the point $(-1,-2,3)$ and the xOy plane.

7. Find a point on the ellipsoid $\frac{x^2}{a^2}+\frac{y^2}{b^2}+\frac{z^2}{c^2}=1$ at which the angles of its normal in three directions are equal.

8. Find the tangent plane of $\frac{x^2}{a^2}+\frac{y^2}{b^2}+\frac{z^2}{c^2}=1$ such that the intercepts on each coordinate are equal.

11.5 The extreme value of the multivariable function and the maximum and minimum

11.5.1 The extreme value of the multivariable function

For some practical problems, the maximum value and minimum value of the multivariable function need to be found. Similar to the one-variable function, the absolute maximum value and absolute minimum value of the multivariable function are closely related to the local maximum value and local minimum value of the multivariable function.

Now let's take the binary function as an example, first, introduce the concept of the extreme value of multivariable functions, and then, study the necessary conditions and sufficient conditions for the existence of extreme values.

Definition A function $z=f(x,y)$ is defined in a neighborhood $U(M_0)$ of point $M_0(x_0,y_0)$, if for any $M\in \overset{\circ}{U}(M_0)$, there is $f(x,y)<f(x_0,y_0)$, $f(x_0,y_0)$ is called the local maximum value of the function $z=f(x,y)$, and $M_0(x_0,y_0)$ is called the local maximum point, and if $f(x,y)>f(x_0,y_0)$, $f(x_0,y_0)$ is called the local minimum value of function $z=f(x,y)$, and $M_0(x_0,y_0)$ is called a local minimum point. The maximum value and the minimum value are both called the extremum, and the point at which the function obtains the extreme value is called the extreme point.

Example 1 Function $z=x^2+y^2$ has a minimum at the point $(0,0)$. The function is positive for all points on the neighborhood of $(0,0)$. And at the point $(0,0)$, the function value is zero, that is,
$$z=x^2+y^2>0, (x,y)\neq(0,0).$$

Geometrically, this is obvious, because the point $(0,0,0)$ is the vertex of the rotation paraboloid $z=x^2+y^2$ open upward.

Example 2 Function $z=\sqrt{1-x^2-y^2}$ has a maximum at the point $(0,0)$, the function value is less than 1 for a sufficiently small neighborhood of the point $(0,0)$. And at the point $(0,0)$, the function value is 1, that is,

$$z = \sqrt{1-x^2-y^2} < 1, (x,y) \neq (0,0).$$

It's geometrically obvious, because the graph of $z = \sqrt{1-x^2-y^2}$ is the upper hemisphere with center at the origin and radius 1.

Example 3 At the point $(0,0)$, the function $z = xy$ obtains neither the maximum nor minimum. Because at the point $(0,0)$, the function value is zero, and in any neighborhood of $(0,0)$, there is always a point that makes the function value positive or negative.

The partial derivative can be used to solve the extremum problem of the binary function. The following two theorems are the conclusion of this problem.

Theorem 1 (Necessary conditions for extreme existence) Suppose that the function $z = f(x,y)$ has partial derivatives at the point $P_0(x_0, y_0)$, and the extreme value is obtained at that point, then

$$f_x(x_0, y_0) = 0, \ f_y(x_0, y_0) = 0. \tag{1}$$

Prove Suppose that the function $f(x,y)$ has a maximum value at point (x_0, y_0).

From the definition of maximum value, for all points (x,y) different from (x_0, y_0) in the neighborhood of (x_0, y_0), there is $f(x,y) < f(x_0, y_0)$. And in particular, take $y = y_0$, $x \neq x_0$ in this neighborhood, there are inequality

$$f(x, y_0) < f(x_0, y_0).$$

It means that the value of $f(x, y_0)$ at $x = x_0$ reaches the maximum value, so

$$\frac{\mathrm{d}}{\mathrm{d}x} f(x, y_0) \bigg|_{x=x_0} = 0,$$

that is $f_x(x_0, y_0) = 0$.

In the same way, we get $f_y(x_0, y_0) = 0$.

Geometrically, the tangent plane equation of the surface $z = f(x,y)$ at point $M_0(x_0, y_0, z_0)$ is

$$z - z_0 = f_x(x_0, y_0)(x - x_0) + f_y(x_0, y_0)(y - y_0). \tag{2}$$

If (x_0, y_0) is the extreme point of the function, by Theorem 1, there is

$$f_x(x_0, y_0) = 0, f_y(x_0, y_0) = 0.$$

The equation of the tangent plane is $z - z_0 = 0$, which shows that the tangent plane of the surface $z = f(x,y)$ at point $M_0(x_0, y_0, z_0)$ is parallel to the xOy coordinate plane.

The point such that $f_x(x,y) = 0$ and $f_y(x,y) = 0$ is called the stationary point of the function $f(x,y)$.

The necessary condition for the existence of extreme values provide a way to find extreme points. For the derivable function, if it has an extreme point, the extreme point must be the stationary point. But the above condition is not sufficient, which means the stationary point of the function is not necessarily the extreme point. For example, the point $(0,0)$ is the resident point of the function $z=xy$, but the function $z=xy$ cannot obtain the extreme value at the point $(0,0)$.

How to determine whether a stationary point is an extreme point or not? This question is answered by the following sufficient condition for the existence of the extreme values.

> **Theorem 2 (Sufficient conditions for extremum existence)** Suppose that $z=f(x,y)$ is continuous in some neighborhood of (x_0, y_0), which has the first and second consecutive partial derivatives, and $f_x(x,y)=0$, $f_y(x,y)=0$, let
> $$A=f_{xx}(x_0,y_0), B=f_{xy}(x_0,y_0), C=f_{yy}(x_0,y_0).$$
> Then
> (1) when $AC-B^2>0$, if $A<0$ (or $C<0$), $f(x_0,y_0)$ is the maximum value of $f(x,y)$, if $A>0$ (or $C>0$), $f(x_0,y_0)$ is the minimum value of $f(x,y)$;
> (2) when $AC-B^2<0$, $f(x_0,y_0)$ is not extreme value;
> (3) when $AC-B^2=0$, it can't be decided, which means $f(x_0,y_0)$ maybe is extreme value, or it is not extreme value. It needs another discussion.

The proof is omitted.

As a result of the Theorem 1 and Theorem 2, the method to find the extreme value of the function with second order continuous partial derivatives can be described as follows.

(1) Solve the following equation set for all real solutions and get all the stationary points.
$$\begin{cases} f_x(x,y)=0, \\ f_y(x,y)=0. \end{cases}$$

(2) Find the second derivatives $f_{xx}(x,y)$, $f_{xy}(x,y)$ and $f_{yy}(x,y)$, and for each stationary point, find the value of the second partial derivative.

(3) For each stationary point, find the symbol of $AC-B^2$. According to the conclusion of Theorem 2, determine whether the stationary point is an extreme point or not? Is it a maximum point or a minimum point?

(4) Find the value of the function at the extreme point, which is the extreme value.

11.5.2 The maximum and minimum of multivariable function

We find maximum and minimum of multivariable functions by extreme values. In the first part of this chapter we have pointed out that, if the function $z=f(x,y)$ is continuous within the bounded region D, the function $f(x,y)$ must have maximum and minimum in the closed region D. The point at which the function has maximum or minimum is either in the closed region D or on the boundary of D.

Now suppose the function $f(x,y)$ is continuous and differentiable in the bounded closed region D, and has finite stationary points. Suppose the function $f(x,y)$ reaches maximum or minimum in the region D, this maximum or minimum must be maxima or minima of $f(x,y)$ in this region. Thus, we get the general method of finding the maximum and minimum.

Find the values of the function $f(x,y)$ at all stationary points in the bounded closed region D or at the boundary curve of D first. And then compare these values, the biggest is the maximum and the smallest is the minimum.

If the function $f(x,y)$ has a single extreme in the region D, it must be the maximum or minimum of the function in the region D, that is, if the extreme is the minima (or maxima), it is the minimum (or maximum).

Example 4 Suppose some factory produces products A and B, their prices are $p_1=12$ and $p_2=18$ (unit: Yuan), the total cost C (unit: ten thousand Yuan) is the function of output of two products x and y (unit: thousand pieces)

$$C(x,y)=2x^2+xy+2y^2.$$

How many is the output of two kinds of product, the profit will be the biggest and how much is the profit?

Solution The function of total cost is

$$R(x,y)=12x+18y.$$

The function of total profit is

$$L(x,y)=R(x,y)-C(x,y)$$
$$=12x+18y-2x^2-xy-2y^2 \quad (x>0,y>0).$$

Let $L_x=12-4x-y=0$ and $L_y=18-x-4y=0$, we can obtain the stationary point $(2,4)$. From this problem we know that the biggest profit must exist, and there is only one stationary point in the region $D: x>0, y>0$, so when $x=2$ and $y=4$, the biggest profit is $L(2,4)=48$.

11.5.3 Conditional extremum Lagrange multiplier

From the extreme we discussed above, the independent variables are in the domain of the function but there is no any other constraint, which is called the unconditional extreme. But in practical problems, there is some other constraint for the independent variables, which is called the conditional extremum.

But in many cases, it is complex to transform conditional extreme into unconditional extreme. The Lagrange multiplier introduced below does not need to transform conditional extreme into unconditional extreme, but find conditional extreme directly.

For the function

$$z=f(x,y) \tag{3}$$

under the condition

$$\varphi(x,y)=0 \tag{4}$$

what is the necessary condition to reach the extreme?

Suppose the point at which the function $z=f(x,y)$ achieves extreme under the condition $\varphi(x,y)=0$ is $P_0(x_0,y_0)$, so

$$\varphi(x_0,y_0)=0. \tag{5}$$

Suppose $f(x,y)$ and $\varphi(x,y)$ have the first partial derivatives in the neighborhood of

Chapter 11 Differentiation of multivariable function and its application

(x_0, y_0), and $\varphi_y(x_0, y_0) \neq 0$, from the existence Theorem 1 of implicit functions we know that the function $\varphi(x, y) = 0$ determines a single-valued derivable implicit function $y = y(x)$ which has continuous derivative, plug it into (3) we obtain

$$z = f[x, y(x)].$$

Because $f(x, y)$ achieves conditional extreme at the point (x_0, y_0), which means $f[x, y(x)]$ achieves extreme at $x = x_0$. From the necessary condition of finding extreme of one variable derivable function, we know that

$$\left.\frac{dz}{dx}\right|_{x=x_0} = f_x(x_0, y_0) + f_y(x_0, y_0) \left.\frac{dy}{dx}\right|_{x=x_0} = 0.$$

From formula of finding derivative of implicit function, we have $\left.\dfrac{dy}{dx}\right|_{x=x_0} = -\dfrac{\varphi_x(x_0, y_0)}{\varphi_y(x_0, y_0)}$, and plug it into the above formula we obtain

$$f_x(x_0, y_0) - \frac{f_y(x_0, y_0) \varphi_x(x_0, y_0)}{\varphi_y(x_0, y_0)} = 0. \tag{6}$$

Formula (5) and (6) are necessary conditions of achieving conditional extreme of $z = f(x, y)$ at (x_0, y_0).

Suppose

$$\lambda = -\frac{f_y(x_0, y_0)}{\varphi_y(x_0, y_0)}, \tag{7}$$

the above necessary condition can be written as

$$\begin{cases} f_x(x_0, y_0) + \lambda \varphi_x(x_0, y_0) = 0, \\ f_y(x_0, y_0) + \lambda \varphi_y(x_0, y_0) = 0, \\ \varphi(x_0, y_0) = 0. \end{cases} \tag{8}$$

According to the above analysis, introduce an auxiliary function

$$L(x, y, \lambda) = f(x, y) + \lambda \varphi(x, y),$$

which is called the Lagrange function, and the parameter λ is called the Lagrange multiplier.

From Formula (8) we can see that (x_0, y_0) matches the equation set $L_x = 0, L_y = 0$, $L_\lambda = \varphi(x, y) = 0$, which means $x = x_0, y = y_0$ are stationary point of the Lagrange function $L(x, y, \lambda)$, and then we will get the Lagrange multiplier method to find conditional extreme.

Lagrange Multiplier Method We can find the possible extreme point of the function $z = f(x, y)$ under attached (constraint) condition $\varphi(x, y) = 0$ following the steps as below.

(1) Construct Lagrange function

$$L(x, y, \lambda) = f(x, y) + \lambda \varphi(x, y), \tag{9}$$

where λ is a constant.

(2) Find the first partial derivative of (9) with respect to x, y, and establish equation set

$$\begin{cases} f_x(x, y) + \lambda \varphi_x(x, y) = 0, \\ f_y(x, y) + \lambda \varphi_y(x, y) = 0, \\ \varphi(x, y) = 0. \end{cases} \tag{10}$$

(3) Find x,y and λ from the Equation(10), x,y are coordinates of stationary point of the possible extreme point.

The above method can be generalized to the situation in which there are more than two variables and more than one conditions. For example, in order to find the extreme of the function $u=f(x,y,z,t)$ under the constraint condition $\varphi(x,y,z,t)=0, \psi(x,y,z,t)=0$, we make the Lagrange function
$$L(x,y,z,t,\lambda,\mu)=f(x,y,z,t)+\lambda\varphi(x,y,z,t)+\mu\psi(x,y,z,t),$$
where λ and μ are parameters. And then solve the equation set, we get
$$L_x=0,\ L_y=0,\ L_z=0,\ L_t=0,\ L_\lambda=0,\ L_\mu=0.$$
Thus, we obtain that x_0, y_0, z_0 are the coordinates of stationary point of possible extreme point.

Example 5 Suppose a company sells the product in two separate markets, demand functions of the two markets are
$$p_1=18-2Q_1,\ p_2=12-Q_2,$$
where p_1 and p_2 are price, Q_1 and Q_2 are the sales, the total cost function is
$$C=2(Q_1+Q_2)+5.$$

(1) If this company carries out price difference strategy, try to determine the sales and price of products in two markets such that the company obtain the biggest profit.

(2) If this company carries out price indifference strategy, try to determine the sales and price of products in two markets such that the company obtain the biggest profit, and compare the total profit in two strategies.

Solution (1) Total profit function is
$$\begin{aligned}P &=R-C=p_1Q_1+p_2Q_2-[2(Q_1+Q_2)+5]\\&=-2Q_1^2-Q_2^2+16Q_1+10Q_2-5.\end{aligned}$$
From
$$\begin{cases}\dfrac{\partial P}{\partial Q_1}=-4Q_1+16=0,\\[2mm]\dfrac{\partial P}{\partial Q_2}=-2Q_2+10=0,\end{cases}$$
we obtain $Q_1=4, Q_2=5$ and $p_1=10, p_2=7$.

This is a practical problem, the maximum must exist, and since the stationary point is unique, at $p_1=10, p_2=7$, the biggest profit is
$$P=-2Q_1^2-Q_2^2+16Q_1+10Q_2-5\Big|_{\substack{Q_1=4\\Q_2=5}}=52.$$

(2) If carry out price indifference strategy, $p_1=p_2$, the constraint condition is
$$2Q_1-Q_2=6.$$
Construct the Lagrange function as follows
$$L(Q_1,Q_2,\lambda)=-2Q_1^2-Q_2^2+16Q_1+10Q_2-5+\lambda(2Q_1-Q_2-6).$$
From

Chapter 11 Differentiation of multivariable function and its application

$$\begin{cases} \dfrac{\partial L}{\partial Q_1} = -4Q_1 + 16 + 2\lambda = 0, \\ \dfrac{\partial L}{\partial Q_2} = -2Q_2 + 10 - \lambda = 0, \\ \dfrac{\partial L}{\partial \lambda} = 2Q_1 - Q_2 - 6 = 0, \end{cases}$$

we obtain $Q_1 = 5, Q_2 = 4, \lambda = 2$ and $p_1 = p_2 = 8$.

The biggest profit is

$$P = -2Q_1^2 - Q_2^2 + 16Q_1 + 10Q_2 - 5 \Big|_{\substack{Q_1=5 \\ Q_2=4}} = 49.$$

Thus, the price difference strategy is more profitable than price indifference strategy.

Exercises 11-5

1. Find the extreme of the function $f(x,y) = x^3 + 3xy^2 - 15x + 12y$.

2. Find the extreme of the function $f(x,y) = xy + \dfrac{50}{x} + \dfrac{20}{y}$ $(x>0, y>0)$.

3. Find the extreme of the function $f(x,y) = e^{x-y}(x^2 - 2y^2)$.

4. Find the extreme of the function $z = x^2 + y^2$ under the condition $\dfrac{x}{a} + \dfrac{y}{b} = 1$ ($a>0$, $b>0$), and prove it is minima according to the characteristic of the figure.

5. Find the shortest distance from the parabola $y = x^2$ to the straight line $x - y - 2 = 0$.

6. Find a point in the first octant of the ellipsoid $\dfrac{x^2}{a^2} + \dfrac{y^2}{b^2} + \dfrac{z^2}{c^2} = 1$, and the volume of the tetrahedron formed by the tangent plane at this point and three coordinate planes.

7. To make a cupped cylindrical lead barrel with volume of 1 m³, how big the barrel is will make the material used least.

8. Find a point on the xOy plane such that the sum of square of the distance from this point to $x=0, y=0$ and $x+2y-16=0$ achieves minimum.

9. Suppose two elements must be put into the production of a product, a and b are inputs of two elements, Q is output. If the production function is $Q = 2a^x b^y$, and constants $x, y > 0, x+y = 1$. Suppose the price of the two elements are p, q, when the output is 12, how much of two elements is put such that the cost is least ?

10. Suppose the production of Cobb-Douglas is

$$L = (x, y) = 100 x^{\frac{3}{4}} y^{\frac{1}{4}},$$

x and y represent the amount of labor force and capital respectively. If the cost of each labor force and per unit capital is 150 Yuan and 250 Yuan, the total budge of this enterprise is 50 000. How to allocate the money to labor and capital, the output will be the largest?

11.6 Taylor's formula of binary function

11.6.1 Taylor's formula of binary function

From the definition of perfect differential of binary function in 11.2, we know that the

function $f(x,y)$ is differentiable at (x_0,y_0), then
$$f(x_0+h,y_0+k)-f(x_0,y_0)=f_x(x_0,y_0)h+f_y(x_0,y_0)k+o(\rho),$$
where $\rho=\sqrt{h^2+k^2}$ and $h=x-x_0$, $k=y-y_0$, which means when $|h|,|k|$ are small enough, $f(x_0+h,y_0+k)$ can be represented approximately by a first order polynomial of h and k in some neighborhood of (x_0,y_0),
$$f(x_0+h,y_0+k)\approx f(x_0,y_0)+f_x(x_0,y_0)h+f_y(x_0,y_0)k.$$

When the degree of approximation of the above formula does not meet the requirement, higher order polynomial of h and k will be considered to replace the function $f(x_0+h,y_0+k)$ of h,k approximately, and the error can be estimated. To solve this problem, Taylor's mean value theorem of one variable function will be generalized to multi-variable function.

Theorem Suppose the binary function $z=f(x,y)$ has the $n+1$ order continuous partial derivatives in some neighborhood of the point (x_0,y_0), (x_0+h,y_0+k) is an arbitrary point in this neighborhood, then
$$f(x_0+h,y_0+k)=f(x_0,y_0)+\left(h\frac{\partial}{\partial x}+k\frac{\partial}{\partial y}\right)f(x_0,y_0)+$$
$$\frac{1}{2!}\left(h\frac{\partial}{\partial x}+k\frac{\partial}{\partial y}\right)^2 f(x_0,y_0)+\cdots+$$
$$\frac{1}{n!}\left(h\frac{\partial}{\partial x}+k\frac{\partial}{\partial y}\right)^n f(x_0,y_0)+R_n, \tag{1}$$
where
$$R_n=\frac{1}{(n+1)!}\left(h\frac{\partial}{\partial x}+k\frac{\partial}{\partial y}\right)^{n+1}f(x_0+\theta h,y_0+\theta k)\ (0<\theta<1). \tag{2}$$
The notations of the formula above,

$\left(h\frac{\partial}{\partial x}+k\frac{\partial}{\partial y}\right)f(x_0,y_0)$ denotes $hf_x(x_0,y_0)+kf_y(x_0,y_0)$;

$\left(h\frac{\partial}{\partial x}+k\frac{\partial}{\partial y}\right)^2 f(x_0,y_0)$ denotes $h^2 f_{xx}(x_0,y_0)+2hk f_{xy}(x_0,y_0)+k^2 f_{yy}(x_0,y_0)$.

Generally, the sign
$$\left(h\frac{\partial}{\partial x}+k\frac{\partial}{\partial y}\right)^m f(x_0,y_0) \text{ denotes } \sum_{r=0}^{m}C_m^r h^r k^{m-r}\frac{\partial^m f(x,y)}{\partial x^r \partial y^{m-r}}\bigg|_{(x_0,y_0)},$$
where $C_m^r=\frac{m!}{r!(m-r)!}$.

Proof In order to use Taylor's mean value theorem of unary function to prove this theorem, let's consider unary function
$$F(t)=f(x_0+th,y_0+tk)\ (0\leqslant t\leqslant 1).$$

Obviously, $F(0)=f(x_0,y_0)$, $F(1)=f(x_0+h,y_0+k)$. From the supposition we know that the function $z=f(x,y)$ has the $n+1$ order continuous derivatives on the interval $[0,1]$. Using the differential method of multi-variable function and letting $x=x_0+th$, $y=$

y_0+tk, we obtain

$$F'(t)=h\frac{\partial f}{\partial x}+k\frac{\partial f}{\partial y}=\left(h\frac{\partial}{\partial x}+k\frac{\partial}{\partial y}\right)f(x_0+th,y_0+tk),$$

$$F''(t)=h^2\frac{\partial^2 f}{\partial x^2}+2hk\frac{\partial^2 f}{\partial x\partial y}+k^2\frac{\partial^2 f}{\partial y^2}=\left(h\frac{\partial}{\partial x}+k\frac{\partial}{\partial y}\right)^2 f(x_0+th,y_0+tk),$$

$$\cdots\cdots\cdots$$

$$F^{(m)}(t)=\left(h\frac{\partial}{\partial x}+k\frac{\partial}{\partial y}\right)^m f(x_0+th,y_0+tk)$$

$$=\sum_{r=0}^{m} C_m^r h^r k^{m-r}\frac{\partial^m}{\partial x^r \partial y^{m-r}}f(x_0+th,y_0+tk).$$

Thus,

$$F'(0)=\left(h\frac{\partial}{\partial x}+k\frac{\partial}{\partial y}\right)f(x_0,y_0),$$

$$F''(0)=\left(h\frac{\partial}{\partial x}+k\frac{\partial}{\partial y}\right)^2 f(x_0,y_0),$$

$$\cdots\cdots\cdots$$

$$F^{(n)}(0)=\left(h\frac{\partial}{\partial x}+k\frac{\partial}{\partial y}\right)^n f(x_0,y_0),$$

$$F^{(n+1)}(\theta)=\left(h\frac{\partial}{\partial x}+k\frac{\partial}{\partial y}\right)^{n+1} f(x_0+\theta h,y_0+\theta k).$$

Using the Maclaurin formula of unary function, we obtain

$$F(t)=F(0)+F'(0)t+\frac{F''(0)}{2!}t^2+\cdots+\frac{F^{(n)}(0)}{n!}t^n+R_n,$$

where

$$R_n=\frac{1}{(n+1)!}F^{(n+1)}(\theta t)\ (0<\theta<1).$$

Let $t=1$, we obtain

$$F(1)=F(0)+F'(0)+\frac{F''(0)}{2!}+\cdots+\frac{F^{(n)}(0)}{n!}+R_n,$$

where

$$R_n=\frac{1}{(n+1)!}F^{(n+1)}(\theta)\ (0<\theta<1).$$

So, we have

$$f(x_0+h,y_0+k)=f(x_0,y_0)+\left(h\frac{\partial}{\partial x}+k\frac{\partial}{\partial y}\right)f(x_0,y_0)+$$

$$\frac{1}{2!}\left(h\frac{\partial}{\partial x}+k\frac{\partial}{\partial y}\right)^2 f(x_0,y_0)+\cdots+$$

$$\frac{1}{n!}\left(h\frac{\partial}{\partial x}+k\frac{\partial}{\partial y}\right)^n f(x_0,y_0)+R_n,$$

where

$$R_n=\frac{1}{(n+1)!}\left(h\frac{\partial}{\partial x}+k\frac{\partial}{\partial y}\right)^{n+1} f(x_0+\theta h,y_0+\theta k)\ (0<\theta<1).$$

The Formula (1) is called the n th order Taylor's formula of the function $f(x,y)$ at (x_0,y_0) and R_n in the Formula (2) is called the Lagrange reminder.

If the distance between $M_0(x_0,y_0)$ and $M(x_0+h,y_0+k)$ is $\rho=\sqrt{h^2+k^2}$, suppose the function $z=f(x,y)$ has the $n+1$ order continuous partial derivative in the neighborhood of the point (x_0,y_0), then its absolute value is no more than some positive number K in the neighborhood of the point (x_0,y_0), so

$$|R_n|=\frac{1}{(n+1)!}\left|\left(h\frac{\partial}{\partial x}+k\frac{\partial}{\partial y}\right)^{n+1}f(x_0+\theta h,y_0+\theta k)\right|$$

$$=\frac{1}{(n+1)!}\rho^{n+1}\left|\left(\frac{h}{\rho}\frac{\partial}{\partial x}+\frac{k}{\rho}\frac{\partial}{\partial y}\right)^{n+1}f(x_0+\theta h,y_0+\theta k)\right|$$

$$\leqslant\frac{K}{(n+1)!}\rho^{n+1}\left(\frac{|h|}{\rho}+\frac{|k|}{\rho}\right)^{n+1}.$$

Because $\frac{|h|}{\rho}\leqslant 1, \frac{|k|}{\rho}\leqslant 1, \left(\frac{|h|}{\rho}+\frac{|k|}{\rho}\right)\leqslant 2$, and

$$|R_n|\leqslant\frac{2^{n+1}K}{(n+1)!}\rho^{n+1}.$$

when $\rho\to 0$, R_n is a higher order infinitesimal than ρ^n.

In the Taylor's Formula (1), if $x_0=0$, $y_0=0$, $h=x, k=y$, the nth order Taylor's formula of $f(x,y)$ is

$$f(x,y)=f(0,0)+\left(x\frac{\partial}{\partial x}+y\frac{\partial}{\partial y}\right)f(0,0)+\frac{1}{2!}\left(x\frac{\partial}{\partial x}+y\frac{\partial}{\partial y}\right)^2 f(0,0)+\cdots+$$

$$\frac{1}{n!}\left(x\frac{\partial}{\partial x}+y\frac{\partial}{\partial y}\right)^n f(0,0)+\frac{1}{(n+1)!}\left(x\frac{\partial}{\partial x}+y\frac{\partial}{\partial y}\right)^{n+1}f(\theta x,\theta y)\ (0<\theta<1). \qquad (3)$$

The Formula (3) is called the n th order Maclaurin formula of the function $f(x,y)$.

Example 1 Find the third order Maclaurin formula of the function
$$f(x,y)=\ln(1+x+y).$$

Solution The function $f(x,y)=\ln(1+x+y)$ has the forth order continuous partial derivatives in some neighborhood of the point (x_0,y_0). Since

$$f_x(x,y)=f_y(x,y)=\frac{1}{1+x+y},$$

$$f_{xx}(x,y)=f_{xy}(x,y)=f_{yy}(x,y)=-\frac{1}{(1+x+y)^2},$$

$$\frac{\partial^3 f}{\partial x^r \partial y^{3-r}}=\frac{2!}{(1+x+y)^3}\ (r=0,1,2,3),$$

$$\frac{\partial^4 f}{\partial x^r \partial y^{4-r}}=-\frac{3!}{(1+x+y)^4}\ (r=0,1,2,3,4),$$

then,

$$\left(x\frac{\partial}{\partial x}+y\frac{\partial}{\partial y}\right)f(0,0)=xf_x(0,0)+yf_y(0,0)=x+y,$$

$$\left(x\frac{\partial}{\partial x}+y\frac{\partial}{\partial y}\right)^2 f(0,0)=x^2 f_{xx}(0,0)+2xyf_{xy}(0,0)+y^2 f_{yy}(0,0)$$

$$=-(x+y)^2,$$

$$\left(x\frac{\partial}{\partial x}+y\frac{\partial}{\partial y}\right)^3 f(0,0)=x^3 f_{xxx}(0,0)+3x^2 y f_{xxy}(0,0)+$$

$$3xy^2 f_{xyy}(0,0)+y^3 f_{yyy}(0,0)$$

$$=2(x+y)^3.$$

Because $f(0,0)=0$,
$$\ln(1+x+y)=x+y-\frac{1}{2}(x+y)^2+\frac{1}{3}(x+y)^3+R_3,$$
and
$$R_3=\frac{1}{4!}\left(x\frac{\partial}{\partial x}+y\frac{\partial}{\partial y}\right)^4 f(\theta x,\theta y)=-\frac{1}{4}\cdot\frac{(x+y)^4}{(1+\theta x+\theta y)^4}\ (0<\theta<1).$$

11.6.2 The proof of sufficient condition of extreme of binary function

Suppose the function $z=f(x,y)$ is continuous in some neighborhood $U_1(M_0)$ of the point $M_0(x_0,y_0)$, $z=f(x,y)$ has the first and second order continuous partial derivatives, and $f_x(x_0,y_0)=0, f_y(x_0,y_0)=0$.

According to the Taylor's formula of binary function $f(x,y)$ at $M_0(x_0,y_0)$, for $(x_0+h, y_0+k)\in U_1(M_0)$, we obtain

$$f(x_0+h,y_0+k)-f(x_0,y_0)$$
$$=f_x(x_0,y_0)h+f_y(x_0,y_0)k+\frac{1}{2!}[f_{xx}(x_0+\theta h,y_0+\theta k)h^2+$$
$$2f_{xy}(x_0+\theta h,y_0+\theta k)hk+f_{yy}(x_0+\theta h,y_0+\theta k)k^2]$$
$$=\frac{1}{2}[h^2 f_{xx}(x_0+\theta h,y_0+\theta k)+2hk f_{xy}(x_0+\theta h,y_0+\theta k)+$$
$$k^2 f_{yy}(x_0+\theta h,y_0+\theta k)]\ (0<\theta<1). \tag{4}$$

(1) Suppose $AC-B^2>0$, which means
$$f_{xx}(x_0,y_0)f_{yy}(x_0,y_0)-[f_{xy}(x_0,y_0)]^2>0. \tag{5}$$

Because the second order partial derivatives of $f(x,y)$ are continuous in $U_1(M_0)$, from the Formula (5), the neighborhood $U_2(M_0)\subset U_1(M_0)$ of the point M_0 exists, such that for any $(x_0+h,y_0+k)\in U_2(M_0)$, we have
$$f_{xx}(x_0+\theta h,y_0+\theta k)f_{yy}(x_0+\theta h,y_0+\theta k)-[f_{xy}(x_0+\theta h,y_0+\theta k)]^2>0. \tag{6}$$

To write easily, we always denote the value of $f_{xx}(x,y), f_{xy}(x,y), f_{yy}(x,y)$ at $(x_0+\theta h, y_0+\theta k)$ by f_{xx}, f_{xy}, f_{yy} in turn. From the Formula (6), when $(x_0+h, y_0+k)\in U_2(M_0)$, f_{xx} and f_{yy} are not zero and have the same sign. So, the Formula (4) can be written as

$$\Delta f=\frac{1}{2f_{xx}}[(hf_{xx}+kf_{xy})^2+k^2(f_{xx}f_{yy}-f_{xy}^2)].$$

When h and k are not equal to zero at the same time, and $(x_0+h,y_0+k)\in U_2(M_0)$, the value in square bracket in the above formula is positive, so Δf is not zero and has the same sign as f_{xx}. From continuity of the second order partial derivatives of $f(x,y)$, we know that f_{xx} and A have the same sign, so Δf has the same sign with A. When $A>0$, $f(x_0,y_0)$ is the minima; when $A<0$, $f(x_0,y_0)$ is the maxima.

(2) Suppose $AC-B^2<0$, which means
$$f_{xx}(x_0,y_0)f_{yy}(x_0,y_0)-[f_{xy}(x_0,y_0)]^2<0. \tag{7}$$

Suppose $f_{xx}(x_0,y_0)=f_{yy}(x_0,y_0)=0$, from the Formula (7) we know $f_{xy}(x_0,y_0)\neq 0$. Let $k=h$ and $k=-h$, from Formula (4) we know

$$\Delta f = \frac{h^2}{2}[f_{xx}(x_0+\theta_1 h, y_0+\theta_1 h)+2f_{xy}(x_0+\theta_1 h, y_0+\theta_1 h)+f_{yy}(x_0+\theta_1 h, y_0+\theta_1 h)],$$

$$\Delta f = \frac{h^2}{2}[f_{xx}(x_0+\theta_2 h, y_0-\theta_2 h)-2f_{xy}(x_0+\theta_2 h, y_0-\theta_2 h)+f_{yy}(x_0+\theta_2 h, y_0-\theta_2 h)],$$

where $0<\theta_1, \theta_2<1$. When $h\to 0$, the formula in square bracket in the above two formulas respectively tends to the limit $2f_{xy}(x_0, y_0)$ and $-2f_{xy}(x_0, y_0)$.

So, when h approaches 0 enough, values in square bracket have opposite sign. Thus Δf can have different signs, $f(x_0, y_0)$ is not extreme.

And then prove the situation when $f_{xx}(x_0, y_0)$ and $f_{yy}(x_0, y_0)$ are not zero at the same time. Suppose $f_{yy}(x_0, y_0)\neq 0$ and let $k=0$, from Formula (4) we obtain

$$\Delta f = \frac{1}{2}h^2 f_{xx}(x_0+\theta h, y_0).$$

Thus, when h approaches 0 enough, Δf has the same sign with $f_{xx}(x_0, y_0)$.

If
$$h=-f_{xy}(x_0, y_0)s, \quad k=f_{xx}(x_0, y_0)s, \tag{8}$$

where s is not zero but approaches to zero enough. When $|s|$ is small enough, Δf and $f_{xx}(x_0, y_0)$ have different signs. In fact, after plugging h and k in Formula (4) with given value into (8), we obtain

$$\Delta f = \frac{1}{2}s^2\{[f_{xy}(x_0, y_0)]^2 f_{xx} - 2f_{xy}(x_0, y_0)f_{xx}(x_0, y_0)f_{xy} + [f_{xx}(x_0, y_0)]^2 f_{yy}\}. \tag{9}$$

When $s\to 0$, the term on the right side of above formula tends to the limit

$$f_{xx}(x_0, y_0)\{f_{xx}(x_0, y_0)f_{yy}(x_0, y_0) - [f_{xy}(x_0, y_0)]^2\}.$$

From Formula (7), the value in bracket above is negative, so when h approaches to 0 enough, the right side of Formula (9) and $f_{xx}(x_0, y_0)$ have opposite sign.

This proves that near the point (x_0, y_0), Δf has values with different signs. Thus $f(x_0, y_0)$ is not extreme.

(3) Suppose $AC-B^2=0$, the value of $f(x_0+h, y_0+k)-f(x_0, y_0)$ need to be discussed, such as the following two functions

$$f(x,y)=x^3 y^3, \quad g(x,y)=x^2+y^4.$$

It is obvious that $O(0,0)$ is their stationary point, and we can verify them satisfy $AC-B^2=0$ easily. However $f(x,y)$ has no extreme at $O(0,0)$, while $g(x,y)$ has.

Exercises 11-6

1. Find Taylor's formula of $f(x,y)=2x^2-xy-y^2-6x-3y+5$ at $(1,-2)$.
2. Find the third order Maclaurin formula of $f(x,y)=e^x \ln(1+y)$.
3. Find the nth order Maclaurin formula of $f(x,y)=e^{x+y}$.

Summary

Differential method of multivariable function is derived and developed from unary function. When we study this chapter, we should compare them. We should pay attention to the basic concept, theorem and method in common between unary function and multi-variable functions, as well as the difference between them.

1. Main contents

(1) Multi-variable function, limit and continuity.

(2) Partial derivatives and perfect differential.

① Partial derivatives of $z=f(x,y)$ to x and y are

$$f_x(x,y)=\lim_{\Delta x \to 0}\frac{f(x+\Delta x,y)-f(x,y)}{\Delta x},$$

$$f_y(x,y)=\lim_{\Delta y \to 0}\frac{f(x,y+\Delta y)-f(x,y)}{\Delta y}.$$

② Higher order partial derivative: suppose the partial derivatives $\frac{\partial z}{\partial x}=f_x(x,y)$, and $\frac{\partial z}{\partial y}=f_y(x,y)$ of the binary function $z=f(x,y)$ also has partial derivatives, these partial derivatives are called the second partial derivative of $z=f(x,y)$, denoted by

$$\frac{\partial}{\partial x}\left(\frac{\partial z}{\partial x}\right)=\frac{\partial^2 z}{\partial x^2}=f_{xx}(x,y), \quad \frac{\partial}{\partial y}\left(\frac{\partial z}{\partial x}\right)=\frac{\partial^2 z}{\partial x \partial y}=f_{xy}(x,y),$$

$$\frac{\partial}{\partial x}\left(\frac{\partial z}{\partial y}\right)=\frac{\partial^2 z}{\partial y \partial x}=f_{yx}(x,y), \quad \frac{\partial}{\partial y}\left(\frac{\partial z}{\partial y}\right)=\frac{\partial^2 z}{\partial y^2}=f_{yy}(x,y).$$

③ Perfect differential: the total increment $\Delta z=f(x+\Delta x, y+\Delta y)-f(x,y)$ of the function $z=f(x,y)$ at $P(x,y)$ can be denoted by $\Delta z=A\Delta x+B\Delta y+o(\rho)$, A, B do not rely on Δx, Δy is only related to x, y, $\rho=\sqrt{(\Delta x)^2+(\Delta y)^2}$, so we call $z=f(x,y)$ is differentiable at $P(x,y)$, $A\Delta x+B\Delta y$ is perfect differential of the function $z=f(x,y)$ at $P(x,y)$, denoted by dz or $df(x,y)$, which means $dz=A\Delta x+B\Delta y$.

(3) Derivation rule of multivariable function.

Suppose partial derivatives of the function $u=\varphi(x,y), v=\psi(x,y)$ at (x,y) exist, and the function $z=f(u,v)$ is differentiable at corresponding point (u,v), so two partial derivative formula of $z=f[\varphi(x,y),\psi(x,y)]$ at (x,y) are

$$\frac{\partial z}{\partial x}=\frac{\partial z}{\partial u}\frac{\partial u}{\partial x}+\frac{\partial z}{\partial v}\frac{\partial v}{\partial x},$$

$$\frac{\partial z}{\partial y}=\frac{\partial z}{\partial u}\frac{\partial u}{\partial y}+\frac{\partial z}{\partial v}\frac{\partial v}{\partial y}.$$

When intermediate variable and independent variable are more than two, the above formula can be derived. For example, if $z=f(u,x,y)$ has continuous partial derivatives, and $u=\varphi(x,y)$ has partial derivatives, partial derivatives of compound function $z=f[\varphi(x,y),x,y]$ is

$$\frac{\partial z}{\partial x}=\frac{\partial f}{\partial u}\frac{\partial u}{\partial x}+\frac{\partial f}{\partial x},$$

$$\frac{\partial z}{\partial y}=\frac{\partial f}{\partial u}\frac{\partial u}{\partial y}+\frac{\partial f}{\partial y}.$$

(4) Derivation formula of implicit function.

Suppose $y=f(x)$ is implicit function determined by $F(x,y)=0$,

$$\frac{\mathrm{d}y}{\mathrm{d}x} = -\frac{F_x(x,y)}{F_y(x,y)}.$$

Suppose $z = f(x,y)$ is implicit function determined by $F(x,y,z) = 0$,

$$\frac{\partial z}{\partial x} = -\frac{F_x(x,y,z)}{F_z(x,y,z)}, \quad \frac{\partial z}{\partial y} = -\frac{F_y(x,y,z)}{F_z(x,y,z)}.$$

(5) Application of differential method in geometry.

① Tangent and normal plane of space curve:

Suppose parametric equation of space curve Γ is $\begin{cases} x = x(t), \\ y = y(t), \\ z = z(t), \end{cases}$ t is parameter, $M_0(x_0, y_0, z_0)$ is a point on the space curve Γ, and its corresponding parameter is t_0, so the tangent equation of space curve Γ at M_0 is

$$\frac{x - x_0}{x'(t_0)} = \frac{y - y_0}{y'(t_0)} = \frac{z - z_0}{z'(t_0)} \quad (x'(t_0), y'(t_0), z'(t_0) \text{ are not all zero}).$$

The normal equation of space curve Γ at M_0 is

$$x'(t_0)(x - x_0) + y'(t_0)(y - y_0) + z'(t_0)(z - z_0) = 0.$$

② Tangent plane and normal line of curve surface.

Suppose the curve surface equation is $F(x,y,z) = 0$, $M_0(x_0, y_0, z_0)$ is a point on curve surface. Suppose partial derivatives of $F(x,y,z)$ at M_0 are continuous and not all zero. So the tangent plane equation of curve surface at M_0 is

$$F_x(x_0, y_0, z_0)(x - x_0) + F_y(x_0, y_0, z_0)(y - y_0) + F_z(x_0, y_0, z_0)(z - z_0) = 0.$$

The normal equation of curve surface at M_0 is

$$\frac{x - x_0}{F_x(x_0, y_0, z_0)} = \frac{y - y_0}{F_y(x_0, y_0, z_0)} = \frac{z - z_0}{F_z(x_0, y_0, z_0)}.$$

If the curve surface equation is $z = f(x,y)$, $M_0(x_0, y_0, z_0)$ is a point on the curve surface, partial derivative $f_x(x,y), f_y(x,y)$ of $z = f(x,y)$ are continuous at (x_0, y_0). So, the tangent plane equation of curve surface at M_0 is

$$f_x(x_0, y_0)(x - x_0) + f_y(x_0, y_0)(y - y_0) - (z - z_0) = 0.$$

The normal equation of curve surface at M_0 is

$$\frac{x - x_0}{f_x(x_0, y_0)} = \frac{y - y_0}{f_y(x_0, y_0)} = \frac{z - z_0}{-1}.$$

(6) Direction derivative and gradient.

① Direction derivative: if binary function $z = f(x,y)$ is differentiable at (x,y),

$$\frac{\partial f}{\partial l} = \frac{\partial f}{\partial x}\cos\alpha + \frac{\partial f}{\partial y}\cos\beta,$$

where $\cos\alpha, \cos\beta$ are direction cosine of the direction l.

If the three variables function $u = f(x,y,z)$ is differentiable at (x,y,z), then

$$\frac{\partial f}{\partial l} = \frac{\partial f}{\partial x}\cos\alpha + \frac{\partial f}{\partial y}\cos\beta + \frac{\partial f}{\partial z}\cos\gamma,$$

where $\cos\alpha, \cos\beta, \cos\gamma$ are direction cosine of the direction l.

② Gradient: if partial derivatives of the binary function $z = f(x,y)$ at (x,y) exist, the gradient of the function $z=f(x,y)$ at (x,y) is

$$\mathbf{grad}\ f(x,y)=\frac{\partial f}{\partial x}\mathbf{i}+\frac{\partial f}{\partial y}\mathbf{j}.$$

If partial derivatives of the three variable function $u=f(x,y,z)$ at (x,y,z) exist, the gradient of the function $u=f(x,y,z)$ at (x,y,z) is

$$\mathbf{grad}\ f(x,y,z)=\frac{\partial f}{\partial x}\mathbf{i}+\frac{\partial f}{\partial y}\mathbf{j}+\frac{\partial f}{\partial z}\mathbf{k}.$$

(7) Extreme of multivariable function.

① The determining method of extreme of binary function.

Suppose $z=f(x,y)$ has the second continuous partial derivatives in some neighborhood of (x_0,y_0). If $f_x(x,y)=0, f_y(x,y)=0$, the condition when the function $f(x,y)$ has extreme at (x_0,y_0) is as follows:

$\Delta=B^2-AC$	$f(x_0,y_0)$
$\Delta<0$	When $A<0$, it has maxima
	When $A>0$, it has minima
$\Delta>0$	Not extreme
$\Delta=0$	Uncertain

where $A=f_{xx}(x_0,y_0), B=f_{xy}(x_0,y_0), C=f_{yy}(x_0,y_0)$.

② Conditional extreme: the method to find possible extreme point of $z=f(x,y)$ under the condition $\varphi(x,y)=0$ is constructing Lagrange's function

$$L(x,y,\lambda)=f(x,y)+\lambda\varphi(x,y),$$

solve the equation set

$$\begin{cases} f_x(x,y)+\lambda\varphi_x(x,y)=0, \\ f_y(x,y)+\lambda\varphi_y(x,y)=0, \\ \varphi(x,y)=0, \end{cases}$$

we can obtain x,y, and x,y are possible extreme point.

2. Basic requirements

(1) Understand the concept of multivariable function.

(2) Study concept of limit and continuity of binary function, and characters of continuous function on bounded closed region.

(3) Understand concept of partial derivative and perfect differential, understand necessary condition and sufficient condition of existence of perfect differential.

(4) Study concept and calculation method of direction derivative and gradient.

(5) Finding calculation method of the first partial derivative of compound function, and be able to find the second partial derivative of compound function.

(6) Be able to find partial derivative of implicit function (include implicit function determined by two equation sets).

(7) Understand tangent of curve surface, tangent plane and normal line of normal

plane and curve plane, be able to find their equation.

(8) Understand concept of extreme and conditional extreme of multivariable function, and find extreme of binary function. Understand Lagrange multiplier method of finding conditional extreme, be able to deal with the application of maximum and minimum.

Quiz

1. Multiple choice.

(1) The function $f(x,y)=\begin{cases} \dfrac{xy}{x^2+y^2}, & (x,y)\neq(0,0), \\ 0, & (x,y)=(0,0) \end{cases}$, at $(0,0)$ ( ).

A. is continuous, partial derivative exists
B. is continuous but partial derivative doesn't exist
C. is not continuous, partial derivative exists
D. is not continuous, partial derivative doesn't exist

(2) Partial derivatives $f_x(x,y), f_y(x,y)$ of the function $f(x,y)$ at the point (x_0, y_0) are ( ) of $f(x,y)$ is differential at this point.

A. necessary condition
B. sufficient condition
C. necessary and sufficient condition
D. neither necessary hor sufficient condition

(3) Suppose $f(x,y)=\begin{cases} \dfrac{1}{xy}\sin x^2 y, & xy\neq 0, \\ 0, & xy=0, \end{cases}$, $f_x(0,1)=($).

A. 0   B. 1   C. 2   D. does not exist

2. Completion.

(1) The domain of the function $z=\ln(x\ln y)$ is _____.

(2) The extreme point of the binary function $z=x^3-y^3-3x^2+3y-9x$ is _____.

(3) The normal plane of the function $\begin{cases} x-y-z=0, \\ 2x+y+z=-2 \end{cases}$ at $(0, 1, -1)$ is _____.

(4) Suppose the function $z=z(x,y)$ is determined by the equation $\sin x+2y-z=e^z$, $\dfrac{\partial z}{\partial x}=$ _____.

3. Solve the following questions.

(1) Suppose $z=ue^v\sin u$ and $u=xy, v=x+y$, find $\dfrac{\partial z}{\partial x}, \dfrac{\partial z}{\partial y}$.

(2) Find the limit $\lim\limits_{(x,y)\to(0,0)} \dfrac{1-\sqrt{x^2 y+1}}{x^3 y^2}\sin(xy)$.

Chapter 11 Differentiation of multivariable function and its application

(3) Find the tangent equation of $\begin{cases} x=t, \\ y=-t^2, \\ z=t^3 \end{cases}$, which is parallel to the plane $x+2y+z=4$.

4. Suppose $z=\sqrt{y}+f(\sqrt{x}-1)$, $x\geqslant 0, y\geqslant 0$, if $y=1$, $z=x$, try to determine the function $f(x)$ and z.

5. Find the limit of $z=x^2+y^2$ under the condition $\dfrac{x}{a}+\dfrac{y}{b}=1$.

6. Find the direction derivative of the function $u=e^z-x+xy$ at $(2, 1, 0)$ along the normal line direction of curve surface $e^z-z+xy=3$.

7. Suppose $f(x,y)=x^2+(x+3)y+ay^2+y^3$, we know $\dfrac{\partial f}{\partial x}=0$ and $\dfrac{\partial f}{\partial y}=0$ are tangent, find a.

8. Suppose $f(u)$ has the second continuous derivatives, and $z=f(e^x \sin y)$ satisfies the equation $\dfrac{\partial^2 z}{\partial x^2}+\dfrac{\partial^2 z}{\partial y^2}=e^{2x}z$, find $f(u)$.

9. Suppose $u=\sqrt{x^2+y^2+z^2}$, prove $\dfrac{\partial^2 u}{\partial x^2}+\dfrac{\partial^2 u}{\partial y^2}+\dfrac{\partial^2 u}{\partial z^2}=\dfrac{2}{u}$.

Exercises

1. Completion.
   (1) Continuous region of the function $z=\ln(x^2+y^2-1)$ is _____.
   (2) The function $f(x,y)$ is differentiable at (x,y) is _____ condition of the continuity of $f(x,y)$ at this point, the continuity of $f(x,y)$ at (x,y) is _____ condition of the differentiability of the function $f(x,y)$ at this point.

2. Multiple-choice.
   (1) Suppose the function $z=1-\sqrt{x^2+y^2}$, so the point $(0, 0)$ is ( ) of the function.
      A. minima point and minimum point
      B. maxima point and maximum point
      C. minima point but not minimum point
      D. maxima point but not maximum point
   (2) $z_x(x_0, y_0)=0$ and $z_y(x_0, y_0)=0$ are ( ) of the obtaining extreme of $z=z(x,y)$ at (x_0, y_0).
      A. necessary but not sufficient condition
      B. sufficient but not necessary condition
      C. sufficient condition
      D. neither necessary nor sufficient condition

3. Find the limit of $\lim\limits_{\substack{x\to 0 \\ y\to 0}} \dfrac{3y^3+2yx^2}{x^2-xy+y^2}$.

4. Prove that $\lim\limits_{\substack{x\to 0 \\ y\to 0}} \dfrac{2x-y}{x+y}$ does not exist.

5. Suppose $f(x,y)=\begin{cases} xy-\dfrac{x^3+y^3}{x^2+y^2}, & (x,y)\neq(0,0), \\ 0, & (x,y)=(0,0), \end{cases}$ find $f_x(0,0), f_y(0,0)$.

6. Find the first and second partial derivatives of the following functions:

(1) $z=xy+\ln\sqrt{x^2+y^2}$; (2) $u=x^y$.

7. Find perfect differential of the following functions:

(1) $z=\dfrac{xy}{x^2-y^2}$; (2) $u=\ln(x^xy^yz^z)$.

8. Suppose $z=F(u,v,x), u=\varphi(x), v=\psi(x)$, find $\dfrac{dz}{dx}$.

9. If $z=\sin y+f(\sin x+\sin y)$, and f is a differentiable function, prove that

$$\sec x \dfrac{\partial z}{\partial x}+\sec y \dfrac{\partial z}{\partial y}=1.$$

10. Suppose $z=xf\left(\dfrac{y}{x}\right)+2y\varphi\left(\dfrac{x}{y}\right)$, and f, φ have the second continuous derivatives, find $\dfrac{\partial^2 z}{\partial x^2}, \dfrac{\partial^2 z}{\partial x \partial y}$.

11. Find the extreme of $z=x^2+xy+y^2-3ax-3by$.

12. Suppose $\varphi(u,v)$ is a differentiable function, prove that tangent plane of any point on the curve plane $\varphi(x-az, y-bz)=0$ is parallel to the straight line $\dfrac{x}{a}=\dfrac{y}{b}=\dfrac{z}{1}$.

13. Find the normal line and normal plane of $x=2t^2, y=\cos(\pi t), z=2\ln t$ at the point which is correspond to $t=2$.

14. Find the maximum and minimum of direction derivative of the function $u=xy^2z^3$ at $(1, 1, 1)$.

15. A semi-cylindrical open container with a rectangle cross section, let its surface area be S, find the length of contain and section radius such that the volume will be the biggest?

16. The sellers make two kinds of advertisement, suppose when promotional cost is x and y (unit: thousand Yuan), the sales S (unit: piece) is the function of x and y

$$S=\dfrac{200x}{5+x}+\dfrac{100y}{10+y}.$$

If the profit is equal to one fifth of sales minus total advertisement cost, how to allocate advertisement cost can make the profit largest? What is the largest profit?

17. Some company produce a product which is put into two markets, price is p_1 and p_2, the sales is q_1 and q_2, demand function is

$$q_1=48-0.4p_1, \quad q_2=20-0.1p_2.$$

Total cost function is

$$C=35+40(q_1+q_2).$$

How to determine the sale in two markets such that the profit will be maximum?

Chapter 12 Multiple integral

Multiple integral is a generalization of the definite integral, whose scope is a bounded area of plane or space. Multiple integral and definite integral, though different forms, but the essence is consistent, which is a type of limits. In this chapter we will introduce the definition of multiple integral (including double integral and triple integral), calculation method, and some of their applications.

12.1 The concept and properties of double integrals

In this section, the concept of double integral is introduced from an example, and the concept of triple integral is only briefly described as the extension of double integral.

12.1.1 Examples

1) The volume of curved roof cylinder

The so-called curved roof cylinder refers to a kind of solid, whose bottom is a bounded closed area D on the xOy surface, the side is a cylinder with the directrix parallel to the z-axis and the boundary of the D as the generatrix, and the top is the curved surface $z = f(x, y)$, $(x, y) \in D$ (where $f(x, y) \geqslant 0$ and it's continuous on D) (Figure 12-1). Now, we will find the volume V of curved roof cylinder.

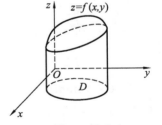

Figure 12-1

We know that the volume of a flat-top cylinder can be found by the formula

$$\text{volume} = \text{height} \times \text{base area}.$$

For the curved roof cylinder, as point (x, y) changes in the area D, the height of the cylinder is a variable, so if you still use the formula above to find the volume, a big error could be generated because of the difference of height. The reason is that there is a big change of height in D. How can we solve this problem? Recall that in Chapter 6, the problem of the area of the curved edge trapezoid, we notice that the volume is additive, which means the large volume can be divided into the sum of small volumes. It's not hard to think of the ways to solve the current problem.

Firstly, I'm going to divide D into n small closed areas arbitrarily, and we use $\Delta\sigma_i$ to denote the area of the i th little region ($i=1,2,\cdots,n$). Then, the solid is divided into n thin curved roof cylinders (Figure 12-2) by the cylinders with the directrix along the boundary of these little regions. Let the volume of the i th thin curved roof cylinder be ΔV_i ($i=1,2,\cdots,n$), then

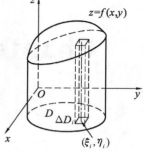

Figure 12-2

$$V = \sum_{i=1}^{n} \Delta V_i.$$

If the diameter of the small region ΔD_i ($i=1,2,\cdots,n$) (Figure 12-2) (diameter is the maximum distance between any two points in the region) is very small, due to the continuity of $f(x,y)$, we know that $f(x,y)$ changes little in ΔD_i. So the curved roof cylinder can be considered as a flat-top cylinder approximately in ΔD_i.

Now, suppose (ξ_i, η_i) is an arbitrary point in ΔD_i, the volume of the flat-top cylinder with height $f(\xi_i, \eta_i)$ and bottom ΔD_i is $f(\xi_i, \eta_i)\Delta\sigma_i$. Then

$$\Delta V_i \approx f(\xi_i, \eta_i)\Delta\sigma_i \quad (i=1,2,\cdots,n).$$

Add up the volumes of these thin flat-top cylinders, we get the approximate volume of the solid,

$$V = \sum_{i=1}^{n} \Delta V_i \approx \sum_{i=1}^{n} f(\xi_i, \eta_i)\Delta\sigma_i.$$

The approximations $\sum_{i=1}^{n} f(\xi_i, \eta_i)\Delta\sigma_i$ is obviously determined by the segmentation method of the region D and the point (ξ_i, η_i). But as the maximum diameter (denoted by λ) of the n small closed area ΔD_i ($i=1,2,\cdots,n$) tends to zero, the limit of the sum above is the exact volume of the solid,

$$V = \lim_{\lambda \to 0} \sum_{i=1}^{n} f(\xi_i, \eta_i)\Delta\sigma_i.$$

And, the volume is not related to the segmentation method of regional D and the selection of the point (ξ_i, η_i).

2) The mass of a slice

There is a slice lying on the xOy plane, which is surrounded by a closed area D.

The area density $\rho(x,y)$ is continuous in D, and $\rho(x,y) > 0$. Now we will find the mass of the slice M.

If the slice is homogeneous, we know that the area density is constant, the mass of the slice can be found by the formula

$$\text{mass} = \text{surface density} \times \text{area}.$$

Now the area density $\rho(x,y)$ is variable, the mass of the slice cannot be found by the formula above directly. But due to the mass is also additive and the area density $\rho(x,y)$ is continuous, the technique to find the curved roof cylinder's volume will be suitable for the mass problem.

We divide the slice into n small pieces and use $\Delta\sigma_i$ to denote the area of the i th piece $\Delta D_i (i=1,2,\cdots,n)$. Since $\rho(x,y)$ is continuous, as long as the diameter of the small piece ΔD_i is small, it can be regarded as uniform approximately. So we choose a point (ξ_i,η_i) in ΔD_i, and take $\rho(\xi_i,\eta_i)$ as the density of this little piece, then we get the approximation $\rho(\xi_i,\eta_i)\Delta\sigma_i$ of ΔM_i, that is,

$$\Delta M_i \approx \rho(\xi_i,\eta_i)\Delta\sigma_i \quad (i=1,2,\cdots,n).$$

We get the approximation of M by adding up these approximations (Figure 12-3),

$$M = \sum_{i=1}^{n}\Delta M_i \approx \sum_{i=1}^{n}\rho(\xi_i,\eta_i)\Delta\sigma_i.$$

The sum $\sum_{i=1}^{n}\rho(\xi_i,\eta_i)\Delta\sigma_i$ is obviously related to the segmentation method of the regional D and the point (ξ_i,η_i).

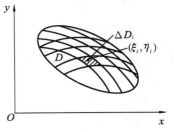

Figure 12-3

Similar to the technique to find the volume above, let $\lambda = \max\{\text{the diameter of }\Delta D_i\}$, the mass of the slice is

$$M = \lim_{\lambda \to 0}\sum_{i=1}^{n}\rho(\xi_i,\eta_i)\Delta\sigma_i.$$

And the mass is also not related to the segmentation method of the regional D and the selection of the point (ξ_i,η_i).

The two examples are different questions, but the technique is exactly the same, which is the limit of a sum. This kind of limit is also used in other questions, such as physics, mechanics, geometry and engineering technology, physical or geometric quantities, then we give the following definition of double integral.

12.1.2 The definition of double integrals

Definition $f(x,y)$ is a bounded function defined on a bounded closed area D. The closed area D is divided into n small closed regions $\Delta D_i (i=1,2,\cdots,n)$, and $\Delta\sigma_i$ is the area of $\Delta D_i (i=1,2,\cdots,n)$. For any point (ξ_i,η_i) in ΔD_i, we have $f(\xi_i,\eta_i)\Delta\sigma_i (i=1,2,\cdots,n)$, and add them up, we get the sum $\sum_{i=1}^{n}f(\xi_i,\eta_i)\Delta\sigma_i$. Let λ be the maximum of the diameters of these small regions, if as λ tends to zero, the limit of the sum exists, which does not depend on the partition method of the region D and the selection of the point (ξ_i,η_i), this limit is called the double integral of the function $f(x,y)$ in the closed region D, and denoted by $\iint\limits_{D} f(x,y)\mathrm{d}\sigma$, that is,

$$\iint\limits_{D} f(x,y)\mathrm{d}\sigma = \lim_{\lambda \to 0}\sum_{i=1}^{n}f(\xi_i,\eta_i)\Delta\sigma_i, \qquad (1)$$

where $f(x,y)$ is called the integrand, $d\sigma$ is called the area element, x and y are called the integral variables, D is called the integral area and $\sum_{i=1}^{n} f(\xi_i, \eta_i) \Delta \sigma_i$ is called the integral sum (or Riemann sum).

Obviously, the double integral is the generalization of the definite integral.

The area element $d\sigma$ in the double integral $\iint_D f(x,y) d\sigma$ is actually the area of a small region of D. According to the definition of double integrals, the limit of the sum does not depend on the partition method of D. So in a rectangular coordinate system, we can divide D by the lines parallel to the two axes. In addition to the small closed regions with boundary points, the remain small closed regions are rectangles. Let the length and width of the rectangle ΔD_i be Δx_i and Δy_i, then the area $\Delta \sigma_i = \Delta x_i \Delta y_i$. So, in a rectangular coordinate system, as $\lambda \to 0$, $\Delta D_i (i=1,2,\cdots,n)$ can be considered as rectangular closed regions. That is, the area element $d\sigma = dxdy$, and the double integral could also be

$$\iint_D f(x,y) dxdy,$$

where $dxdy$ is called the area element in the rectangular coordinate system.

We point out without proof that if $f(x,y)$ is continuous on the closed region D, the limit of the right sum in Formula (1) always exists. That is to say, if $f(x,y)$ is continuous on the region D, the double integral on D exists. And it can be further proved that if we use some piecewise smooth curves (the smooth curve means that the tangent exists at any point on the curve, and the tangent line moves along the curve continuously) to divide D into a finite number of small regions, and $f(x,y)$ is continuous on each region, the double integral of $f(x,y)$ on D also exists.

According to the definition of double integral, the volume of the column with curved top is the double integral of $f(x,y)$ on the bottom D,

$$V = \iint_D f(x,y) d\sigma.$$

The mass of the slice is the double integral of its area density $\rho(x,y)$ in the closed region D,

$$M = \iint_D \rho(x,y) d\sigma.$$

For the curved top cylinder, in general, if $f(x,y) \geq 0$, the integrand $f(x,y)$ can be interpreted as the vertical coordinates of the curved top at the point (x,y). So the geometric meaning of the double integral is the volume of the cylinder. If $f(x,y)$ is negative, the cylinder is just below the xOy plane. And the absolute value of the double integral is still equal to the volume of the cylinder, though the value of the double integral is negative. If $f(x,y)$ is positive in some parts of D and negative in other parts, we take the volume of the cylinder above the xOy plane to be positive, and the volume below the xOy plane be negative. The double integral of $f(x,y)$ on D is equal to the algebraic sum of these volumes.

12.1.3 The property of double integrals

Notice that the double integral is the limit of a sum which is similar to the definite integral, so it has the following properties.

Property 1 If the function $f(x,y), g(x,y)$ are both integrable on the bounded closed region D, then the function $kf(x,y)+lg(x,y)$ is integrable on the bounded closed region D, where k and l are arbitrary constants, that is,

$$\iint_D [kf(x,y)+lg(x,y)] dxdy = k\iint_D f(x,y) dxdy + l\iint_D g(x,y) dxdy.$$

This property is called the linearity of the double integral.

Property 2 If the function $f(x,y)$ is integrable on the region D, divide D into two closed regions D_1 and D_2, then $f(x,y)$ is integrable on both regions D_1 and D_2, that is,

$$\iint_D f(x,y) dxdy = \iint_{D_1} f(x,y) dxdy + \iint_{D_2} f(x,y) dxdy.$$

This property can be extended by dividing D into a finite number of regions $D_i (i=1, 2, \cdots, n)$, then

$$\iint_D f(x,y) dxdy = \sum_{i=1}^{n} \iint_{D_i} f(x,y) dxdy.$$

This property is called that the double integral is additive on the region.

Property 3 If $f(x,y)=1$ on D, let σ be the area of D, then

$$\iint_D 1 \cdot d\sigma = \iint_D d\sigma = \sigma.$$

The geometrical significance of this property is obvious, which means the volume of the flat-topped column with height 1 is equal to the base area of the cylinder.

Property 4 If the function $f(x,y)$ is integrable on the region D, and $f(x,y) \geq 0$ on D, then

$$\iint_D f(x,y) dxdy \geq 0.$$

This property is called the sign-preserving property of the double integral.

Corollary 1 If the functions $f(x,y)$ and $g(x,y)$ are both integrable on the region D, and $f(x,y) \leqslant g(x,y)$ on D, then $\iint_D f(x,y) \mathrm{d}x \mathrm{d}y \leqslant \iint_D g(x,y) \mathrm{d}x \mathrm{d}y$.

Corollary 2 If the function $f(x,y)$ and $|f(x,y)|$ are both integrable on the region D, then $\left| \iint_D f(x,y) \mathrm{d}x \mathrm{d}y \right| \leqslant \iint_D |f(x,y)| \mathrm{d}x \mathrm{d}y$.

Property 5 Let M and m are the maximum and minimum values of $f(x,y)$ in the closed region D, and σ be the area of D, then

$$m\sigma \leqslant \iint_D f(x,y) \mathrm{d}\sigma \leqslant M\sigma.$$

This property is called the estimate theorem of the double integral.

Property 6 If the function $f(x,y)$ is integrable on the region D, there exists at least one point (ξ, η) on D, such that

$$\iint_D f(x,y) \mathrm{d}x \mathrm{d}y = f(\xi, \eta)\sigma,$$

where σ is the area of D.

This is called the mean value theorem of the double integral.

Example 1 Estimate the value of the double integral

$$\iint_D e^{\sin x \cos y} \mathrm{d}x \mathrm{d}y,$$

where D is the circular region $x^2 + y^2 \leqslant 4$.

Solution For arbitrary point $(x,y) \in \mathbf{R}^2$, because $-1 \leqslant \sin x \cos y \leqslant 1$, we have

$$\frac{1}{e} \leqslant e^{\sin x \cos y} \leqslant e.$$

The area of D is $\sigma = 4\pi$, then

$$\frac{4\pi}{e} \leqslant \iint_D e^{\sin x \cos y} \mathrm{d}x \mathrm{d}y \leqslant 4\pi e.$$

Exercises 12-1

1. Use the definition of the double integral to prove the following equations:

(1) $\iint_D \mathrm{d}\sigma = S$ (where S is the area of D);

(2) $\iint_D \sqrt{R^2 - x^2 - y^2} \mathrm{d}\sigma = \frac{2}{3}\pi R^3$ (D is the circular region with center at the origin and

radius R).

2. Compare the following integrals by the properties of the double integral:

(1) $\iint\limits_D (x+y)^2 d\sigma$ and $\iint\limits_D (x+y)^3 d\sigma$, where the region D is surrounded by the x-axis, y-axis and the line $x+y=1$;

(2) $\iint\limits_D \ln(x+y)d\sigma$ and $\iint\limits_D [\ln(x+y)]^2 d\sigma$, where D is the rectangular region $3 \leqslant x \leqslant 5, 0 \leqslant y \leqslant 1$.

3. Determine the sign of the double integral $\iint\limits_D \ln(x^2+y^2)dxdy$, where D is the closed area $|x|+|y| \leqslant 1$.

4. Estimate the following integrals by the properties of the double integral.

(1) $I = \iint\limits_D \sin^2 x \sin^2 y d\sigma$, where D is the rectangular region, $0 \leqslant x \leqslant \pi, 0 \leqslant y \leqslant \pi$;

(2) $I = \iint\limits_D (x^2+4y^2+9)dxdy$, where D is the circular region $x^2+y^2 \leqslant 4$.

12.2 Calculation of double integrals

Similar to the definite integral, it is difficult to find a double integral by the definition. The technique we will give is to write a double integral into an iterated integral.

12.2.1 Calculate double integrals in rectangular coordinate system

We discuss the formula to find a double integral first.

Suppose the area of the cross section of a solid is $S(x), x \in [a,b]$, we know that the volume of the solid is

$$V = \int_a^b S(x)dx.$$

And then, we will find the integral.

Suppose $f(x,y) \geqslant 0$, and regard $\iint\limits_D f(x,y)dxdy$ as the volume of the curved top cylinder in the region D. Suppose the boundary curve of D is divided into two curves $y = \varphi_1(x)$ and $y = \varphi_2(x)$ by two straight lines $x=a$ and $x=b(a<b)$ and $\varphi_1(x) \leqslant \varphi_2(x)$. Let a straight line on the plane xOy which is parallel to y-axis to pass through the region D. Suppose the intersection point is not more than two. Then, the region D can be denoted by $\varphi_1(x) \leqslant y \leqslant \varphi_2(x), a \leqslant x \leqslant b$. We call this kind of region the X-type region (Figure 12-4). Then find the volume of the curved top cylinder (Figure 12-5).

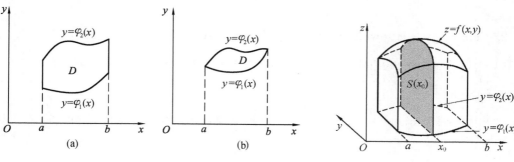

Figure 12-4   Figure 12-5

Suppose there is a set of planes $x = x_0 \ (a \leqslant x_0 \leqslant b)$ intercept the cylinder. The cross section at point $(x_0, 0, 0)$ is a curved edge trapezoid with the base $[\varphi_1(x_0), \varphi_2(x_0)]$ and curved edge $z = f(x_0, y)$ (see the shaded section in Figure 12-5). So the area of the cross section is

$$S(x_0) = \int_{\varphi_1(x_0)}^{\varphi_2(x_0)} f(x_0, y) \mathrm{d}y.$$

In general, for any point x on $[a, b]$, the area of the cross section is

$$S(x) = \int_{\varphi_1(x)}^{\varphi_2(x)} f(x, y) \mathrm{d}y.$$

By the formula of "the cubic volume with the known area of the parallel cross section" $V = \int_a^b S(x) \mathrm{d}x$, we get the volume of the curved top cylinder

$$V = \int_a^b S(x) \mathrm{d}x = \int_a^b \left[\int_{\varphi_1(x)}^{\varphi_2(x)} f(x, y) \mathrm{d}y \right] \mathrm{d}x.$$

And the volume is also equal to the double integral, so

$$\iint_D f(x, y) \mathrm{d}x \mathrm{d}y = \int_a^b \left[\int_{\varphi_1(x)}^{\varphi_2(x)} f(x, y) \mathrm{d}y \right] \mathrm{d}x. \tag{1}$$

The right term of the Equation (1) is called the iterated integral. For the inner integral, x is deemed as a constant, and $f(x, y)$ is just a function of y. We can integrate y from $\varphi_1(x)$ to $\varphi_2(x)$, and then we get a function of x which is the area of the cross section $S(x)$. Then we find the integral of the variable x over the interval $[a, b]$.

This example gives the technique to find the double integral, that is, write the double integral into a cumulative integral, like the right term of the Formula (1), the cumulative integral of first integrating over x then over y could be written by

$$\int_a^b \mathrm{d}x \int_{\varphi_1(x)}^{\varphi_2(x)} f(x, y) \mathrm{d}y,$$

that is,

$$\iint_D f(x, y) \mathrm{d}x \mathrm{d}y = \int_a^b \mathrm{d}x \int_{\varphi_1(x)}^{\varphi_2(x)} f(x, y) \mathrm{d}y.$$

Similarly, if the boundary curve of the region D is divided into two curves $x = \psi_1(y)$ and $x = \psi_2(y)$, $\psi_1(y) \leqslant \psi_2(y)$ by the straight lines $y = c$ and $y = d (c < d)$. In the xOy plane, suppose the intersection point of the lines parallel to the x-axis and the boundary of the region is less than 2. Then the region D can be represented by $\psi_1(y) \leqslant x \leqslant \psi_2(y)$, $c \leqslant y \leqslant d$. This region is called the Y region (Figure 12-6).

And then, we have

$$\iint_D f(x,y)\,dxdy = \int_c^d \left[\int_{\psi_1(y)}^{\psi_2(y)} f(x,y)\,dx\right]dy. \qquad (2)$$

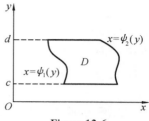

Figure 12-6

We can also write Formula (2) as

$$\iint_D f(x,y)\,dxdy = \int_c^d dy \int_{\psi_1(y)}^{\psi_2(y)} f(x,y)\,dx.$$

That is, write the double integral into a cumulative integral formula, first integrate over x then over y.

Formula (1) and Formula (2) give two calculation methods of double integrals in rectangular coordinates. In the derivation above, $f(x,y)$ is actually the arbitrary continuous function on the bounded closed region D, where it is assumed that $f(x,y) \geqslant 1$, Formula (1) and Formula (2) are all valid.

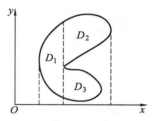

Figure 12-7

Both of the two methods require that the intersection point of the boundary curve and the lines parallel to the x-axis or y-axis is not more than two. If area D does not meet the above requirement, for example, the region D in Figure 12-7, we can divide D into D_1, D_2, and D_3 by a line that is parallel to the y-axis such that every section meets the requirement. And the double integral on the region D is equal to the sum of the double integrals on D_1, D_2 and D_3. Similarly, a line parallel to the x-axis can also divide a region.

To find the double integral $\iint_D f(x,y)\,dxdy$, we should determine whether the region D is X-type or Y-type first. For the integrals, we have to choose a proper method, otherwise it will complicate the calculation, or even can't work out.

Especially when the integral region D is rectangular region

$$D = \{(x,y) \mid c \leqslant y \leqslant d, a \leqslant x \leqslant b\}$$

and the function is $f(x,y) = f_1(x) f_2(y)$, we have

$$\iint_D f(x,y)\,dxdy = \int_a^b f_1(x)\,dx \cdot \int_c^d f_2(y)\,dy.$$

Example 1 Find the double integral

$$\iint_D \left(1 - \frac{x}{3} - \frac{y}{4}\right)dxdy,$$

where D is the rectangular region $D = \{(x,y) \mid -2 \leqslant y \leqslant 2, -1 \leqslant x \leqslant 1\}$.

Solution (Method 1) Draw the graph of the integral region D first (Figure 12-8). Because D is a rectangular region, it can be deemed as both X-type region and Y-type region. If it is a X-type, we can obtain

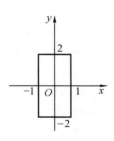

Figure 12-8

$$\iint_D \left(1 - \frac{x}{3} - \frac{y}{4}\right) dx dy = \int_{-1}^{1} dx \int_{-2}^{2} \left(1 - \frac{x}{3} - \frac{y}{4}\right) dy$$

$$= \int_{-1}^{1} \left[y - \frac{x}{3}y - \frac{1}{8}y^2\right]_{-2}^{2} dx$$

$$= \int_{-1}^{1} \left(4 - \frac{4}{3}x\right) dx = 8.$$

(**Method 2**) If D is Y-type, we have

$$\iint_D \left(1 - \frac{x}{3} - \frac{y}{4}\right) dx dy = \int_{-2}^{2} dy \int_{-1}^{1} \left(1 - \frac{x}{3} - \frac{y}{4}\right) dx$$

$$= \int_{-2}^{2} \left[x - \frac{1}{6}x^2 - \frac{y}{4}x\right]_{-1}^{1} dy$$

$$= \int_{-2}^{2} \left(2 - \frac{1}{2}y\right) dy = 8.$$

Example 2 Find

$$\iint_D xy \, dx dy,$$

where D is the region surrounded by the parabola $y^2 = x$ and the straight line $y = x - 2$.

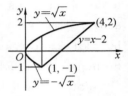

Figure 12-9

Solution Draw the graph of the region (Figure 12-9).

Find the intersection point of the two curves first. Solve the equation set $\begin{cases} y^2 = x, \\ y = x - 2, \end{cases}$ and then we can find the intersection points $(1, -1)$ and $(4, 2)$.

(1) Choose D as Y-type, since $D: \begin{cases} -1 \leqslant y \leqslant 2, \\ y^2 \leqslant x \leqslant y+2, \end{cases}$ we obtain

$$\iint_D xy \, dx dy = \int_{-1}^{2} dy \int_{y^2}^{y+2} xy \, dx = \int_{-1}^{2} y\left(\frac{1}{2}x^2\right)\bigg|_{y^2}^{y+2} dy$$

$$= \int_{-1}^{2} \frac{1}{2}[(y+2)^2 - y^5] dy = \frac{45}{8}.$$

(2) Choose D as X-type, D is divided into D_1 and D_2 by $x = 1$,

$$D_1: \begin{cases} 0 \leqslant x \leqslant 1, \\ -\sqrt{x} \leqslant y \leqslant \sqrt{x}, \end{cases} \quad D_2: \begin{cases} 1 \leqslant x \leqslant 4, \\ x - 2 \leqslant y \leqslant \sqrt{x}. \end{cases}$$

So we have

$$\iint_D xy \, dx dy = \iint_{D_1} xy \, dx dy + \iint_{D_2} xy \, dx dy = \int_0^1 x dx \int_{-\sqrt{x}}^{\sqrt{x}} y \, dy + \int_1^4 x dx \int_{x-2}^{\sqrt{x}} y \, dy = \frac{45}{8}.$$

Comparing two solutions above, (1) is better than (2).

Example 3 Find the double integral

$$\iint_D x^2 e^{-y^2} dx dy,$$

where D is the region surrounded by the straight lines $y = x$, $y = 1$ and $x = 0, x = 1$.

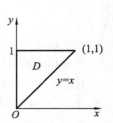

Figure 12-10

Solution Draw the graph of region D first (Figure 12-10).

The region D is deemed as a Y-type region, in which $0 \leqslant y \leqslant 1, 0 \leqslant x \leqslant y$. We choose integrate over x first and then over y, that is

$$\iint_D x^2 e^{-y^2} dx dy = \int_0^1 dy \int_0^y x^2 e^{-y^2} dx = \frac{1}{3} \int_0^1 y^3 e^{-y^2} dy$$

$$= \frac{1}{6} \int_0^1 y^2 e^{-y^2} dy^2 \xrightarrow{\diamondsuit u = y^2} \frac{1}{6} \int_0^1 u e^{-u} du$$

$$= \frac{1}{6} \left[-ue^{-u} \Big|_0^1 + \int_0^1 e^{-u} du \right] = \frac{1}{6} - \frac{1}{3e}.$$

In this example, if we denote the region D by X-type, the double integral will be integrate over y first and then over x. Because we can't find the antiderivative of e^{-y^2}, it is difficult to integrate. The Example 2 and 3 show that when we write a double integral into the iterated integral, the choice of the integrate order will make the solution easy or difficult, or even cannot be calculated. So when we find a double integral, we should consider the characteristic of the integrand and the integral region to determine the proper order of the iterated integral.

Example 4 Find the double integral

$$\iint_D \sqrt{|y - x^2|} \, dx dy,$$

the region D is rectangular region $-1 \leqslant x \leqslant 1, 0 \leqslant y \leqslant 2$.

Solution Draw the graph of the region D (Figure 12-11).

Because

$$|y - x^2| = \begin{cases} y - x^2, & y \geqslant x^2, \\ x^2 - y, & y < x^2, \end{cases}$$

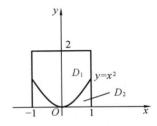

Figure 12-11

in the region D, D is divided into two parts D_1 and D_2 by the parabola $y = x^2$, and they are both X-type region, we have

$$D_1 : \begin{cases} -1 \leqslant x \leqslant 1, \\ x^2 \leqslant y \leqslant 2, \end{cases} \quad D_2 : \begin{cases} -1 \leqslant x \leqslant 1, \\ 0 \leqslant y \leqslant x^2. \end{cases}$$

Thus

$$\iint_D \sqrt{|y - x^2|} \, dx dy = \iint_{D_1} \sqrt{y - x^2} \, dx dy + \iint_{D_2} \sqrt{x^2 - y} \, dx dy$$

$$= \int_{-1}^1 dx \int_{x^2}^2 \sqrt{y - x^2} \, dy + \int_{-1}^1 dx \int_0^{x^2} \sqrt{x^2 - y} \, dy$$

$$= \int_{-1}^1 \frac{2}{3} (2 - x^2)^{\frac{3}{2}} dx + \int_{-1}^1 \frac{2}{3} |x|^3 dx = \frac{\pi}{2} + \frac{1}{3}.$$

Because when we write double integral into iterated integral, there are two orders. Sometimes for the given iterated integral, in order to calculate easily, the integrate order will be exchanged.

Example 5 Suppose $f(x, y)$ is continuous, change the integrate order of the iterated integral

$$I = \int_0^1 dx \int_0^{3\sqrt{x}} f(x, y) dy + \int_1^{\sqrt{10}} dx \int_0^{\sqrt{10 - x^2}} f(x, y) dy.$$

215

Solution The given iterated integrals are equal to the double integrals of the function $f(x,y)$ in the regions D_1 and D_2, where $D_1 = \{(x,y) \mid 0 \leqslant y \leqslant 3\sqrt{x}, 0 \leqslant x \leqslant 1\}$ and $D_2 = \{(x,y) \mid 0 \leqslant y \leqslant \sqrt{10-x^2}, 1 \leqslant x \leqslant \sqrt{10}\}$. Draw the graph of D_1 and D_2 in the same coordinate (Figure 12-12).

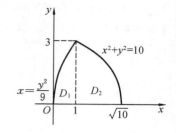

Figure 12-12

Then, we change the integral order, that is, integrate over x first and then over y, we have

$$I = \iint_D f(x,y)\,dxdy = \int_0^3 dy \int_{\frac{y^2}{9}}^{\sqrt{10-y^2}} f(x,y)\,dx.$$

Example 6 Find the volume of the solid surrounded by two orthogonal cylindrical surfaces with bottom radius R (Figure 12-13a).

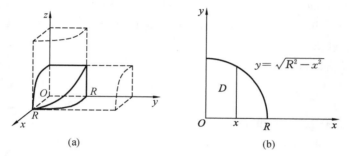

Figure 12-13

Solution From figure 12-13, the equations of two cylindrical surfaces are

$$x^2 + y^2 = R^2, \quad x^2 + z^2 = R^2.$$

By the symmetry of the coordinate plane and the solid, we just need to find the volume V_1 in the first octant (Figure 12-13a), and then multiplied by 8.

The solid in the first octant is a cylinder with curved top $x^2 + z^2 = R^2$ and bottom D_{xy} that is surrounded by the circle $x^2 + y^2 = R^2$, the x-axis and the y-axis. That is, the region D_{xy} is

$$\begin{cases} 0 \leqslant x \leqslant R, \\ 0 \leqslant y \leqslant \sqrt{R^2 - x^2} \end{cases} \text{(Figure 12-13b)}.$$

Therefore, we have

$$V_1 = \iint_D \sqrt{R^2 - x^2}\,d\sigma = \int_0^R \left[\int_0^{\sqrt{R^2-x^2}} \sqrt{R^2 - x^2}\,dy \right] dx$$

$$= \int_0^R \left[y\sqrt{R^2 - x^2} \right]_0^{\sqrt{R^2-x^2}} dx = \int_0^R (R^2 - x^2)\,dx = \frac{2}{3}R^3.$$

The volume of the solid is

$$V = 8V_1 = \frac{16}{3}R^3.$$

12.2.2 Calculate double integral in the polar coordinate system

In this section, we will focus on the solution of double integral in the polar coordinate system.

1) Expression

For some integrals, it is difficult to find the values in the rectangular coordinate system. In this case, the polar coordinate system maybe a better option. Now, let's begin with the notations of the polar coordinate system.

Since the relationship between the rectangular coordinates and the polar coordinates is
$$\begin{cases} x = r\cos\theta, \\ y = r\sin\theta, \end{cases}$$
in the polar coordinate system, we have $f(x,y) = f(r\cos\theta, r\sin\theta)$. Then, we will find the expression of the integral in the polar coordinate system. First of all, suppose the region of integration D is the area confined by two radials and a curve (Figure 12-14a). Except for the two radials, there are not more than two intersection points of a radial coming from O and passing through the boundary of D.

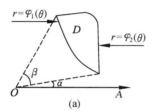
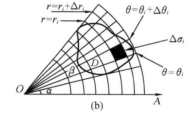

(a)   (b)

Figure 12-14

Let r and θ be constants, the two coordinate lines will divide the area D into many tiny regions and the area of each small region can be denoted by $\Delta\sigma_i$. Some small regions cover the boundary, and for the rest regions, we can find the area by the following way,

$$\Delta\sigma_i = \frac{1}{2}(r_i + \Delta r_i)^2 \Delta\theta_i - \frac{1}{2}r_i^2 \Delta\theta_i = \left(r_i + \frac{1}{2}\Delta r_i\right)\Delta r_i \Delta\theta_i$$
$$= r_i \Delta r_i \Delta\theta_i + \frac{1}{2}\Delta r_i^2 \Delta\theta.$$

So, $\frac{1}{2}\Delta r_i^2 \Delta\theta$ is the higher order infinitesimal of $r_i \Delta r_i \Delta\theta_i$. By the definition of the differential, we have $d\sigma = r dr d\theta$, which is the area element in the polar coordinate system. Therefore, the expression formula of the double integral in the polar coordinate system is

$$\iint_D f(x,y) dx dy = \iint_D f(r\cos\theta, r\sin\theta) r dr d\theta. \tag{3}$$

From the Equation (3), to switch a double integral from the rectangular coordinates to the polar coordinates, we just need to change $f(x,y)$ into $f(r\cos\theta, r\sin\theta)$ and change $dxdy$ to $rdrd\theta$.

The definition of double integral reveals that the double integral is independent on how the domain of integration divides. In other words, we find the value of a double integral in the rectangular coordinate system or the polar coordinate system, and then, we will get the

2) Iterated integrals in the polar coordinate system

Similar to the double integrals in the rectangular coordinate system, the computation of the double integrals in the polar coordinate system is also a matter of iterated integrals. And we always integrate over r first and then over θ. Suppose there are not more than two intersection points of the bounds of the region (denoted by D) and any radials.

(1) If the polar point O is in the region D(Figure 12-15), we have
$$0 \leqslant r \leqslant r(\theta), 0 \leqslant \theta \leqslant 2\pi.$$
So
$$\iint_D f(r\cos\theta, r\sin\theta) r dr d\theta = \int_0^{2\pi} d\theta \int_0^{r(\theta)} f(r\cos\theta, r\sin\theta) r dr.$$

(2) If the polar point is on the bound of the region D(Figure 12-16), we have
$$0 \leqslant r \leqslant r(\theta),\ \alpha \leqslant \theta \leqslant \beta.$$
So
$$\iint_D f(r\cos\theta, r\sin\theta) r dr d\theta = \int_\alpha^\beta d\theta \int_0^{r(\theta)} f(r\cos\theta, r\sin\theta) r dr.$$

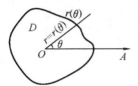
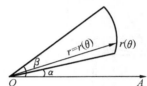

Figure 12-15   Figure 12-16

(3) If the polar point is outside the region D (Figure 12-17), we have
$$r_1(\theta) \leqslant r \leqslant r_2(\theta),\ \alpha \leqslant \theta \leqslant \beta.$$

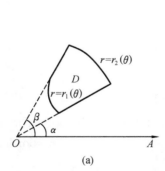
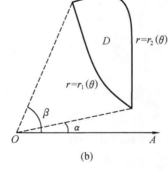

(a)   (b)

Figure 12-17

So
$$\iint_D f(r\cos\theta, r\sin\theta) r dr d\theta = \int_\alpha^\beta d\theta \int_{r_1(\theta)}^{r_2(\theta)} f(r\cos\theta, r\sin\theta) r dr. \tag{4}$$

Example 7 Find the double integral in the polar coordinate system,
$$\iint_D (1 - x^2 - y^2) dx dy,$$

where D is the region $x^2+y^2\leqslant 1$.

Solution In the polar coordinate, the region can be expressed by
$$0\leqslant r\leqslant 1, 0\leqslant \theta\leqslant 2\pi,$$
and we also have
$$x=r(\theta)\cos\theta,\ y=r(\theta)\sin\theta,\ dxdy=rdrd\theta.$$
So
$$\iint_D (1-x^2-y^2)dxdy = \iint_D (1-r^2)rdrd\theta = \int_0^{2\pi} d\theta \int_0^1 (1-r^2)rdr$$
$$=2\pi\cdot\left[\frac{1}{2}r^2-\frac{1}{4}r^4\right]_0^1 = \frac{\pi}{2}.$$

Example 8 Find $\iint_D (x+2y)\sqrt{x^2+y^2}dxdy$, where D is the region surrounded by the circle $x^2+y^2-2ax=0(a>0)$.

Solution In the polar coordinate system, the circle is $r^2-2ar\cos\theta=0$ or $r=2a\cos\theta$ (Figure 12-18). The region D can be expressed by
$$\begin{cases} -\dfrac{\pi}{2}<\theta<\dfrac{\pi}{2}, \\ 0\leqslant r\leqslant 2a\cos\theta. \end{cases}$$

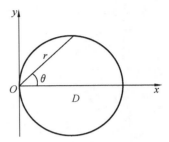

Figure 12-18

Therefore,
$$\iint_D (x+2y)\sqrt{x^2+y^2}dxdy = \iint_D (r\cos\theta+2r\sin\theta)r^2 drd\theta$$
$$=\int_{-\frac{\pi}{2}}^{\frac{\pi}{2}} d\theta \int_0^{2a\cos\theta} (\cos\theta+2\sin\theta)r^3 dr$$
$$=4a^4 \int_{-\frac{\pi}{2}}^{\frac{\pi}{2}} (\cos\theta+2\sin\theta)\cos^4\theta d\theta$$
$$=4a^4\left[2\int_0^{\frac{\pi}{2}}\cos^5\theta d\theta+2\int_{-\frac{\pi}{2}}^{\frac{\pi}{2}}\sin\theta\cos^4\theta d\theta\right]$$
$$=8a^4\left(\frac{4\times 2}{5\times 3}+0\right)=\frac{64}{15}a^4.$$

Example 9 Find the double integral
$$\iint_D e^{-x^2-y^2}dxdy,$$
where $D=\{(x,y)\,|\,x^2+y^2\leqslant a^2\}\ (a>0)$.

Solution Since D is a circular region, we will find the integral in the polar coordinate system. The region is $D=\{(r,\theta)\,|\,0\leqslant r\leqslant a, 0\leqslant\theta\leqslant 2\pi\}$, so we have
$$\iint_D e^{-x^2-y^2}dxdy = \iint_D e^{-r^2}rdrd\theta = \int_0^{2\pi} d\theta \int_0^a e^{-r^2}rdr$$
$$=2\pi\cdot\left[-\frac{1}{2}e^{-r^2}\right]_0^a = \pi(1-e^{-a^2}).$$

Note In the Example 9, rectangular coordinate system doesn't work because we cannot find the antiderivative of e^{-x^2}.

Example 10 Find the value of the generalized integral
$$\int_0^{+\infty} e^{-x^2} dx.$$

Solution Since we cannot find the antiderivative of e^{-x^2}, it is difficult to find the integral directly. By the result of the Example 9 and the properties of double integrals, we let

$D_1 = \{(x,y) \mid x^2 + y^2 \leqslant R^2, x \geqslant 0, y \geqslant 0\}$ $(R > 0)$,
$D = \{(x,y) \mid 0 \leqslant x \leqslant R, 0 \leqslant y \leqslant R\}$,
$D_2 = \{(x,y) \mid x^2 + y^2 \leqslant 2R^2, x \geqslant 0, y \geqslant 0\}$.

As shown in Figure 12-19, we have $D_1 \subset D \subset D_2$. And since $e^{-x^2-y^2} > 0$, we get

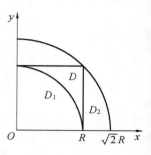

Figure 12-19

$$\iint_{D_1} e^{-x^2-y^2} dxdy \leqslant \iint_D e^{-x^2-y^2} dxdy \leqslant \iint_{D_2} e^{-x^2-y^2} dxdy,$$

that is,

$$\frac{\pi}{4}(1 - e^{-R^2}) \leqslant \int_0^R e^{-x^2} dx \int_0^R e^{-y^2} dy \leqslant \frac{\pi}{4}(1 - e^{-2R^2}).$$

Let $R \to +\infty$, since the limit of the left term and the right term are both $\frac{\pi}{4}$, by the squeeze theorem, we have

$$\int_0^\infty e^{-x^2} dx \int_0^\infty e^{-y^2} dy = \left(\int_0^\infty e^{-x^2} dx\right)^2 = \frac{\pi}{4}.$$

Therefore,

$$\int_0^{+\infty} e^{-x^2} dx = \frac{\sqrt{\pi}}{2}.$$

The result above is very useful for the normal distribution. From the example above, we also know that it is more convenient to find the double integral in the polar coordinate system if the integrand is a function of $x^2 + y^2$ and the region is surrounded by a circle.

12.2.3 Application in the economic management

The double integral is defined by

$$\iint_D f(x,y) d\sigma = \lim_{d \to 0} \sum_{i=1}^n f(x_i, y_i) \Delta\sigma_i.$$

That is to say, the double integral is actually the limit of a sum. And we get the limit by the following steps.

(1) Divide the region D into n small regions.

(2) Multiply the area of each small region and the value of the function at some point in this region.

(3) Add each product we get in step 2.

(4) Find the limit of the sum we get in step 3 as the diameter of each small region approaches 0.

We know that if the integrand is greater than 0, the double integral equals the volume of corresponding curved roof cylinder. It means that we can use the double integral to solve the economic question if there is a binary function of the known economic variables, that is, $f(x,y)$ is known, and the unknown value is additive.

Take the demographic model as an example, suppose the function of population density is given, (x,y) is a point in the rectangular coordinate system whose origin is viewed as a certain city. With the fact that the population is the product of the area and the density, we can find the population in region D by double integrals. Now, let's solve it.

Example 11 In 2008, the population density is
$$P(x,y)=\frac{20}{\sqrt{x^2+y^2+96}},$$
where (x,y) is the coordinates of some point. Find the population on the region within two kilometers from the city center. (The unit of the population is ten thousand, while the unit of the distance is kilometer.)

Solution The population is
$$\iint_D P(x,y)\,d\sigma = \iint_{x^2+y^2 \leqslant 4} \frac{20}{\sqrt{x^2+y^2+96}}\,dxdy$$
$$= 20 \times \int_0^{2\pi}\left[\int_0^2 \frac{1}{\sqrt{r^2+96}}r\,dr\right]d\theta$$
$$= 20 \times \int_0^{2\pi}\left[\int_0^2 \frac{1}{2\sqrt{r^2+96}}d(r^2+96)\right]d\theta$$
$$= 20 \times \int_0^{2\pi}\left(\sqrt{r^2+96}\,\Big|_0^2\right)d\theta$$
$$= 20 \times \int_0^{2\pi}(10-4\sqrt{6})d\theta$$
$$= 20 \times 2\pi \times (10-4\sqrt{6}) \approx 25.4.$$

Therefore, the population of this area amounts to 254 thousand.

Note It is always difficult to get the population density at some point (x,y), so we usually find the density $p(r)$, where r is the distance between the point (x,y) and the center of the city.

This is a function of one variable, how can we figure out the population according to this function? Is it valid that we regard $p(r)$ as the integrand and r as the integral variable?

This is not a question of definite integration of a simple function, since the region is an area instead of an interval. Thus, we have to consider multiplying the area and the density.

Example 12 Suppose the population density is
$$p(r)=12e^{-0.2r},$$
which is the population density r kilometers away from the city center. Find the population on the region within two kilometers from the city center. (The unit is the same as the Example 11.)

Solution
$$\iint_D 12e^{-0.2\sqrt{x^2+y^2}} dxdy = 12 \times \int_0^{2\pi}\left[\int_0^2 e^{-0.2r} r dr\right]d\theta = 12 \times \int_0^{2\pi}\left[\int_0^2 -5r de^{-0.2r}\right]d\theta$$

$$= 12 \times \int_0^{2\pi}\left[-5re^{-0.2r}\Big|_0^2 - \int_0^2 (-5)e^{-0.2r} dr\right]d\theta$$

$$= 12 \times \int_0^{2\pi}\left[-10e^{-0.4} + 5(-5e^{-0.2r})\Big|_0^2\right]d\theta$$

$$= 12 \times \int_0^{2\pi}(-35e^{-0.4} + 25)d\theta$$

$$= 12 \times 2\pi \times (-35e^{-0.4} + 25)$$

$$= -840\pi e^{-0.4} + 600\pi \approx 116.02.$$

Thus, we've roughly got a population of 1 160 200.

12.2.4 Variable substitution

We know how to substitute the double integral in the rectangular coordinate system for that in the polar coordinate system just by the formula $\begin{cases} x=r\cos\theta, \\ y=r\sin\theta, \end{cases}$ and the equation is

$$\iint_D f(x,y) dxdy = \iint_D f(r\cos\theta, r\sin\theta) r dr d\theta.$$

Next, we will give another explanation of the above equation. The set $\begin{cases} x=r\cos\theta, \\ y=r\sin\theta \end{cases}$ can be deemed as a substitution between the plane $rO\theta$ and the plane xOy, that is,

$$M'(r,\theta) \xrightleftharpoons[]{\begin{cases} x=r\cos\theta \\ y=r\sin\theta \end{cases}} M(x,y).$$

The substitution is to establish a corresponding relationship between two points in the two planes respectively and then use it to help us solve the problems.

Now we give the general formula of the substitution.

Theorem Suppose $f(x,y)$ is a continuous function defined on a bounded area D, let $x=x(u,v), y=y(u,v)$, then substitute the region D in the plane xOy for the bounded region D' in the plane uOv. Suppose $x=x(u,v), y=y(u,v)$ have continuous partial derivatives in the region D', and the Jacobi determinant

$$J = \frac{\partial(x,y)}{\partial(u,v)} = \begin{vmatrix} \dfrac{\partial x}{\partial u} & \dfrac{\partial x}{\partial v} \\ \dfrac{\partial y}{\partial u} & \dfrac{\partial y}{\partial v} \end{vmatrix} \neq 0.$$

Then, we have

$$\iint_D f(x,y) dxdy = \iint_{D'} f[x(u,v), y(u,v)] |J| dudv. \tag{5}$$

The use of substitution should conform to the following three principles.

(1) The function has continuous first-order partial derivative, and $J \neq 0$ (sometimes, J could be 0 at some specific points or some curves).

(2) For the new variables, it is easy to confirm the limit of integration.

(3) The integral of u, v is much easier to find than the previous one.

Example 13 Find the relationship of the double integral in the rectangular coordinate system and in the polar coordinate system by substitution theorem.

Solution Since $x = r\cos\theta, y = r\sin\theta$, and

$$J = \frac{\partial(x,y)}{\partial(r,\theta)} = \begin{vmatrix} \frac{\partial x}{\partial r} & \frac{\partial x}{\partial \theta} \\ \frac{\partial y}{\partial r} & \frac{\partial y}{\partial \theta} \end{vmatrix} = \begin{vmatrix} \cos\theta & -r\sin\theta \\ \sin\theta & r\cos\theta \end{vmatrix} = r.$$

If $r \neq 0$, we have

$$\iint_D f(x,y)\,dx\,dy = \iint_{D'} f(r\cos\theta, r\sin\theta)\, r\, dr\, d\theta.$$

Example 14 Find the area of the region D surrounded by the parabolas $y^2 = x$, $y^2 = 2x$ and the hyperbolas $xy = 2$, $xy = 3$ (Figure 12-20a).

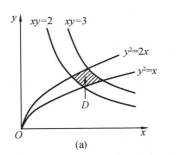
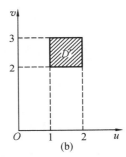

Figure 12-20

Solution By the substitution $\begin{cases} u = \frac{y^2}{x}, \\ v = xy, \end{cases}$ the region D in the plane xOy could be changed to the region D' in the plane uOv (Figure 12-20b).

The region D' is $1 \leqslant u \leqslant 2, 2 \leqslant v \leqslant 3$, and

$$J = \frac{\partial(x,y)}{\partial(u,v)} = \frac{1}{\frac{\partial(u,v)}{\partial(x,y)}} = \frac{1}{-\frac{3y^2}{x}} = -\frac{1}{3u} \neq 0.$$

So the area is

$$A = \iint_D 1\,dx\,dy = \iint_{D'} |J|\,du\,dv = \frac{1}{3}\int_2^3 dv \int_1^2 \frac{1}{u}du = \frac{1}{3}\ln 2.$$

Example 15 Find

$$\iint_D \sqrt{1 - \left(\frac{x^2}{a^2} + \frac{y^2}{b^2}\right)}\,dx\,dy,$$

where the region D is surrounded by $\frac{x^2}{a^2} + \frac{y^2}{b^2} = 1 (a > 0, b > 0)$.

Solution Let $\begin{cases} x = ra\cos\theta, \\ y = rb\sin\theta, \end{cases}$ where $r \geqslant 0, a > 0, b > 0, 0 \leqslant \theta \leqslant 2\pi$.

By this substitution, we have $D: \frac{x^2}{a^2}+\frac{y^2}{b^2}\leqslant 1 \Leftrightarrow D': 0\leqslant r\leqslant 1, 0\leqslant \theta\leqslant 2\pi$, and

$$J(r,\theta)=\begin{vmatrix} \frac{\partial x}{\partial r} & \frac{\partial y}{\partial \theta} \\ \frac{\partial y}{\partial r} & \frac{\partial y}{\partial \theta} \end{vmatrix}=abr.$$

Since $J(r,\theta)=0$ if $r=0$, we have

$$\iint_D \sqrt{1-\left(\frac{x^2}{a^2}+\frac{y^2}{b^2}\right)}\,dxdy = \iint_{D'} \sqrt{1-r^2}\,abr\,drd\theta$$
$$= \int_0^{2\pi} d\theta \int_0^1 \sqrt{1-r^2}\,r\,dr = \frac{2}{3}\pi ab.$$

Exercises 12-2

1. Find the following double integrals:

(1) $I = \iint_D xy\,dxdy$, where D is $0\leqslant x\leqslant 1, 0\leqslant y\leqslant \pi$.

(2) $I = \iint_D e^{x+y}\,d\sigma$, where D is the region surrounded by $0\leqslant x\leqslant 1, 0\leqslant y\leqslant 1$.

(3) $\iint_D \cos(x+y)\,d\sigma$, where D is the region surrounded by $x=0, y=\pi, y=x$.

(4) $\iint_D \frac{x}{y+1}\,dxdy$, where D is the region surrounded by $y=x^2+1, y=2x$ and $x=0$.

2. Draw the integral region, and evaluate the following double integrals:

(1) $\iint_D xy(x-y)\,d\sigma$, where D is the closed region surrounded by $0\leqslant x\leqslant a, 0\leqslant y\leqslant b$.

(2) $\iint_D \frac{x^2}{y^3}\,d\sigma$, where D is the region surrounded by $x=2, y=x, xy=1$.

(3) $\iint_D xy^2\,dxdy$, where D is the region surrounded by $y^2=2px$ and $x=\frac{p}{2}(p>0)$.

(4) $\iint_D xy^2\,d\sigma$, where D is the right region surrounded by the circle $x^2+y^2=4$ and y-axis.

3. Write the double integral $\iint_D f(x,y)\,dxdy$ into the iterated integral (two orders), where the region D is as follows:

(1) D is the region surrounded by $x+y=1, x-y=1, x=0$.
(2) D is the region surrounded by $y=x^2, y=4-x^2$.
(3) D is the region surrounded by $y=x, y=3x, x=1, x=3$.
(4) D is the region surrounded by $\frac{x^2}{a^2}+\frac{y^2}{b^2}=1$.

4. Prove $\int_a^b dx \int_a^x f(y)\,dy = \int_a^b f(x)(b-x)\,dx$.

5. If $f(x)$ is a continuous function and greater than zero on $[a,b]$, show that
$$\int_a^b f(x)\,dx \int_a^b \frac{1}{f(x)}\,dx \geqslant (b-a)^2.$$

6. Draw the graph of the integral region correspond to the following iterated integral, and exchange the order of the iterated integral:

(1) $\int_0^1 dy \int_y^{\sqrt{y}} f(x,y)\,dx$;

(2) $\int_{-a}^0 dx \int_{-x}^a f(x,y)\,dy + \int_0^{\sqrt{a}} dx \int_{x^2}^a f(x,y)\,dy \ (a>0)$;

(3) $\int_0^1 dx \int_{\sqrt{2+x^2}}^{\sqrt{4-x^2}} f(x,y)\,dy$;

(4) $\int_1^e dx \int_0^{\ln x} f(x,y)\,dy$;

(5) $\int_{-1}^1 dx \int_{x^2+x}^{x+1} f(x,y)\,dy$;

(6) $\int_0^1 dx \int_0^x f(x,y)\,dy + \int_1^2 dx \int_0^{2-x} f(x,y)\,dy$.

7. Find the plane slice surrounded by the curve $y=x^2$, $y=x+2$, and the area density at each point $u=1+x^2$, find the weight of this slice.

8. Find the volume of the solid surrounded by the curved plane $z=x^2+2y^2$ and $z=6-2x^2-y^2$.

9. Find the volume of the solid surrounded by the rotating paraboloid $z=x^2+y^2$ and the plane $z=a^2\ (a>0)$.

10. Write the following iterated integral into the iterated integral in polar coordinate system:

(1) $\int_0^R dx \int_0^{\sqrt{R^2-x^2}} f(\sqrt{x^2+y^2})\,dy$;

(2) $\int_0^{2R} dy \int_0^{\sqrt{2Ry-y^2}} f(x,y)\,dx$;

(3) $\int_0^2 dx \int_x^{\sqrt{3}x} f(\sqrt{x^2+y^2})\,dy$;

(4) $\int_0^1 dx \int_0^{1-x} f(x,y)\,dy$;

(5) $\int_0^1 dx \int_0^{x^2} f(x,y)\,dy$;

(6) $\int_0^1 dx \int_{1-x}^{\sqrt{1-x^2}} f(x,y)\,dy$.

11. Find the following double integrals by the polar coordinates:

(1) $\iint_D e^{x^2+y^2}\,d\sigma$, where D is $x^2+y^2 \leqslant R^2$.

(2) $\iint_D \sin\sqrt{x^2+y^2}\,d\sigma$, where D is $\pi^2 \leqslant x^2+y^2 \leqslant 4\pi^2$.

(3) $\iint_D \arctan\frac{y}{x}\,d\sigma$, where D is $x^2+y^2 \leqslant a^2$.

(4) $\iint_D \ln(1+x^2+y^2)\,d\sigma$, where D is the closed region in the first quadrant surrounded by the circle $x^2+y^2=1$ and the axis.

(5) $\iint_D \sqrt{x^2+y^2}\,dxdy$, where D is the closed region surrounded by $y=x$ and $y=x^2$.

(6) $\iint_D (x^2+y^2)\,d\sigma$, where D is the region surrounded by the upper half of the circle $(x-a)^2+y^2=a^2$.

12. Choose appropriate coordinate to find the following integrals:

(1) $\iint_D \sqrt{\dfrac{1-x^2-y^2}{1+x^2+y^2}}\,dxdy$, where D is the region in the first quadrant surrounded by $x^2+y^2=1$ and $x=0, y=0$.

(2) $\iint_D \arctan\dfrac{y}{x}\,dxdy$, where D is the region surrounded by $x^2+y^2\geq 1, x^2+y^2\leq 9, y\geq \dfrac{x}{\sqrt{3}}$ and $y\leq \sqrt{3}x$.

(3) $\iint_D (|x|+|y|)\,dxdy$, where D is $x^2+y^2\leq 1$.

13. Suppose the closed region D of a plane slice is surrounded by an arc on the spiral $l=2\theta$ and the straight line $\theta=\dfrac{\pi}{2}$ $\left(0\leq \theta\leq \dfrac{\pi}{2}\right)$, its plane density is $\rho(x,y)=x^2+y^2$, find the weight of this slice.

14. Find the volume of the solid surrounded by the cone $z=\sqrt{x^2+y^2}$, the cylinder $x^2+y^2=1$, and $z=0$.

*15. Exchange appropriately to evaluate the following double integrals:

(1) $\iint_D x^2 y^2\,dxdy$, where D is the closed region in the first quadrant surrounded by two hyperbolas $xy=1$ and $xy=2$, and the straight lines $y=x$ and $y=4x$.

(2) $\iint_D e^{\frac{y}{x+y}}\,dxdy$, where D is the closed region surrounded by the x-axis, y-axis and the straight line $x+y=1$.

(3) $\iint_D \left(\dfrac{x^2}{a^2}+\dfrac{y^2}{b^2}\right)dxdy$, where $D=\left\{(x,y)\Big|\dfrac{x^2}{a^2}+\dfrac{y^2}{b^2}\leq 1\right\}$.

(4) $\iint_D \cos\dfrac{x-y}{x+y}\,dxdy$, where D is surrounded by $x+y=1, x=0, y=0$.

*16. Choose appropriate exchange to prove the following equations:

(1) $\iint_D f(x+y)\,dxdy = \int_{-1}^{1} f(u)\,du$, where $D=\{(x,y)\,|\,|x|+|y|\leq 1\}$.

(2) $\iint_D f(ax+by+c)\,dxdy = 2\int_{-1}^{1}\sqrt{1-u^2}f(u\sqrt{a^2+b^2}+c)\,du$, $D=\{(x,y)\,|\,x^2+y^2\leq 1\}$, and $a^2+b^2\neq 0$.

12.3 Triple integral and its calculation

The concept of double integral can be directly extended to the triple integral.

12.3.1 The definition and property of triple integral

Definition Suppose that $f(x,y,z)$ is a bounded function defined in a bounded closed space region Ω, then divide Ω into numerous small regions Δv_i arbitrarily, and Δv_i also shows its volume ($i=1,2,\cdots,n$). If we denote the diameter of Δv_i ($i=1,2,\cdots,n$), the maximum distance between any two points, by λ_i, and $\lambda = \max\{\lambda_1, \lambda_2, \cdots, \lambda_n\}$, then take a point ($\xi_i, \eta_i, \zeta_i$) arbitrarily in each small region Δv_i. Then make multiplication $f(\xi_i, \eta_i, \zeta_i)\Delta v_i$ ($i=1,2,\cdots,n$) and add these products up, that is, $\sum_{i=1}^{n} f(\xi_i, \eta_i, \zeta_i)\Delta v_i$. When $\lambda \to 0$, if the limit of the above sum exists and independent of the segmentation method of Ω as well as the method of taking $f(\xi_i, \eta_i, \zeta_i)$, the limit of the sum is called triple integral of the function $f(x,y,z)$ on Ω, and denote it by $\iiint_\Omega f(x,y,z)\mathrm{d}v$, that is

$$\iiint_\Omega f(x,y,z)\mathrm{d}v = \lim_{\lambda \to 0} \sum_{i=1}^{n} f(\xi_i, \eta_i, \zeta_i)\Delta v_i, \tag{1}$$

where $f(x,y,z)$ is called integrand, $\mathrm{d}v$ is called volume element, $f(x,y,z)\mathrm{d}v$ is called integral expression and Ω is called the space region of integration.

Since the limit is independent of the segmentation method of Ω, we can use the planes parallel to the coordinate planes to segment Ω in space right angle coordinate system. In addition to some irregular small closed regions including the boundary points of Ω, the small closed region is cuboid. We assume that the length of sides of the i th small cuboid Δv_i is Δx_i, Δy_i, Δz_i, and then $\Delta v_i = \Delta x_i \Delta y_i \Delta z_i$. Thus, in space right angle coordinate system, when $\lambda \to 0$, all Δv_i can be considered as small cuboid, so the volume elements $\mathrm{d}v$ is written as $\mathrm{d}x\mathrm{d}y\mathrm{d}z$, consequently, the triple integral is written as $\iiint_\Omega f(x,y,z)\mathrm{d}x\mathrm{d}y\mathrm{d}z$, where $\mathrm{d}x\mathrm{d}y\mathrm{d}z$ is called the volume element in space right angle coordinate system.

If the function $f(x,y,z)$ is continuous in the closed space region Ω, the limit of the right side of Equation (1) exists. In other words, the triple integral of $f(x,y,z)$ in closed region Ω exists. Afterwards, we will always assume that the function $f(x,y,z)$ is continuous in closed dominant Ω. Some terms about the double integral can be also used in triple integral. The properties of triple integral and double integral are similar, so it is not going to repeat it here.

If $f(x,y,z)$ represents the density of an object at the point (x,y,z), Ω represents the space region occupied by the object, and $f(x,y,z)$ is continuous on Ω, then

$$\sum_{i=1}^{n} f(\xi_i, \eta_i, \zeta_i)\Delta v_i$$

is the approximate mass of the object, so as $\lambda \to 0$, the limit of the approximate mass is the exact mass M of the object, that is,

$$M = \iiint_\Omega f(x,y,z) dv.$$

12.3.2 Evaluate the triple integral by space right angle coordinate

The basic calculation method of the triple integral $\iiint_\Omega f(x,y,z) dv$ is to turn triple integral into simple integrals three times and then calculate it. In the space right angle coordinate system, we can turn triple integral into a double integral and a definite integral. Then, let's discuss the calculation methods of triple integral with different types of integral region.

If the projected region of the integral region Ω on the xOy plane is denoted by D_{xy}, Ω can be shown as

$$\Omega = \{(x,y,z) \mid z_1(x,y) \leqslant z \leqslant z_2(x,y), (x,y) \in D_{xy}\},$$

and $z_1(x,y)$ and $z_2(x,y)$ are both continuous functions in D_{xy}, then, Ω is called space region of XY. The characteristic of this kind of region is that there is no more than two intersection points of any line perpendicular to the xOy plane and the edge surface Σ of Ω (Figure 12-21).

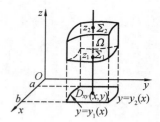

Figure 12-21

As shown in Figure 12-21, the projecting cylinder of Ω towards the xOy plane divides the edge surface Σ into two parts, the lower edge surface Σ_1 and the upper edge surface Σ_2. We assume that the equations are $\Sigma_1: z = z_1(x,y)$ and $\Sigma_2: z = z_2(x,y)$, so

$$z_1(x,y) \leqslant z_2(x,y).$$

Then we draw a straight line parallel to the z-axis passing through any point (x,y) in D_{xy}. This straight line passes through Σ_1 and Σ_2, and the intersection points are $z_1(x,y)$ and $z_2(x,y)$ respectively. Thus, we can make definite integral $\int_{z_1(x,y)}^{z_2(x,y)} f(x,y,z) dz$ (integral variable is z) at the point $(x,y) \in D_{xy}$. When the point (x,y) changes on $D(x,y)$, the definite integral is a binary function on D_{xy}, that is,

$$\varphi(x,y) = \int_{z_1(x,y)}^{z_2(x,y)} f(x,y,z) dz.$$

Then, we make double integral of $\varphi(x,y)$ on D_{xy}

$$\iint_{D_{xy}} \varphi(x,y) dx dy = \iint_{D_{xy}} \left[\int_{z_1(x,y)}^{z_2(x,y)} f(x,y,z) dz \right] dx dy.$$

Thus, we have

$$\iiint_\Omega f(x,y,z) dx dy dz = \iint_{D_{xy}} \left[\int_{z_1(x,y)}^{z_2(x,y)} f(x,y,z) dz \right] dx dy. \quad (2)$$

If the triple integral is written into the form (2), that is, a definite integral first and then a double integral, we will write the double integral into an iterated integral. For example, if D_{xy} can be shown as

$$D_{xy} = \{(x,y) \mid y_1(x) \leqslant y \leqslant y_2(x), a \leqslant x \leqslant b\},$$

the triple integral can be written into
$$\iiint_\Omega f(x,y,z)\mathrm{d}x\mathrm{d}y\mathrm{d}z = \int_a^b \mathrm{d}x \int_{y_1(x)}^{y_2(x)} \mathrm{d}y \int_{z_1(x,y)}^{z_2(x,y)} f(x,y,z)\mathrm{d}z.$$
That is, evaluating a triple integral is the same as integral for three times.

Similarly, there are two other types of the space region Ω, project Ω on the yOz plane or the zOx plane. For the two types, the triple integral is evaluated by the steps of "single integral" and "double integral". Because this method means that projecting the region Ω on the coordinate plane, and the region of double integral is the projection region of Ω, this method is always called Coordinate Surface Projection.

Example 1 Evaluate
$$\iiint_\Omega xyz\,\mathrm{d}x\mathrm{d}y\mathrm{d}z,$$
where Ω is the place region of first octant enclosed by curve surface $x^2+y^2+z^2=1$ and three coordinates (Figure 12-22).

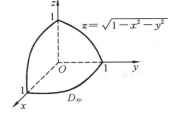

Figure 12-22

Solution For the region Ω, firstly, project the region Ω on the xOy plane, we get the range of z is $0 \leqslant z \leqslant \sqrt{1-x^2-y^2}$ and D_{xy}, which is the region enclosed by $x^2+y^2=1$ in the first quadrant. So D_{xy} can be shown as
$$D_{xy}: \begin{cases} 0 \leqslant x \leqslant 1, \\ 0 \leqslant y \leqslant \sqrt{1-x^2}. \end{cases}$$
And then, Ω can be shown as
$$\begin{cases} 0 \leqslant z \leqslant \sqrt{1-x^2-y^2}, \\ 0 \leqslant y \leqslant \sqrt{1-x^2}, \\ 0 \leqslant x \leqslant 1. \end{cases}$$
As a result,
$$\iiint_\Omega xyz\,\mathrm{d}x\mathrm{d}y\mathrm{d}z = \iint_D \mathrm{d}x\mathrm{d}y \int_0^{\sqrt{1-x^2-y^2}} xyz\,\mathrm{d}z = \int_0^1 \mathrm{d}x \int_0^{\sqrt{1-x^2}} \mathrm{d}y \int_0^{\sqrt{1-x^2-y^2}} xyz\,\mathrm{d}z$$
$$= \frac{1}{2} \int_0^1 \mathrm{d}x \int_0^{\sqrt{1-x^2}} xy(1-x^2-y^2)\mathrm{d}y = \frac{1}{8} \int_0^1 x(1-x^2)^2 \mathrm{d}x = \frac{1}{48}.$$

Sometimes, when we need to evaluate a triple integral, we can firstly find a double integral, and then find a definite integral.

If we project the region Ω to the z-axis, suppose that we get the projection interval $[c_1,c_2]$, and then, Ω can be shown as $\Omega = \{(x,y,z) \mid (x,y) \in D_z, c_1 \leqslant z \leqslant c_2\}$, where D_z is the cross section region of Ω and the plane vertical to the z-axis and pass through the point $(0,0,z)$ (Figure 12-23). Thus, we get
$$\iiint_\Omega f(x,y,z)\mathrm{d}x\mathrm{d}y\mathrm{d}z = \int_{c_1}^{c_2} \mathrm{d}z \iint_{D_z} f(x,y,z)\mathrm{d}x\mathrm{d}y, \qquad (3)$$
and Ω is called the space region of type Z.

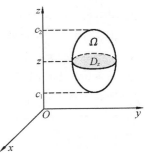

Figure 12-23

Note We use the form (3) to evaluate a triple integral if the double integral $\iint\limits_{D_z} f(x,y,z)\mathrm{d}x\mathrm{d}y$ can be found easily.

Similarly, there are two other types of the space region. If Ω is the X-type or the Y-type, the triple integral can be found by the steps of "double integral" first and then "single integral". Because this method is to project the region Ω to the axis and the double integral is on the cross section region of Ω, this method is called coordinate axis projection or method of section.

Example 2 Evaluate
$$\iiint\limits_{\Omega} z^2 \mathrm{d}x\mathrm{d}y\mathrm{d}z,$$
where Ω is the ellipsoid $\dfrac{x^2}{a^2}+\dfrac{y^2}{b^2}+\dfrac{z^2}{c^2}\leqslant 1$.

Solution By the Equation (2), $\Omega=\{(x,y,z)\mid (x,y)\in D_z, -c\leqslant z\leqslant c\}$, and in this question, we get $D_z=\left\{(x,y)\,\Big|\,\dfrac{x^2}{a^2}+\dfrac{y^2}{b^2}\leqslant 1-\dfrac{z^2}{c^2}\right\}\ (-c\leqslant z\leqslant c)$, so
$$\iiint\limits_{\Omega} z^2 \mathrm{d}x\mathrm{d}y\mathrm{d}z = \int_{-c}^{c}\mathrm{d}z\iint\limits_{D_z} z^2 \mathrm{d}x\mathrm{d}y = \int_{-c}^{c} z^2 \sigma(D_z)\mathrm{d}z,$$
where $\sigma(D_z)$ is the area of D_z. Since the area of the ellipsoid is
$$\sigma(D_z)=\pi\left(a\sqrt{1-\dfrac{z^2}{c^2}}\right)\left(b\sqrt{1-\dfrac{z^2}{c^2}}\right)=\pi ab\left(1-\dfrac{z^2}{c^2}\right),$$
we get
$$\iiint\limits_{\Omega} z^2 \mathrm{d}x\mathrm{d}y\mathrm{d}z = \int_{-c}^{c}\pi ab\left(1-\dfrac{z^2}{c^2}\right)z^2\mathrm{d}z = \dfrac{4}{15}\pi abc^3.$$

If we change the integrand z^2 to the constant "1", we can get the volume of the ellipsoid with semi-axis a, b and c,
$$V=\iiint\limits_{\Omega}\mathrm{d}x\mathrm{d}y\mathrm{d}z = \int_{-c}^{c}\mathrm{d}z\iint\limits_{D_z}\mathrm{d}x\mathrm{d}y = \int_{-c}^{c}\pi ab\left(1-\dfrac{z^2}{c^2}\right)\mathrm{d}z = \dfrac{4}{3}\pi abc.$$

In general, $\iiint\limits_{\Omega}\mathrm{d}x\mathrm{d}y\mathrm{d}z$ is the volume of the solid enclosed by Ω.

The calculations of triple integral and double integral are actually same. For some triple integral, it is difficult to find it by the space right angle coordinates. However, cylindrical coordinate and spherical coordinate is often applied in calculating triple integral.

12.3.3 Calculate the triple integral by cylindrical coordinate

1) Cylindrical coordinate

Assume that $M(x,y,z)$ is a point and its projection point on the xOy plane is $P(x,y,0)$. In a space right angle coordinate system, we regard x-axis as polar axis and construct a polar coordinate system on the xOy plane, and then regard z-axis as vertical axis. It is called cylindrical coordinate system (Figure 12-24) and these three order numbers, r, θ, z, are the cylindrical coordinates of point M, that is, $M(r,\theta,z)$. And r is called polar radius,

θ is polar angle, z is vertical scale and the variable range of r, θ, z is

$$\begin{cases} 0 \leqslant r < +\infty, \\ 0 \leqslant \theta \leqslant 2\pi, \\ -\infty < z < +\infty. \end{cases}$$

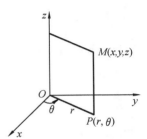

Figure 12-24

Three coordinate planes are $r = r_0$, $\theta = \theta_0$ and $z = z_0$, where $r = r_0$ is a cylindrical surface with the z-axis as center axis, $\theta = \theta_0$ is a half-plane with the z-axis as a side, $z = z_0$ is the plane parallel to the xOy plane.

Obviously, the relationship of the space right angle coordinates and the cylindrical coordinates of point M is

$$\begin{cases} x = r\cos \theta, \\ y = r\sin \theta, \\ z = z. \end{cases} \qquad (4)$$

2) Computing method

Now we discuss how to write $\iiint_\Omega f(x, y, z) \mathrm{d}x\mathrm{d}y\mathrm{d}z$ into a triple integral in the cylindrical coordinate system. Firstly, we can write $f(x, y, z)$ into $f(r\cos \theta, r\sin \theta, z)$ by the Equation (4). And then, the most important is how to represent the integral region in the cylindrical coordinate system. We can use the three coordinate planes to divide Ω into many small closed regions. Except for the small regions including boundary planes, most of

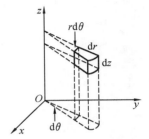

Figure 12-25

the small regions are the fan-shaped cylinder as shown in Figure 12-25 and the volume $\mathrm{d}v$ is the product of the base area and the height $\mathrm{d}z$, that is, $\mathrm{d}v = r\mathrm{d}r\mathrm{d}\theta\mathrm{d}z$. This is the expression of the volume element in the cylindrical coordinate system.

Thus we get the formula for transforming a triple integral from the form of space right angle coordinates to the cylindrical coordinates,

$$\iiint_\Omega f(x, y, z) \mathrm{d}x\mathrm{d}y\mathrm{d}z = \iiint_\Omega f(r\cos \theta, r\sin \theta, z) r\mathrm{d}r\mathrm{d}\theta\mathrm{d}z.$$

As for the calculation method of triple integral on the cylindrical coordinates, we still need to write the triple integral into an iterated integral. And the limit of each integral can be determined by the variation range of r, θ, z in integral region Ω. Now look at the following example.

Example 3 Evaluate $\iiint_\Omega z\mathrm{d}x\mathrm{d}y\mathrm{d}z$ by the cylindrical coordinates and Ω is the space region enclosed by the spherical surface $x^2 + y^2 + z^2 = 4$.

Solution Because the projection of the spherical surface on the xOy plane can be shown as $D_{xy}: x^2 + y^2 \leqslant 4$ and its expression in polar coordinates is

$$D_{r\theta}: \begin{cases} 0 \leqslant \theta \leqslant 2\pi, \\ 0 \leqslant r \leqslant 2. \end{cases}$$

Then the expression of Ω is

$$\begin{cases} 0 \leqslant \theta \leqslant 2\pi, \\ 0 \leqslant r \leqslant 2, \\ 0 \leqslant z \leqslant \sqrt{4-r^2}. \end{cases}$$

Thus, we have

$$\iiint_\Omega z\,\mathrm{d}x\mathrm{d}y\mathrm{d}z = \iiint_\Omega zr\,\mathrm{d}r\mathrm{d}\theta\mathrm{d}z = \int_0^{2\pi}\mathrm{d}\theta \int_0^2 r\,\mathrm{d}r \int_0^{\sqrt{4-r^2}} z\,\mathrm{d}z$$

$$= 2\pi \int_0^2 r\left[\frac{1}{2}(4-r^2)\right]\mathrm{d}r = 4\pi.$$

Example 4 Find

$$\iiint_\Omega (x^2+y^2)\,\mathrm{d}v,$$

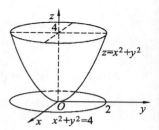

Figure 12-26

where Ω is the closed region enclosed by $\{(x,y,z) \mid x^2+y^2=z, z=4\}$ (Figure 12-26).

Solution Because the projection of Ω on the xOy plane is $D_{xy}: x^2+y^2 \leqslant 4$, in the cylindrical coordinate system, Ω can be shown as

$$\Omega: \begin{cases} 0 \leqslant \theta \leqslant 2\pi, \\ 0 \leqslant r \leqslant 2, \\ \dfrac{r^2}{2} \leqslant z \leqslant 2. \end{cases}$$

So we have,

$$\iiint_\Omega (x^2+y^2)\,\mathrm{d}v = \int_0^{2\pi}\mathrm{d}\theta \int_0^2 r^3\,\mathrm{d}r \int_{\frac{r^2}{2}}^2 \mathrm{d}z = 2\pi \int_0^2 r^3\left(2-\frac{r^2}{2}\right)\mathrm{d}r = \frac{16}{3}\pi.$$

12.3.4 Calculate the triple integral by spherical coordinate

Assume that $M(x,y,z)$ is a point, r is the distance between the point M and the origin point O, φ is the included angle between OM and the positive direction of z-axis, P is the projection of point M on the xOy plane, and θ is the included angle between the positive direction of x-axis and OP-ray (Figure 12-27). Thus, the point M can be determined by the three ordered numbers, r, φ, θ, that is, the point M can be expressed by the spherical coordinates $M(r, \varphi, \theta)$. By the definition above, the variation range of r, φ, θ is

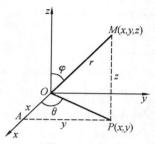

Figure 12-27

$$0 \leqslant r < +\infty, \; 0 \leqslant \varphi \leqslant \pi, \; 0 \leqslant \theta < 2\pi.$$

The three coordinate planes of the spherical coordinate system are $r=r_0, \varphi=\varphi_0$ and $\theta=\theta_0$, where $r=r_0$ is the spherical surface with the center on origin point O, $\varphi=\varphi_0$ is the circular conical surface with vertex at the origin point and the central shaft z-axis, $\theta=\theta_0$ is the half-plane passing through the z-axis.

If we assume that P is the projection of point M on the xOy plane and A is the projection of point P on the x-axis, $OA=x, AP=y, PM=z$, then we get $x=OP\cos\theta, y=OP\sin\theta, z=OM\cos\varphi$ from the Figure 12-27. Since $OM=r, OP=r\sin\varphi$, we get the relationship between the space right angle coordinate and the spherical coordinate,

$$\begin{cases} x=r\sin\varphi\cos\theta, \\ y=r\sin\varphi\sin\theta, \\ z=r\cos\varphi. \end{cases}$$

In order to get the volume element in spherical coordinates, we use three groups of coordinate planes to divide the space region Ω into many small closed regions. Except for the small closed regions containing the edge surface, most of these small closed regions are all hexahedral shape (Figure 12-28). Since the volume of hexahedron consisting of the tiny increments dr, $d\varphi$, $d\theta$, this hexahedron can be regarded as a cuboid with the length to the warp direction $rd\varphi$, the width to the weft direction $r\sin\varphi d\theta$ and the height to the radial line direction dr. Thus, the volume element in the spherical coordinates system is

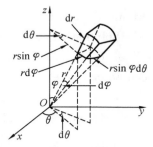

Figure 12-28

$$dv=r^2\sin\varphi dr d\varphi d\theta.$$

As a result, we have

$$\iiint_\Omega f(x,y,z)dxdydz = \iiint_\Omega f(r\sin\varphi\cos\theta, r\sin\varphi\sin\theta, r\cos\varphi)r^2\sin\varphi dr d\varphi d\theta.$$

By this formula, we can transform a triple integral from the space right angle coordinates to the spherical coordinates. And then, we write the triple integral into an iterated integral of r, φ and θ.

If the origin is in the region Ω and the edge surface of Ω can be expressed by the formula $r=(\varphi,\theta)$, that is, Ω is

$$\Omega: \begin{cases} 0 \leqslant \theta \leqslant 2\pi, \\ 0 \leqslant \varphi \leqslant \pi, \\ 0 \leqslant r \leqslant r(\varphi,\theta), \end{cases}$$

then the triple integral in spherical coordinate will be transformed to the iterated integral

$$\iiint_\Omega f(r\sin\varphi\cos\theta, r\sin\varphi\sin\theta, r\cos\varphi)r^2\sin\varphi dr d\varphi d\theta$$

$$= \int_0^{2\pi} d\theta \int_0^\pi \sin\varphi d\varphi \int_0^{r(\varphi,\theta)} f(r\sin\varphi\cos\theta, r\sin\varphi\sin\theta, r\cos\varphi)r^2 dr.$$

If the region Ω is enclosed by the spherical surface $r=a$, then

$$\iiint_\Omega f(r\sin\varphi\cos\theta, r\sin\varphi\sin\theta, r\cos\varphi)r^2\sin\varphi dr d\varphi d\theta$$

$$= \int_0^{2\pi} d\theta \int_0^\pi \sin\varphi d\varphi \int_0^a f(r\sin\varphi\cos\theta, r\sin\varphi\sin\theta, r\cos\varphi)r^2 dr.$$

Especially, if $f(x,y,z)=1$, the formula above is
$$\int_0^{2\pi} d\theta \int_0^{\pi} \sin\varphi d\varphi \int_0^a r^2 dr = 2\pi \cdot 2 \cdot \frac{a^3}{3} = \frac{4}{3}\pi a^3,$$
which is the volume of sphere $r=a$.

Example 5 Evaluate
$$\iiint_{\Omega} dxdydz,$$
where Ω is enclosed by the spherical surface $x^2+y^2+(z-R)^2=R^2$ and the circular conical surface $z=\sqrt{x^2+y^2}$ (Figure 12-29).

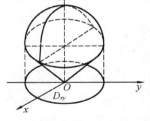

Figure 12-29

Solution In spherical coordinates, the spherical equation
$$x^2+y^2+z^2-2Rz=0$$
can be expressed by
$$r^2\sin^2\varphi\cos^2\theta+r^2\sin^2\varphi\sin^2\theta+r^2\cos^2\varphi-2Rr\cos\varphi=0,$$
or $r^2-2Rr\cos\varphi=0$. Thus the center of the sphere is $(0,0,R)$, the radius is $r=2R\cos\varphi$ and the circular conical surface with vertex at the origin is $\varphi=\frac{\pi}{4}$. So, the region Ω can be shown as
$$\Omega: \begin{cases} 0\leqslant\theta\leqslant 2\pi, \\ 0\leqslant\varphi\leqslant\frac{\pi}{4}, \\ 0\leqslant r\leqslant 2R\cos\varphi. \end{cases}$$
Then
$$\iiint_{\Omega} dxdydz = \iiint_{\Omega} r^2\sin\varphi dr d\varphi d\theta = \int_0^{2\pi} d\theta \int_0^{\frac{\pi}{4}} \sin\varphi d\varphi \int_0^{2R\cos\varphi} r^2 dr$$
$$= \int_0^{2\pi} d\theta \int_0^{\frac{\pi}{4}} \sin\varphi \cdot \frac{r^3}{3}\Big|_0^{2R\cos\varphi} d\varphi = \frac{16}{3}\pi R^3 \int_0^{\frac{\pi}{4}} \cos^3\varphi\sin\varphi d\varphi$$
$$= \frac{4}{3}\pi R^3\left(1-\cos^4\frac{\pi}{4}\right) = \pi R^3.$$

Exercises 12-3

1. Transform the triple integral $z = \iiint_{\Omega} f(x,y,z)dxdydz$ into the iterated integral and there are four different definitions of Ω.

   (1) The region Ω is enclosed by the curve surface $z=x^2+y^2$, $y=x^2$ and the plane $y=1$ and $z=0$.

   (2) The region Ω is enclosed by the hyperbolic paraboloid $xy=z$ and the plane $x+y-1=0$ and $z=0$.

   (3) The region Ω is enclosed by the spherical surface $x^2+y^2+z^2=1$ and in the first octant.

   (4) The region Ω is enclosed by the curve surfaces $z=x^2+2y^2$ and $z=2-x^2$.

2. Evaluate the following triple integrals by Cartesian coordinates.

(1) $\iiint_\Omega x\,dxdydz$, where Ω is enclosed by three coordinate planes and the plane $x+2y+z=1$.

(2) $\iiint_\Omega \dfrac{x\,dxdydz}{(1+x+y+z)^3}$, where Ω is the tetrahedron enclosed by the plane of $x=0$, $z=0$ and $x+y+z=1$.

(3) $\iiint_\Omega z\,dxdydz$, where Ω is the upper part of an spherical surface enclosed by edge surface of $x^2+y^2+z^2=1$ and $z=0$.

(4) $\iiint_\Omega xz\,dxdydz$, where Ω is enclosed by the planes $z=0$, $z=y$, $y=1$ and the paraboloid $y=x^2$.

(5) $\iiint_\Omega z^2\,dxdydz$, where Ω is the common part of the sphere $x^2+y^2+z^2 \leqslant R^2$ and $x^2+y^2+z^2 \leqslant 2Rz (R>0)$.

(6) $\iiint_\Omega x^3 y^2 \,dv$, where Ω is enclosed by the hyperbolic paraboloid $z=xy$, the coordinate planes $z=0, y=x$ and $x=a$.

3. If an object covers the space region Ω: $0\leqslant x\leqslant 1, 0\leqslant y\leqslant 1, 0\leqslant z\leqslant 1$ and the density of the point (x,y,z) is $\rho(x,y,z)=x+y+z$, find the weight of this object.

4. For the triple integral $\iiint_\Omega f(x,y,z)\,dxdydz$, if $f(x,y,z)$ is the product of the functions $f_1(x), f_2(y), f_3(z)$, that is, $f(x,y,z)=f_1(x)f_2(y)f_3(z)$, the region Ω is $a\leqslant x\leqslant b, c\leqslant y\leqslant d, l\leqslant z\leqslant m$, prove the following equality
$$\iiint_\Omega f_1(x)f_2(y)f_3(z)\,dxdydz = \int_a^b f_1(x)dx \int_c^d f_2(y)dy \int_l^m f_3(z)dz.$$

5. Evaluate the following triple integrals by the cylindrical coordinates.

(1) $\iiint_\Omega (x^2+y^2)\,dv$, where Ω is the closed region enclosed by the circular conical surface $x^2+y^2=z^2$ and the plane $z=h(h>0)$.

(2) $\iiint_\Omega z\,dv$, where Ω is the closed region enclosed by the curved surface $x^2+y^2+z^2=2$ and $x^2+y^2=z$.

6. Evaluate the following triple integrals by the spherical coordinates.

(1) $\iiint_\Omega z\sqrt{x^2+y^2}\,dv$, where Ω is enclosed by the cylindrical surface $x^2+y^2=2x(y\geqslant 0)$ and the plane $z=0, z=a, y=0$.

(2) $\iiint_\Omega z\,dv$, where Ω is enclosed by $x^2+y^2+(z-a)^2 \leqslant a^2$ and $x^2+y^2 \leqslant z^2$.

7. Evaluate the following triple integrals by the proper coordinates.

(1) $\iiint_\Omega (x+y+z)\mathrm{d}v$, where Ω is $x\geqslant 0, y\geqslant 0, z\geqslant 0, x^2+y^2+z^2\leqslant R^2$.

(2) $\iiint_\Omega (x^2+y^2)\mathrm{d}v$, where Ω is enclosed by the paraboloid $x^2+y^2=2z$ and the plane $z=1, z=2$.

(3) $\iiint_\Omega \sqrt{x^2+y^2+z^2}\,\mathrm{d}v$, where Ω is the space region enclosed by $x^2+y^2+(z-1)^2\leqslant 1$.

(4) $\iiint_\Omega (x^2+y^2)\mathrm{d}v$, where Ω is enclosed by the semi-spherical surface of $z=\sqrt{A^2-x^2-y^2}$, $z=\sqrt{a^2-x^2-y^2}\,(A>a>0)$ and the plane $z=0$.

8. Find the volume of the solid enclosed by the following curved surfaces by the triple integrals.

(1) The solid is enclosed by the curved surface $x^2+y^2=az$ and $z=2a-\sqrt{x^2+y^2}$ $(a>0)$.

(2) The solid is enclosed by the curved surface $z=\sqrt{x^2+y^2}$ and $z=x^2+y^2$.

(3) The solid is enclosed by the curved surface $z=\sqrt{5-x^2+y^2}$ and $x^2+y^2=4z$.

9. Find the volume of the two parts of the sphere of $x^2+y^2+z^2\leqslant 4z$ separated by the curved surface $z=4-x^2-y^2$.

12.4 Application of multiple integral

We have previously pointed out that the mass of the plane lamina and the volume of the curved roof cylinder can be calculated by the double integral, and now we will solve the geometrical problems and the physical problems by the multiple integral.

12.4.1 Applications in geometry

1) The volume of a solid enclosed by a closed surface

Example 1 Find the volume of the tetrahedron Ω, which is enclosed by three coordinate planes and the plane $x+y+z=1$.

Solution As shown in Figure 12-30, the projection of tetrahedron Ω on the xOy plane is

$$D_{xy}:\begin{cases} 0\leqslant x\leqslant 1,\\ 0\leqslant y\leqslant 1-x.\end{cases}$$

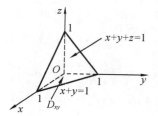

Figure 12-30

So Ω can be expressed by $0\leqslant x\leqslant 1, 0\leqslant y\leqslant 1-x$ and $0\leqslant z\leqslant 1-x-y$. Then we will get the volume of Ω,

$$V=\int_0^1 \mathrm{d}x \int_0^{1-x} \mathrm{d}y \int_0^{1-x-y} \mathrm{d}z = \int_0^1 \mathrm{d}x \int_0^{1-x}(1-x-y)\mathrm{d}y$$
$$=\int_0^1 \left[(1-x)y-\frac{y^2}{2}\right]_0^{1-x}\mathrm{d}x = \int_0^1 \left[(1-x)^2-\frac{1}{2}(1-x)^2\right]\mathrm{d}x$$
$$=\frac{1}{2}\int_0^1 (1-x)^2\mathrm{d}x = -\frac{1}{6}(1-x)^3\Big|_0^1 = \frac{1}{6}.$$

2) The area of hook face

Assume that the hook face Σ is determined by the equation of
$$z = z(x, y) \tag{1}$$
the projection of the hook face Σ on the xOy plane is D_{xy} and the partial derivative of $z(x,y)$, z_x and z_y, is continuous in D_{xy}. Find the area A of the hook face Σ.

For any closed region dσ, which is in the region D_{xy}, with a very small diameter (d is also the area of the closed region), and any point $P(x,y)$ in dσ, there is a corresponding point $M(x, y, z(x, y))$ on the hook face Σ. The projection of the point M on the xOy plane is the point P and the tangent plane of the hook face Σ at the point M is T (Figure 12-31a).

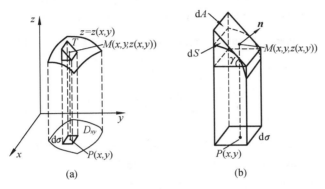

Figure 12-31

Draw a cylindrical surface with the generatrix parallel to the z-axis and the boundary line of dσ as directrix. This cylindrical surface cuts a small hook face ds in the hook face Σ. It also cuts a small plane dA in tangent plane T. Because the diameter of dσ is very small, we can nearly think the area of the small plane dA can replace the area of ds. Assume that the angle of the normal vector $\boldsymbol{n}$ at the point M of hook face Σ and the z-axis is γ, and the normal vector $\boldsymbol{n} = (-z_x(x,y), -z_y(x,y), 1)$. Thus
$$\cos\gamma = \frac{1}{\sqrt{1 + z_x(x,y) + z_y(x,y)}}. \tag{2}$$

From Figure 12-31, since dσ = d$A\cos\gamma$, we have
$$dA = \frac{d\sigma}{\cos\gamma}. \tag{3}$$

As a result,
$$dA = \sqrt{1 + z_x^2(x,y) + z_y^2(x,y)}\, d\sigma. \tag{4}$$

This is the area element ds of hook face Σ. The area of the hook face Σ is
$$A = \iint_{D_{xy}} \sqrt{1 + z_x^2(x,y) + z_y^2(x,y)}\, d\sigma \tag{5}$$

or
$$A = \iint_{D_{xy}} \sqrt{1 + \left(\frac{\partial z}{\partial x}\right)^2 + \left(\frac{\partial z}{\partial y}\right)^2}\, dx dy. \tag{5'}$$

Similarly, we can project hook face Σ on the yOz plane (or xOz plane) if we assume the equation of hook face to be $x=x(y,z)$ or $y=y(x,z)$, and denote the projection region by D_{yz} (or D_{xz}). Then we will get the formula

$$A = \iint_{D_{yz}} \sqrt{1+\left(\frac{\partial x}{\partial y}\right)^2 + \left(\frac{\partial x}{\partial z}\right)^2}\, dydz \tag{6}$$

or

$$A = \iint_{D_{xz}} \sqrt{1+\left(\frac{\partial y}{\partial x}\right)^2 + \left(\frac{\partial y}{\partial z}\right)^2}\, dxdz. \tag{7}$$

Example 2 Find the superficial area of a sphere with the radius R.

Solution Take the origin as the center of the sphere, so the equation of the sphere is $x^2+y^2+z^2=R^2$. We know the superficial area of the whole sphere is twice as much as the area of upper hemispherical surface. The equation of upper hemispherical surface is $z=\sqrt{R^2-x^2-y^2}$. The closed projection region of the sphere on the xOy plane is

$$D_{xy} = \{(x,y) \mid x^2+y^2 \leqslant R^2\}.$$

And

$$\sqrt{1+\left(\frac{\partial z}{\partial x}\right)^2 + \left(\frac{\partial z}{\partial y}\right)^2} = \frac{R}{\sqrt{R^2-x^2-y^2}}$$

is not continuous on the boundary $x^2+y^2=R^2$ of D_{xy}, so we cannot use the Formula (5) to find the superficial area directly. We take the closed region $D_1 = \{(x,y) \mid x^2+y^2 \leqslant \rho^2\}$ $(0<\rho<R)$ as integral region, then we find the area A_1 of upper hemispherical surface in D_1. We can make $\rho \to R$ and use the ultimate value of A_1. Finally, the area of upper hemispherical surface is the limit of A_1 as $\rho \to R$. In polar coordinate system, the expression of D_{xy} is $\begin{cases} 0 \leqslant \theta \leqslant 2\pi, \\ 0 \leqslant r \leqslant R, \end{cases}$ so

$$A_1 = \iint_{D_1} \frac{R}{\sqrt{R^2-x^2-y^2}} dxdy = R\int_0^{2\pi} d\theta \int_0^{\rho} \frac{rdr}{\sqrt{R^2-r^2}}$$

$$= 2\pi R \int_0^{\rho} \frac{r}{\sqrt{R^2-r^2}} dr = 2\pi R(R - \sqrt{R^2-\rho^2}).$$

Thus

$$A = 2\lim_{\rho \to R} A_1 = 2\lim_{\rho \to R} 2\pi R(R-\sqrt{R^2-\rho^2}) = 2 \cdot 2\pi R^2 = 4\pi R^2.$$

Example 3 Find the area A of the plane $z=1-x$ in the first octant cut by two cylinders $x=y^2$ and $2x=y^2$ (Figure 12-32a).

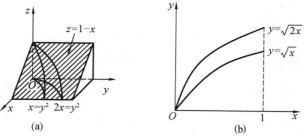

Figure 12-32

Solution The closed projection region on the plane xOy is
$$D_{xy} = \{(x,y) \mid \sqrt{x} \leqslant y \leqslant \sqrt{2x}, 0 \leqslant x \leqslant 1\} \text{ (Figure 12-32b)}.$$
Because of $\sqrt{1+\left(\dfrac{\partial z}{\partial x}\right)^2+\left(\dfrac{\partial z}{\partial y}\right)^2}=\sqrt{2}$, we have

$$\begin{aligned}
A &= \iint_{D_{xy}} \sqrt{1+\left(\frac{\partial z}{\partial x}\right)^2+\left(\frac{\partial z}{\partial y}\right)^2}\,dxdy = \iint_{D_{xy}} \sqrt{2}\,dxdy \\
&= \sqrt{2}\int_0^1 dx \int_{\sqrt{x}}^{\sqrt{2x}} dy = \sqrt{2}\int_0^1 (\sqrt{2}-1)\sqrt{x}\,dx \\
&= \frac{2(2-\sqrt{2})}{3}.
\end{aligned}$$

12.4.2 Applications in physics

We can solve the quality, gravity, moment of inertia and other issues of the plane objects by triple integrals. By the triple integral, we can also solve the same issues of space objects. The following highlights will be the solution of plane issues and the way to find the equality, gravity, moment of inertia on the space objects.

1) Quality of an object

The double integral can find the quality M of plane slice D with areal density $\rho=\rho(x,y)$, and the formula is

$$M = \iint_D \rho(x,y)\,d\sigma. \tag{8}$$

The triple integral can find the quality M of space solid Ω with volume density $\rho(x,y,z)$, and the formula is

$$M = \iiint_\Omega \rho(x,y,z)\,dv. \tag{9}$$

2) Center of gravity of an object

We will discuss the coordinates of the center of gravity of plane slice.

Assume that there are n particles on the xOy plane with the coordinates (x_1, y_1), (x_2, y_2), $\cdots$, (x_n, y_n) and the quality of them $m_1, m_2, \cdots, m_n$. From the Mechanical knowledge, we get the center of gravity of the system of particles,

$$\overline{x} = \frac{\sum_{i=1}^n m_i x_i}{\sum_{i=1}^n m_i} = \frac{m_y}{m}, \quad \overline{y} = \frac{\sum_{i=1}^n m_i y_i}{\sum_{i=1}^n m_i} = \frac{m_x}{m}. \tag{10}$$

In the formula above, $m_y = \sum_{i=1}^n m_i x_i$, $m_x = \sum_{i=1}^n m_i y_i$ are called the static moment of the system of particles on the y-axis and the x-axis. $M = \sum_{i=1}^n m_i$ is called the total mass of the system of particles.

Now, we can find the center of gravity of the plane slice. Assume that the plane slice on the xOy plane is enclosed by the bounded closed region D and the areal density at the point (x,y) is $\rho=\rho(x,y)$, then we can find the center of gravity of this plane slice if $\rho(x,y)$ is continuous on D.

For a closed region $d\sigma$ with a very small diameter on the closed region D ($d\sigma$ also is the area of the small region), (x,y) is a point in $d\sigma$. Since its diameter is very small and $\rho(x,y)$ is continuous on D, the quality of $d\sigma$ is nearly equal to $\rho(x,y)d\sigma$. It can be thought that the quality of $d\sigma$ could be nearly focus on the point (x,y). Then we get that
$$dm_y = x\rho(x,y)d\sigma, \ dm_x = y\rho(x,y)d\sigma.$$
Make these elements as integral expression, and then make integral in the closed region D, that is,
$$m_y = \iint_D x\rho(x,y)d\sigma, \ m_x = \iint_D y\rho(x,y)d\sigma. \tag{11}$$
Thus the coordinates of the particle of the plane lamina are
$$\bar{x} = \frac{m_y}{m} = \frac{\iint_D x\rho(x,y)d\sigma}{\iint_D \rho(x,y)d\sigma}, \ \bar{y} = \frac{m_x}{m} = \frac{\iint_D y\rho(x,y)d\sigma}{\iint_D \rho(x,y)d\sigma}. \tag{12}$$

If the slice is well-proportioned, the areal density is constant. Thus we can take the ρ of Equation (12) out of the sign of integration and then cancel it, as a result, we get the center of gravity of the well-proportioned slice,
$$\bar{x} = \frac{1}{A}\iint_D x d\sigma, \ \bar{y} = \frac{1}{A}\iint_D y d\sigma, \tag{13}$$
where $A = \iint_D d\sigma$ represents the area of the closed region D.

Example 4 Find the center of gravity of the well-proportioned slice located between the circles of
$x^2+(y-1)^2=1$ and $x^2+(y-2)^2=4$ (Figure 12-33).

Solution From the figure, we know that the region D is symmetrical to the y-axis, so the center of gravity must be on the y-axis, that is, $\bar{x}=0$.

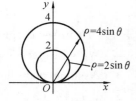

Figure 12-33

The area of the slice is $A=3\pi$.

Because the polar equations of $x^2+(y-1)^2=1$ and $x^2+(y-2)^2=4$ are respectively $r=2\sin\theta$ and $r=4\sin\theta$ where $0\leqslant\theta\leqslant\pi$, we have
$$\iint_D y d\sigma = \iint_D r^2 \sin\theta dr d\theta = \int_0^\pi \sin\theta d\theta \int_{2\sin\theta}^{4\sin\theta} r^2 dr$$
$$= \frac{56}{3}\int_0^\pi \sin^4\theta d\theta = 7\pi.$$

Then we get $\bar{y}=\dfrac{7\pi}{3\pi}=\dfrac{7}{3}$, so the center of gravity is $\left(0,\dfrac{7}{3}\right)$.

Similarly, we can find the center of gravity of a space object. Assume that there is an object in a closed region Ω and its body density is $\rho=\rho(x,y,z)$, we can get the coordinates of the center of gravity of the space object,

$$\overline{x} = \frac{\iiint\limits_{\Omega} x\rho(x,y,z)\mathrm{d}v}{\iiint\limits_{\Omega} \rho(x,y,z)\mathrm{d}v}, \quad \overline{y} = \frac{\iiint\limits_{\Omega} y\rho(x,y,z)\mathrm{d}v}{\iiint\limits_{\Omega} \rho(x,y,z)\mathrm{d}v}, \quad \overline{z} = \frac{\iiint\limits_{\Omega} z\rho(x,y,z)\mathrm{d}v}{\iiint\limits_{\Omega} \rho(x,y,z)\mathrm{d}v}, \quad (14)$$

where $\iiint\limits_{\Omega} \rho(x,y,z)\mathrm{d}v$ is the quality of the object.

If $\rho = \rho(x,y,z)$ is a constant, we can find the coordinates of the center of gravity directly from the Formula (14).

Example 5 Find the center of gravity of a well-proportioned object Ω enclosed by rotating paraboloid $z = x^2 + y^2$ and plane $z = 1$.

Solution Draw the graph of Ω firstly (Figure 12-34).

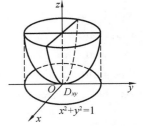

Figure 12-34

Since the object Ω is symmetrical to the yOz plane and the zOx plane, and the object is uniform, the center of gravity is on the z-axis, that is, $\overline{x} = \overline{y} = 0$.

Because ρ is a constant, we can get

$$\overline{z} = \frac{\iiint\limits_{\Omega} z\mathrm{d}v}{\iiint\limits_{\Omega} \mathrm{d}v}.$$

Project the edge surface of Ω on the xOy plane, we get a projection region D_{xy} surrounded by $x^2 + y^2 = 1$, so the expression of Ω in the cylindrical coordinates is

$$\begin{cases} 0 \leqslant \theta \leqslant 2\pi, \\ 0 \leqslant r \leqslant 1, \\ r^2 \leqslant z \leqslant 1. \end{cases}$$

Since

$$\iiint\limits_{\Omega} z\mathrm{d}v = \iint\limits_{D} r\mathrm{d}r\mathrm{d}\theta \int_{r^2}^{1} z\mathrm{d}z = \int_{0}^{2\pi} \mathrm{d}\theta \int_{0}^{1} \frac{1}{2}(r - r^5)\mathrm{d}r = \frac{\pi}{3},$$

$$\iiint\limits_{\Omega} \mathrm{d}v = \iint\limits_{D} r\mathrm{d}r\mathrm{d}\theta \int_{r^2}^{1} \mathrm{d}z = \int_{0}^{2\pi} \mathrm{d}\theta \int_{0}^{1} (r - r^3)\mathrm{d}r = \frac{\pi}{2},$$

by Formula (14), we have

$$\overline{z} = \frac{\iiint\limits_{\Omega} z\mathrm{d}v}{\iiint\limits_{\Omega} \mathrm{d}v} = \frac{\frac{\pi}{3}}{\frac{\pi}{2}} = \frac{2}{3}.$$

Therefore, the center of gravity is $\left(0, 0, \frac{2}{3}\right)$.

3) The rotational inertia of an object

By Mechanics, the rotational inertia of a particle to an axis is the product of the gravity of the particle and the square of the distance from the particle to the axis.

Assume that there are n particles on the xOy plane and their quality are $m_1, m_2, \cdots, m_n$, from Mechanics, the rotational inertia of the system of particles to the x-axis, the y-

axis and the origin point are

$$I_x = \sum_{i=1}^{n} y_i^2 m_i, \quad I_y = \sum_{i=1}^{n} x_i^2 m_i, \quad I_O = \sum_{i=1}^{n} (x_i^2 + y_i^2) m_i. \qquad (15)$$

Then we will find the rotational inertia of a plane slice.

Assume that the slice is enclosed by a closed region D on the xOy plane, the areal density of point (x,y) is $\rho(x,y)$ and $\rho(x,y)$ is continuous on D. For a closed region $d\sigma$ with a very small diameter in the closed region D ($d\sigma$ is also the area of this small region) and a point (x,y) in the region $d\sigma$, since the diameter of $d\sigma$ is very small and $\rho(x,y)$ is continuous on D, the quality of $d\sigma$ is nearly equal to $\rho(x,y)d\sigma$. Because $d\sigma$ is very small, the quality of $d\sigma$ can be thought that it nearly focuses on the point (x,y). Then, we can get the rotational inertia of $d\sigma$ to the x-axis, the y-axis and the origin O,

$$dI_x = y^2 \rho(x,y) d\sigma, \quad dI_y = x^2 \rho(x,y) d\sigma, \quad dI_O = (x^2 + y^2) \rho(x,y) d\sigma.$$

Make these elements above as integral expression, and then make integral in the closed region D, we get

$$I_x = \iint_D y^2 \rho(x,y) d\sigma, \quad I_y = \iint_D x^2 \rho(x,y) d\sigma, \quad I_O = \iint_D (x^2 + y^2) \rho(x,y) d\sigma. \qquad (16)$$

Example 6 Find the rotational inertia of a well-proportioned plane slice D enclosed by cardioid $r = a(1 + \cos \theta)$ ($a > 0$) to the x-axis and the y-axis (Figure 12-35).

Solution Because the slice D is well-proportioned, we could assume the density $\rho = \rho(x,y) = 1$, and the region enclosed by the cardioid is $D = \{(r,\theta) \mid 0 \leqslant r \leqslant a(1 + \cos \theta), -\pi \leqslant \theta \leqslant \pi\}$, then we get

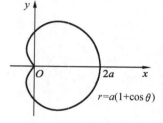

Figure 12-35

$$I_x = \iint_D y^2 d\sigma = \iint_D r^2 \sin^2 \theta \cdot r dr d\theta = \int_{-\pi}^{\pi} \sin^2 \theta d\theta \int_0^{a(1+\cos\theta)} r^3 dr$$

$$= \frac{a^4}{4} \int_{-\pi}^{\pi} (1 + \cos \theta)^4 \sin^2 \theta d\theta = \frac{a^4}{2} \int_0^{\pi} (1 + \cos \theta)^4 \sin^2 \theta d\theta$$

$$= 2^6 a^4 \int_0^{\pi} \cos^{10} \frac{\theta}{2} \sin^2 \frac{\theta}{2} d\left(\frac{\theta}{2}\right).$$

Let $t = \frac{\theta}{2}$, we have

$$I_x = 2^6 a^4 \int_0^{\frac{\pi}{2}} \cos^{10} t \sin^2 t dt = 2^6 a^4 \int_0^{\frac{\pi}{2}} \cos^{10} t (1 - \cos^2 t) dt$$

$$= 2^6 \cdot a^4 \cdot \frac{9 \cdot 7 \cdot 5 \cdot 3 \cdot 1}{10 \cdot 8 \cdot 6 \cdot 4 \cdot 2} \left(1 - \frac{11}{12}\right) \cdot \frac{\pi}{2} = \frac{21}{32} \pi a^4.$$

And $\quad I_y = \iint_D x^2 d\sigma = \iint_D r^2 \cos^2 \theta \cdot r dr d\theta = \int_{-\pi}^{\pi} \cos^2 \theta d\theta \int_0^{a(1+\cos\theta)} r^3 dr$

$$= \frac{a^4}{2} \int_0^{\pi} (1 + \cos \theta)^4 \cos^2 \theta d\theta = \frac{a^4}{2} \int_0^{\pi} (1 + \cos \theta)^4 d\theta - \frac{21}{32} \pi a^4.$$

Let $t = \frac{\theta}{2}$, we have

$$I_y = 2^4 a^4 \int_0^{\frac{\pi}{2}} \cos^8 t\, dt - \frac{21}{32}\pi a^4 = \frac{70}{32}\pi a^4 - \frac{21}{32}\pi a^4 = \frac{49}{32}\pi a^4.$$

Similarly, if $\rho(x,y,z)$ is continuous on Ω, the rotational inertia of an object which occupies the space bounded closed region Ω with the density $\rho(x,y,z)$, to the x-axis, the y-axis, the z-axis and the origin point are

$$I_x = \iiint_\Omega (y^2 + z^2)\rho(x,y,z)\,dv, \quad I_y = \iiint_\Omega (z^2 + x^2)\rho(x,y,z)\,dv,$$

$$I_z = \iiint_\Omega (x^2 + y^2)\rho(x,y,z)\,dv, \quad I_O = \iiint_\Omega (x^2 + y^2 + z^2)\rho(x,y,z)\,dv. \tag{17}$$

Example 7 Find the rotational inertia of an uniform sphere $\Omega = \{(x,y,z) \mid x^2 + y^2 + z^2 \leqslant 1\}$ with density 1 to the x-axis, the y-axis and the z-axis.

Solution From the formula above, we have

$$I_x = \iiint_\Omega (y^2 + z^2)\,dv, \quad I_y = \iiint_\Omega (z^2 + x^2)\,dv, \quad I_z = \iiint_\Omega (x^2 + y^2)\,dv.$$

From the symmetry of the sphere, we have $I_x = I_y = I_z = I$. Add them up, we have

$$3I = \iiint_\Omega 2(x^2 + y^2 + z^2)\,dv.$$

In the spherical coordinates system, the integral region is $\Omega = \{(r,\varphi,\theta) \mid 0 \leqslant r \leqslant 1, 0 \leqslant \varphi \leqslant \pi, 0 \leqslant \theta \leqslant 2\pi\}$.

Then, we get

$$I = \frac{2}{3}\iiint_\Omega (x^2 + y^2 + z^2)\,dv = \frac{2}{3}\iiint_\Omega r^2 \cdot r^2 \sin\varphi\, dr\, d\varphi\, d\theta$$

$$= \frac{2}{3}\int_0^{2\pi} d\theta \int_0^\pi \sin\varphi\, d\varphi \int_0^1 r^4\, dr = \frac{2}{3} \cdot 2\pi \cdot 2 \cdot \frac{1}{5} = \frac{8}{15}\pi.$$

So, we get the rotational inertia is

$$I_x = I_y = I_z = \frac{8}{15}\pi.$$

4) Gravitation

Assume that a particle with mass 1 located at point $P_0(x_0, y_0, z_0)$ and another particle with mass m located at point $P(x,y,z)$, we get the gravitation of this particle to the unit mass particle, from the law of gravitation of Mechanics, is

$$\boldsymbol{F} = \frac{Gm}{r^3}(x - x_0, y - y_0, z - z_0) = \frac{Gm}{r^3}\boldsymbol{r}, \tag{18}$$

where G is the gravitational constant and

$$\boldsymbol{r} = \overrightarrow{P_0 P} = (x - x_0, y - y_0, z - z_0),$$

$$r = |\boldsymbol{r}| = \sqrt{(x-x_0)^2 + (y-y_0)^2 + (z-z_0)^2}.$$

Then we discuss the gravitation of a space object to an unit mass particle which located at the point $P_0(x_0, y_0, z_0)$ outside of the object.

Assume that this object is enclosed by a space closed region Ω and its density is $\rho(x,y,z)$, and suppose that $\rho(x,y,z)$ is continuous on Ω. Take a closed region $d\sigma$ with a very small diameter on the closed region D ($d\sigma$ is also the volume of the small region). The point

(x,y,z) is on this small region dv. The mass of the small object $\rho(x,y,z)dv$ could be thought to be nearly focus on the point (x,y,z). By the gravitation equation between two particles, we can get the gravitation of the small object to the unit mass particle

$$d\mathbf{F} = (dF_x, dF_y, dF_z)$$
$$= \left(G\frac{\rho(x,y,z)(x-x_0)}{r^3}dv, G\frac{\rho(x,y,z)(y-y_0)}{r^3}dv, \right.$$
$$\left. G\frac{\rho(x,y,z)(z-z_0)}{r^3}dv \right),$$

where dF_x, dF_y, dF_z are projection of $d\mathbf{F}$ on three axes,

$$r = \sqrt{(x-x_0)^2 + (y-y_0)^2 + (z-z_0)^2},$$

and G is gravity constant. Then make integral of dF_x, dF_y, dF_z, we get

$$\mathbf{F} = (F_x, F_y, F_z)$$
$$= \left(\iiint_\Omega \frac{G\rho(x,y,z)(x-x_0)}{r^3}dv, \iiint_\Omega \frac{G\rho(x,y,z)(y-y_0)}{r^3}dv, \right.$$
$$\left. \iiint_\Omega \frac{G\rho(x,y,z)(z-z_0)}{r^3}dv \right). \tag{19}$$

In the calculation of gravity, it is not always necessary to work out the three projections by integral. We can directly get the projection on some axis is zero from the symmetry of the object.

To find the gravity of a plane slice to an unit mass particle located at the point $P_0(x_0, y_0, z_0)$ outside of the slice, we assume that the slice is enclosed by a closed region D on the xOy plane and its areal density is $\rho(x,y)$. Then turn the triple integral on Ω into a double integral, we will get a corresponding calculation formula.

Example 8 Evaluate the gravitation of an well-proportioned sphere $x^2+y^2+z^2 \leqslant R^2$ with the radius R and constant volume density ρ, to an unit mass particle located at point $(0,0,a)(a>R)$.

Solution By the symmetry and uniformity of the well-proportioned sphere, we get

$$F_x = F_y = 0.$$

The gravitation component on the z-axis is

$$F_z = \iiint_\Omega \frac{G(z-a)\rho}{[x^2+y^2+(z-a)^2]^{\frac{3}{2}}}dxdydz$$
$$= G\rho \int_{-R}^{R} dz \iint_{D_z} \frac{z-a}{[x^2+y^2+(z-a)^2]^{\frac{3}{2}}}dxdy,$$

where $D_z = \{(x,y) \mid x^2+y^2 \leqslant R^2-z^2\}$. And since

$$\iint_{D_z} \frac{z-a}{[x^2+y^2+(z-a)^2]^{\frac{3}{2}}}dxdy = (z-a)\int_0^{2\pi} d\varphi \int_0^{\sqrt{R^2-z^2}} \frac{r}{[r^2+(z-a)^2]^{\frac{3}{2}}}dr$$
$$= 2\pi\left(-1 - \frac{z-a}{\sqrt{R^2+a^2-2az}} \right),$$

we have

$$F_z = 2\pi G\rho \int_{-R}^{R} \left(-1 - \frac{z-a}{\sqrt{R^2+a^2-2az}}\right) dz$$

$$= 2\pi G\rho \left[-2R + \frac{1}{a}\int_{-R}^{R}(z-a)\,d\sqrt{R^2+a^2-2az}\right]$$

$$= 2\pi G\rho\left[-2R + 2R - \frac{2R^3}{3a^2}\right] = -G\frac{M}{a^2},$$

where the mass of this well-proportioned sphere is $M = \frac{4\pi R^3}{3}\rho$.

Exercises 12-4

1. Find the area of a semi-spherical surface with the radius R.

2. Find the area of the spherical surface $x^2+y^2+z^2=a^2$ included in the cylinder $x^2+y^2=ax\,(a>0)$.

3. Find the surface area of the solid enclosed by the semi-spherical surface of $z=\sqrt{3a^2-x^2-y^2}$ and the paraboloid of revolution $x^2+y^2=2az$.

4. Find the center of gravity of a well-proportioned plane enclosed by the graph of $\rho=\cos\theta$ and $\rho=2\cos\theta$.

5. Find the center of gravity of the slice which occupies the closed region D enclosed by the parabola $y=x^2$ and the straight line $y=x$, and the areal density $\rho(x,y)=x^2y$.

6. Find the coordinates of the center of gravity of a well-proportioned semi-ellipsoid $\frac{x^2}{a^2}+\frac{y^2}{b^2}+\frac{z^2}{c^2}\leqslant 1, z\geqslant 0$.

7. Find the center of gravity of the bodies enclosed by the following edge surfaces by triple integral if density $\rho=1$.
   (1) $z^2=x^2+y^2, z=1$;
   (2) $z=x^2+y^2, x+y=a, x=0, y=0, z=0$;
   (3) $z=\sqrt{A^2-x^2-y^2}, z=\sqrt{a^2-x^2-y^2}\;(A>a>0), z=0$.

8. Find the center of gravity of an object which occupies the space region $\Omega=\{(x,y,z)\mid 0\leqslant x\leqslant 1, 0\leqslant y\leqslant 1, 0\leqslant z\leqslant 1\}$ and its density is $\rho(x,y,z)=x+y+z$.

9. Find the rotational inertia of an uniform sphere with the radius R and the density 1.

10. Find the rotational inertia of the figure enclosed by $y^2=ax$ and a straight line $x=a\,(a>0)$ over another straight line $y=-a$.

11. A uniform object that occupies the closed region Ω which is enclosed by an edge surface $z=x^2+y^2$ and the planes $z=0, |x|=a, |y|=a$, and its density is a constant.
    (1) Find the volume of the object.
    (2) Find the centroid of the object.
    (3) Find the rotational inertia to the z-axis.

12. Find the gravitation of an uniform cylinder with the density ρ and its closed region $\Omega=\{(x,y,z)\mid x^2+y^2\leqslant R^2, 0\leqslant z\leqslant h\}$ find the gravitational force on the particle per unit mass at $M_0(0,0,a)\,(a>b)$.

*12.5 Integral with parameters

Assume that $f(x,y)$ is a continuous function defined on a rectangular closed region $R=[a,b]\times[\alpha,\beta]$. For any number x on $[a,b]$, $f(x,y)$ is a continuous function of the variable y on $[\alpha,\beta]$, the integral $\int_\alpha^\beta f(x,y)\mathrm{d}y$ is existent, and the integral value is dependent on the value of x. As the value of x changes, the integral value will change. This integral determines a function of x on $[a,b]$ and we denote the integral by $\varphi(x)$, that is,

$$\varphi(x) = \int_\alpha^\beta f(x,y)\mathrm{d}y \ (a \leqslant x \leqslant b). \tag{1}$$

To find the integral in Equation (1), the variable x can be regarded as a constant, and it is often called parameter. Thus the right term of equality (1) is an integral having a parameter x. Now we will discuss more about the characters of $\varphi(x)$.

Theorem 1 If function $f(x,y)$ is continuous on the rectangle $R=[a,b]\times[\alpha,\beta]$, the function $\varphi(x)$ determined by equality (1) is continuous on $[a,b]$.

Proof Assume x and $x+\Delta x$ are two points on $[a,b]$, then

$$\varphi(x+\Delta x) - \varphi(x) = \int_\alpha^\beta [f(x+\Delta x,y) - f(x,y)]\mathrm{d}y. \tag{2}$$

Since $f(x,y)$ is continuous on the closed region R. For any $\varepsilon>0$, there exists $\delta>0$ such that for any two points (x_1,y_1) and (x_2,y_2) on R, if the distance between them is less than δ, that is, $\sqrt{(x_2-x_1)^2+(y_2-y_1)^2}<\delta$, then there will be $|f(x+\Delta x,y)-f(x,y)|<\varepsilon$, so

$$|\varphi(x+\Delta x) - \varphi(x)| \leqslant \int_\alpha^\beta |f(x+\Delta x,y) - f(x,y)| \mathrm{d}y < \varepsilon(\beta-\alpha).$$

So $\varphi(x)$ is continuous on $[a,b]$.

Since $\varphi(x)$ is continuous on $[a,b]$, the integral on $[a,b]$ exists, and its integral can be denoted by

$$\int_a^b \varphi(x)\mathrm{d}x = \int_a^b \left[\int_\alpha^\beta f(x,y)\mathrm{d}y\right]\mathrm{d}x = \int_a^b \mathrm{d}x \int_\alpha^\beta f(x,y)\mathrm{d}y.$$

If $f(x,y)$ is continuous on the rectangular R, the integral above exists. The integral on the right side is the integral of function $f(x,y)$ over y first and then over x, and we also could get the iterated integral over x first and then over y. As a result, we will get Theorem 2.

Theorem 2 If the function $f(x,y)$ is continuous on the rectangle $R=[a,b]\times[\alpha,\beta]$, then

$$\int_a^b \left[\int_\alpha^\beta f(x,y)\mathrm{d}y\right]\mathrm{d}x = \int_\alpha^\beta \left[\int_a^b f(x,y)\mathrm{d}x\right]\mathrm{d}y, \tag{3}$$

or

$$\int_a^b \mathrm{d}x \int_\alpha^\beta f(x,y)\mathrm{d}y = \int_\alpha^\beta \mathrm{d}y \int_a^b f(x,y)\mathrm{d}x.$$

Now we will consider the differential of the function $\varphi(x)$ determined by equality (1).

Theorem 3 If the function $f(x,y)$ and its partial derivation $\dfrac{\partial f(x,y)}{\partial x}$ are both continuous on rectangle $R=[a,b]\times[\alpha,\beta]$, then the function $\varphi(x)$ determined by equality (1) will be derivable in $[a,b]$ and

$$\varphi'(x) = \frac{d}{dx}\int_\alpha^\beta f(x,y)dy = \int_\alpha^\beta \frac{\partial f(x,y)}{\partial x}dy. \qquad (4)$$

Proof If $\varphi'(x)$ exists, we know $\varphi'(x)=\lim\limits_{\Delta x\to 0}\dfrac{\varphi(x+\Delta x)-\varphi(x)}{\Delta x}$. By equality (2), the quotient of difference is

$$\frac{\varphi(x+\Delta x)-\varphi(x)}{\Delta x} = \int_\alpha^\beta \frac{f(x+\Delta x,y)-f(x,y)}{\Delta x}dy.$$

By the Lagrange's mean value theorem and uniform continuity of $\dfrac{\partial f}{\partial x}$, we get

$$\frac{f(x+\Delta x,y)-f(x,y)}{\Delta x} = \frac{\partial f(x+\theta\Delta x,y)}{\partial x} = \frac{\partial f(x,y)}{\partial x}+\eta(x,y,\Delta x),$$

where $0<\theta<1$ and $\lim\limits_{\Delta x\to 0}\eta(x,y,\Delta x)=0$. So we get

$$\frac{\varphi(x+\Delta x,y)-\varphi(x)}{\Delta x} = \int_\alpha^\beta \frac{\partial f(x,y)}{\partial x}dy + \int_\alpha^\beta \eta(x,y,\Delta x)dy.$$

Let $\Delta x\to 0$ and take the limit of formula above, then we can obtain the formula we need to prove.

Example 1 Find the integral

$$\int_0^{\frac{\pi}{2}} \ln(a^2\sin^2 x + b^2\cos^2 x)dx.$$

Solution Because

$$\int_0^{\frac{\pi}{2}} \ln(a^2\sin^2 x + b^2\cos^2 x)dx = \int_0^{\frac{\pi}{2}} \ln[a^2 + (b^2-a^2)\cos^2 x]dx$$

$$= \int_0^{\frac{\pi}{2}} \ln(a^2 + c\cos^2 x)dx = F(c),$$

then we have

$$F'(c) = \int_0^{\frac{\pi}{2}} \frac{\cos^2 x}{a^2+c\cos^2 x}dx = \frac{1}{c}\int_0^{\frac{\pi}{2}}\left(1-\frac{a^2}{a^2+c\cos^2 x}\right)dx$$

$$= \frac{1}{c}\left[\frac{\pi}{2} - \frac{|a|}{\sqrt{a^2+c}}\arctan(\tan x)\Big|_0^{\frac{\pi}{2}}\right]$$

$$= \frac{\pi}{2c}\left(1-\frac{|a|}{\sqrt{a^2+c}}\right) = \frac{\pi}{2\sqrt{a^2+c}}\cdot\frac{1}{\sqrt{a^2+c}+|a|},$$

$$F(c)-F(0) = \int_0^c F'(c)dc = \pi\int_0^c \frac{d(\sqrt{a^2+c}+|a|)}{\sqrt{a^2+c}+|a|}$$

$$= \pi\ln(\sqrt{a^2+c}+|a|) - \pi\ln|2a|.$$

Because $c=b^2-a^2$, $\sqrt{a^2+c}=|b|$, and $F(0)=\pi\ln|a|$, we have

$$F(c) = \int_0^{\frac{\pi}{2}} \ln(a^2 \sin^2 x + b^2 \cos^2 x) \mathrm{d}x = \pi \ln \frac{|a|+|b|}{2}.$$

In equality (1), the integral limits α and β are both constants, but in practical applications, there are different values of the parameter x and different integral limits. The integral limit is also a function of the parameter x, then the integral could be denoted by

$$\Phi(x) = \int_{\alpha(x)}^{\beta(x)} f(x,y) \mathrm{d}y. \tag{5}$$

Then, we discuss the characters of these integrals depending on the parameters.

Theorem 4 If function $f(x,y)$ is continuous on the rectangle $R=[a,b]\times[\alpha,\beta]$, the functions $\alpha(x)$ and $\beta(x)$ are continuous on $[a,b]$, $\alpha \leqslant \alpha(x) \leqslant \beta$, $\alpha < \beta(x) \leqslant \beta$ ($a \leqslant x \leqslant b$), then the function $\Phi(x)$ in (5) is continuous on $[a,b]$.

Proof Assume x and $x+\Delta x$ are two points on $[a,b]$, then

$$\Phi(x+\Delta x) - \Phi(x) = \int_{\alpha(x+\Delta x)}^{\beta(x+\Delta x)} f(x+\Delta x, y) \mathrm{d}y - \int_{\alpha(x)}^{\beta(x)} f(x,y) \mathrm{d}y.$$

And

$$\int_{\alpha(x+\Delta x)}^{\beta(x+\Delta x)} f(x+\Delta x, y) \mathrm{d}y = \int_{\alpha(x+\Delta x)}^{\alpha(x)} f(x+\Delta x, y) \mathrm{d}y + \int_{\alpha(x)}^{\beta(x)} f(x+\Delta x, y) \mathrm{d}y + \int_{\beta(x)}^{\beta(x+\Delta x)} f(x+\Delta x, y) \mathrm{d}y.$$

Thus

$$\Phi(x+\Delta x) - \Phi(x) = \int_{\beta(x+\Delta x)}^{\alpha(x)} f(x+\Delta x, y) \mathrm{d}y + \int_{\beta(x)}^{\beta(x+\Delta x)} f(x+\Delta x, y) \mathrm{d}y + \int_{\alpha(x)}^{\beta(x)} [f(x+\Delta x, y) - f(x,y)] \mathrm{d}y. \tag{6}$$

As $\Delta x \to 0$, the integral limits of the right integral of right side of equality (6) is fixed. By the same method of proving Theorem 1, the integral above tends to zero and

$$\left| \int_{\alpha(x+\Delta x)}^{\alpha(x)} f(x+\Delta x, y) \mathrm{d}y \right| \leqslant M |\alpha(x+\Delta x) - \alpha(x)|,$$

$$\left| \int_{\beta(x)}^{\beta(x+\Delta x)} f(x+\Delta x, y) \mathrm{d}y \right| \leqslant M |\beta(x+\Delta x) - \beta(x)|,$$

where M is the maximum value of $|f(x,y)|$ on the rectangle R. Since $\alpha(x)$ and $\beta(x)$ are continuous on $[a,b]$ and we get, from two formulas above, as $\Delta x \to 0$, the left two integrals on the right side of (6) will tend to zero, thus when $a \leqslant x \leqslant b$,

$$\lim_{\Delta x \to 0} [\Phi(x+\Delta x) - \Phi(x)] = 0.$$

So $\varphi(x)$ is continuous on $[a,b]$.

We have the following conclusions about the derivative of the function $\varphi(x)$.

Theorem 5 If function $f(x,y)$ and its partial derivation $\dfrac{\partial f(x,y)}{\partial x}$ are both continuous on the rectangle $R=[a,b]\times[\alpha,\beta]$, the functions $\alpha(x)$ and $\beta(x)$ are derivable on $[a,b]$ and $\alpha \leqslant \alpha(x) \leqslant \beta$, $\alpha \leqslant \beta(x) \leqslant \beta$ ($a \leqslant x \leqslant b$), then, the function $\varphi(x)$ determined by Equation (5) is derivable on $[a,b]$ and

$$\Phi'(x) = \frac{d}{dx}\int_{\alpha(x)}^{\beta(x)} f(x,y)dy$$
$$= \int_{\alpha(x)}^{\beta(x)} \frac{\partial f(x,y)}{\partial x}dy + f[x,\beta(x)]\beta'(x) - f[x,\alpha(x)]\alpha'(x). \quad (7)$$

Proof By equality (6), we have
$$\frac{\Phi(x+\Delta x) - \Phi(x)}{\Delta x} = \int_{\alpha(x)}^{\beta(x)} \frac{f(x+\Delta x, y) - f(x,y)}{\Delta x}dy +$$
$$\frac{1}{\Delta x}\int_{\beta(x)}^{\beta(x+\Delta x)} f(x+\Delta x, y)dy - \frac{1}{\Delta x}\int_{\alpha(x)}^{\alpha(x+\Delta x)} f(x+\Delta x, y)dy. \quad (8)$$

As $\Delta x \to 0$, the limits of the left integral on the right side of the equality (8) will not change. Because of the same reason of proving Theorem 3, we get
$$\int_{\alpha(x)}^{\beta(x)} \frac{f[(x+\Delta x, y) - f(x,y)]}{\Delta x}dy \to \int_{\alpha(x)}^{\beta(x)} \frac{\partial f(x,y)}{\partial x}dy \ (\Delta x \to 0).$$

As for the second integral on the right side of equality (8), by the mean value theorem, we get
$$\frac{1}{\Delta x}\int_{\beta(x)}^{\beta(x+\Delta x)} f(x+\Delta x, y)dy = \frac{1}{\Delta x}[\beta(x+\Delta x) - \beta(x)]f(x+\Delta x, \eta).$$

In the equality above, η is between $\beta(x)$ and $\beta(x+\Delta x)$, so we obtain
$$\lim_{\Delta x \to 0}\frac{1}{\Delta x}[\beta(x+\Delta x) - \beta(x)] = \beta'(x),$$
$$\lim_{\Delta x \to 0} f(x+\Delta x, \eta) = f(x, \beta(x)).$$

Then
$$\lim_{\Delta x \to 0} \frac{1}{\Delta x}\int_{\beta(x)}^{\beta(x+\Delta x)} f(x+\Delta x, y)dy = f(x, \beta(x))\beta'(x).$$

Similarly, we can prove
$$\lim_{\Delta x \to 0} \frac{1}{\Delta x}\int_{\alpha(x)}^{\alpha(x+\Delta x)} f(x+\Delta x, y)dy = f(x, \alpha(x))\alpha'(x).$$

Thus, as $\Delta x \to 0$, take the limit equality (8), we can get equality (7). Sometimes, the equality (7) is also called Leibniz formula.

Example 2 Find $\Phi'(x)$, if $\Phi(x) = \int_{x}^{x^2} \frac{\cos(xy)}{y}dy$.

Solution By the Leibniz formula, we get
$$\Phi'(x) = \int_{x}^{x^2} (-1)\sin(xy)dy + \frac{\cos x^3}{x^2} \cdot 2x - \frac{\cos x^2}{x} \cdot 1$$
$$= \left[\frac{\cos(xy)}{x}\right]_{x}^{x^2} + \frac{2\cos x^3}{x} - \frac{\cos x^2}{x} = \frac{3\cos x^3 - 2\cos x^2}{x}.$$

Example 3 Find $I = \int_{0}^{1} \frac{x^b - x^a}{\ln x}dx \ (0 < a < b)$.

Solution Because $\int_{a}^{b} x^y dy = \left[\frac{x^y}{\ln x}\right]_{a}^{b} = \frac{x^b - x^a}{\ln x}$, we have

$$I = \int_0^1 dx \int_a^b x^y dy.$$

Then we get $f(x,y)=x^y$ is continuous on the rectangle $R=[0,1]\times[a,b]$ from Theorem 2. By exchanging the order of integrals, we get

$$I = \int_a^b dy \int_0^1 x^y dx = \int_a^b \left[\frac{x^{y+1}}{y+1}\right]_0^1 dy = \int_a^b \frac{1}{y+1} dy = \ln\frac{b+1}{a+1}.$$

*Exercises 12-5

1. Find the following limits:

   (1) $\lim\limits_{x\to 0}\int_x^{1+x} \dfrac{dy}{1+x^2+y^2}$;

   (2) $\lim\limits_{x\to 0}\int_{-1}^1 \sqrt{x^2+y^2}\, dy$.

2. Find the derivatives of the following integrals:

   (1) $\Phi(x) = \int_{\sin x}^{\cos x}(y^2\sin x - y^3)dy$;

   (2) $\Phi(x) = \int_0^x \dfrac{\ln(1+xy)}{y} dy$;

   (3) $\Phi(x) = \int_{x^2}^{x^3} \arctan\dfrac{y}{x}\, dy$;

   (4) $\Phi(x) = \int_x^{x^2} e^{-xy^2} dy$.

3. Find $F''(x)$, if $F(x) = \int_0^x (x+y)f(y)dy$ and $f(y)$ is a differentiable function.

4. Find the integral $I = \int_0^{\frac{\pi}{2}} \ln(\cos^2 x + a^2 \sin^2 x)dx\,(a>0)$.

5. Find the following integrals:

   (1) $\int_0^1 \dfrac{\arctan x}{x}\cdot\dfrac{dx}{\sqrt{1-x^2}}$;

   (2) $\int_0^1 \sin\left(\ln\dfrac{1}{x}\right)\dfrac{x^b-x^a}{\ln x} dx\,(0<a<b)$.

Summary

1. Main contents

(1) The concept, property and calculation method of the double integral (transform it into the iterated integral).

① The methods of calculation in the plane right angle coordinate system.

The method is the cumulative method, that is, regard x as a constant firstly, and take integral over y first and then over x, or regard y as a constant firstly, and take integral over x first and then over y. The integral formula is

$$\iint_D f(x,y)dxdy = \int_a^b \left[\int_{\varphi_1(x)}^{\varphi_2(x)} f(x,y)dy\right]dx = \int_a^b dx\int_{\varphi_1(x)}^{\varphi_2(x)} f(x,y)dy,$$

or

$$\iint_D f(x,y)dxdy = \int_c^d \left[\int_{\psi_1(y)}^{\psi_2(y)} f(x,y)dx\right]dy = \int_c^d dy\int_{\psi_1(y)}^{\psi_2(y)} f(x,y)dx.$$

② The methods of calculation in the polar coordinate system.

If the boundary equation of integrand of the double integral and integral region D will be more easily expressed in polar coordinate system, we can use the methods of calculation in polar coordinate system.

If the pole O is out of D, the region D can be expressed by inequalities $r_1(\theta)\leqslant r\leqslant$

$r_2(\theta)$ and $\alpha \leqslant \theta \leqslant \beta$, the integral formula is

$$\iint_D f(r\cos\theta, r\sin\theta) r dr d\theta = \int_\alpha^\beta d\theta \int_{r_1(\theta)}^{r_2(\theta)} f(r\cos\theta, r\sin\theta) r dr.$$

If the pole O is in the D and the boundary equation of region is $0 \leqslant r \leqslant r(\theta)$ and $0 \leqslant \theta \leqslant 2\pi$, the integral formula is

$$\iint_D f(r\cos\theta, r\sin\theta) r dr d\theta = \int_0^{2\pi} d\theta \int_0^{r(\theta)} f(r\cos\theta, r\sin\theta) r dr.$$

If the pole O is on the boundary of D and the boundary equation of dominate is $0 \leqslant r \leqslant r(\theta)$ and $\theta_1 \leqslant \theta \leqslant \theta_2$, the integral formula is

$$\iint_D f(r\cos\theta, r\sin\theta) r dr d\theta = \int_{\theta_1}^{\theta_2} d\theta \int_0^{r(\theta)} f(r\cos\theta, r\sin\theta) r dr.$$

(2) The concept and methods of calculation of the triple integral (transform it into three-time integral or a iterated integral and a simple integral).

There are two conditions in coordinate system.

① In polar coordinate system.

If the integral region is $\Omega = \{(x,y,z) \mid z_1(x,y) \leqslant z \leqslant z_2(x,y), (x,y) \in D_{xy}\}$, the integral formula is

$$\iiint_\Omega f(x,y,z) dx dy dz = \iint_{D_{xy}} \left[\int_{z_1(x,y)}^{z_2(x,y)} f(x,y,z) dz \right] dx dy.$$

If the integral region is $\Omega = \{(x,y,z) \mid (x,y) \in D_z, c_1 \leqslant z \leqslant c_2\}$, the integral formula is

$$\iiint_\Omega f(x,y,z) dx dy dz = \int_{c_1}^{c_2} dz \iint_{D_z} f(x,y,z) dx dy.$$

② In the cylindrical coordinate system, the integral formula is

$$\iiint_\Omega f(x,y,z) dx dy dz = \iiint_\Omega f(r\cos\theta, r\sin\theta, z) r dr d\theta dz.$$

③ In the spherical coordinate system, the integral formula is

$$\iiint_\Omega f(x,y,z) dx dy dz = \iiint_\Omega f(r\sin\varphi\cos\theta, r\sin\varphi\sin\theta, r\cos\varphi) r^2 \sin\varphi dr d\varphi d\theta.$$

(3) Application of the double integral and the triple integral.

① Application of double integral

Area Suppose the hook face Σ is determined by $z = z(x,y)$, the projection region of Σ on the xOy plane is D and function $z(x,y)$ is the first-order continuous partial derivative of z_x and z_y on D, the area of hook face Σ is

$$A = \iint_D \sqrt{1 + z_x^2(x,y) + z_y^2(x,y)} d\sigma.$$

Volume Suppose there is a curved roof cylinder which the bottom is the closed dominate D of surface xOy and the top is hook face $z = f(x,y) (f(x,y) \geqslant 0)$, its volume is

$$V = \iint_D f(x,y) d\sigma.$$

Quality If a plane slice covered the closed region D on the xoy plane and the areal density

at point (x,y) is $\rho=\rho(x,y)$, the quality of this plane slice is
$$M = \iint_D \rho(x,y)\,d\sigma.$$

Center of gravity If a plane slice covers the closed region D on the xOy plane and the areal density at point (x,y) is $\rho(x,y)$, the coordinates of the center of gravity of this plane slice is

$$\bar{x} = \frac{M_y}{M} = \frac{\iint_D x\rho(x,y)\,d\sigma}{\iint_D \rho(x,y)\,d\sigma},\quad \bar{y} = \frac{M_x}{M} = \frac{\iint_D y\rho(x,y)\,d\sigma}{\iint_D \rho(x,y)\,d\sigma}.$$

Moment of inertia If a plane slice covers the closed region D on the xOy plane and the areal density at point (x,y) is $\rho=\rho(x,y)$, the moments of inertia of this plane slice to the x-axis, the y-axis and the origin point are

$$I_x = \iint_D y^2\rho(x,y)\,d\sigma,\ I_y = \iint_D x^2\rho(x,y)\,d\sigma,\ I_O = \iint_D (x^2+y^2)\rho(x,y)\,d\sigma.$$

② Application of triple integrals

Volume The volume of space closed region Ω is
$$V = \iiint_\Omega dv.$$

Quality If the object covers the space closed region Ω and the volume density $\rho(x,y,z)$ is continuous on Ω, the quality of this object is
$$M = \iiint_\Omega \rho(x,y,z)\,dv.$$

Center of gravity If the object covers the space closed region Ω and the volume density $\rho(x,y,z)$ is continuous on Ω, the center of gravity of this object is

$$\bar{x} = \frac{\iiint_\Omega x\rho(x,y,z)\,dv}{\iiint_\Omega \rho(x,y,z)\,dv},\quad \bar{y} = \frac{\iiint_\Omega y\rho(x,y,z)\,dv}{\iiint_\Omega \rho(x,y,z)\,dv},\quad \bar{z} = \frac{\iiint_\Omega z\rho(x,y,z)\,dv}{\iiint_\Omega \rho(x,y,z)\,dv}.$$

Moment of inertia If the object covers the space closed region Ω and the volume density $\rho(x,y,z)$ is continuous on Ω, the moments of inertia of this object to every coordinate plane and origin are

$$I_x = \iiint_\Omega (y^2+z^2)\rho(x,y,z)\,dv,\ I_y = \iiint_\Omega (z^2+x^2)\rho(x,y,z)\,dv,$$

$$I_z = \iiint_\Omega (x^2+y^2)\rho(x,y,z)\,dv,\ I_O = \iiint_\Omega (x^2+y^2+z^2)\rho(x,y,z)\,dv.$$

2. Basic requirements

(1) Understand the concept of the double integral, triple integral and know the nature of multiple integral.

(2) Master the calculation method of double integral (plane right angle coordinates and polar coordinates), understand the calculation method of triple integral (space right

angle coordinates, cylindrical coordinates and spherical coordinates).

Quiz

1. Choice questions.

(1) Suppose the plane region D is surrounded by straight lines $x=0, y=0, x+y=\frac{1}{2}$, $x+y=1$, if $I_1 = \iint_D [\ln(x+y)]^7 dxdy$, $I_2 = \iint_D \ln(x+y)^7 dxdy$, $I_3 = \iint_D \sin(x+y)^7 dxdy$, the relationship between I_1, I_2, I_3 is (   ).

   A. $I_1 < I_2 < I_3$ B. $I_3 < I_2 < I_1$ C. $I_1 < I_3 < I_2$ D. $I_3 < I_1 < I_2$

(2) Suppose $I = \iint_D |xy| \, dxdy$, where D is $x^2 + y^2 \leqslant a^2$, so $I = ($ $)$.

   A. $\dfrac{a^4}{4}$ B. $\dfrac{a^4}{3}$ C. $\dfrac{a^4}{2}$ D. a^4

(3) Suppose $I = \iiint_\Omega (x^2 + y^2 + z^2) dv$, where Ω is $x^2 + y^2 + z^2 \leqslant 1$, so $I = ($ $)$.

   A. $\iiint_\Omega dv$ B. $\int_0^{2\pi} d\theta \int_0^{2\pi} d\varphi \int_0^1 r^4 \sin\theta d\theta$

   C. $\int_0^{2\pi} d\theta \int_0^\pi d\varphi \int_0^1 r^4 \sin\varphi dr$ D. $\int_0^\pi d\varphi \int_0^{2\pi} d\theta \int_0^1 r^4 \sin\theta dr$

(4) We know $\int_0^1 f(x) dx = \int_0^1 xf(x) dx$, so $\iint_D f(x) dxdy = ($ $)$, $D: x+y \leqslant 1, x \geqslant 0, y \geqslant 0$.

   A. 2   B. 0   C. $\dfrac{1}{2}$ D. 1

2. Completion.

(1) Exchange the integral order of the iterated integral $\int_{-1}^0 dy \int_{1-y}^2 f(x,y) dx =$ _____.

(2) Turn the integral $I = \int_0^1 dy \int_0^y f(x^2 + y^2) dx$ into the iterated integral in polar coordinates _____.

(3) $\iiint_{x^2+y^2+z^2 \leqslant 1} f(x) dxdydz$ can be denoted by repeated integral in spherical coordinate as _____.

(4) The volume of the solid surrounded by the elliptic paraboloid $z = x^2 + 2y^2$ and the parabolic cylinder $z = 2 - x^2$ is _____.

3. Solve the following questions.

(1) Find $\iint_D x d\sigma$, D is the plane region surrounded by the curve $y = x^2 - 1$ and the straight line $y = -x + 1$.

(2) Find $\iiint_\Omega (x+z)dv$, Ω is the solid surrounded by the curved plane $z=x^2+y^2$ and the plane $z=1$.

(3) Find $I = \int_0^1 dx \int_x^1 f(x)dy$, $f(x)$ is continuous on $[0, 1]$ and $\int_0^1 f(x)dx = A$.

4. Find $\iiint_\Omega \left(\dfrac{x^2}{a^2} + \dfrac{y^2}{b^2} + \dfrac{z^2}{c^2}\right)dv$ where Ω is $\dfrac{x^2}{a^2}+\dfrac{y^2}{b^2}+\dfrac{z^2}{c^2} \leqslant 1$.

5. $I = \iint_D (|x|+|y|)dxdy$, and $D: x^2+y^2 \leqslant 1$.

6. Suppose $f(x)$ is continuous, $\Omega: 0 \leqslant z \leqslant h, x^2+y^2 \leqslant t^2 (t \geqslant 0)$, but $F(t) = \iiint_\Omega [z^2 + f(x^2+y^2)]dv$, find $\lim\limits_{t \to 0^+} \dfrac{F(t)}{t^2}$.

7. Find $\iint_D e^{(|x|+|y|)} dxdy$, $D: |x|+|y| \leqslant 1$.

8. Suppose Ω is the closed region surrounded by the curved surface formed by the rotation of the curve $\begin{cases} y^2=2z, \\ x=0 \end{cases}$ around the z-axis and the plane $z=4$, find the triple integral $I = \iiint_\Omega (x^2+y^2+z^2)dv$.

9. Suppose $f(x)$ is a continuous function on the closed interval $[a, b]$, prove that $\left[\int_a^b f(x)dx\right]^2 \leqslant (b-a)\int_a^b f^2(x)dx$.

Exercises

1. Choose a correct conclusion in the four conclusions given in the following questions.

(1) Suppose $D: 1 \leqslant x^2+y^2 \leqslant 4$, f is a continuous function on D, so the double integral is equal to (   ) in the polar coordinate.

A. $2\pi \int_1^2 rf(r^2)dr$

B. $2\pi \left[\int_0^2 rf(r)dr - \int_0^1 rf(r)dr\right]$

C. $2\pi \int_1^2 rf(r)dr$

D. $2\pi \left[\int_0^2 rf(r^2)dr - \int_0^1 rf(r^2)dr\right]$

(2) Suppose the plane region D is surrounded by $x=0, y=0, x+y=\dfrac{1}{4}$ and $x+y=1$, if $I_1 = \iint_D [\ln(x+y)]^3 dxdy, I_2 = \iint_D (x+y)^3 dxdy, I_3 = \iint_D [\sin(x+y)]^3 dxdy$, the order of I_1, I_2, I_3 is (   ).

A. $I_1 < I_2 < I_3$
B. $I_3 < I_2 < I_1$
C. $I_1 < I_3 < I_2$
D. $I_3 < I_1 < I_2$

2. Find the following double integrals:

(1) $\int_0^1 dx \int_x^1 e^{-y^2} dy$;

(2) $\iint_D xy\, d\sigma$, D is surrounded by $y=x, y=0, x=1$;

(3) $\iint_D yx^2 e^{xy}\, d\sigma$, $D: 0 \leqslant x \leqslant 1, 0 \leqslant y \leqslant 1$;

(4) $\iint_D (y^2 + 3x - 6y + 9)\, d\sigma$, $D = \{(x,y) \mid x^2 + y^2 \leqslant R^2\}$.

3. Exchange the order of the following iterated integrals:

(1) $\int_0^a dx \int_x^{2ax-x^2} f(x,y)\, dx$;

(2) $\int_{-1}^1 dx \int_{-\sqrt{1-x^2}}^{1-x^2} f(x,y)\, dydx$;

(3) $\int_0^1 dy \int_{\frac{y^2}{3}}^{\sqrt{3-y^2}} f(x,y)\, dx$.

4. Turn the iterated integral $\int_0^a dx \int_0^x \sqrt{x^2 + y^2}\, dy$ into the iterated integral in polar coordinates, and find the integral.

5. If $f(x)$ is continuous on $[a,b]$ and always greater than zero, prove that
$\int_a^b f(x)\, dx \int_a^b \frac{1}{f(x)} \geqslant (b-a)^2$.

6. Turn the integral $\iiint_\Omega f(x,y,z)\, dxdydz$ into the repeated integral, Ω is the closed region surrounded by the curved plane $z = x^2 + y^2$, $y = x^2$ and the plane $y = 1, z = 0$.

7. Find the following triple integrals.

(1) $\iiint_\Omega xy\, dv$, Ω is the region in the first octant surrounded by cylinder $x^2 + y^2 = 1$ and the plane $z = 1, z = 0, x = 0, y = 0$.

(2) $\iiint_\Omega z\sqrt{x^2 + y^2}\, dv$, Ω is surrounded by the curve $y = \sqrt{2x - x^2}$ and the plane $z = 1, z = a\ (a > 0), y = 0$.

(3) $\iiint_\Omega (x^2 + y^2)\, dv$, Ω is the region surrounded by the curved surface formed by the rotation of the curve $\begin{cases} y^2 = 2z, \\ x = 0 \end{cases}$ around the z-axis and the plane $z = 2, z = 8$.

8. Find the area of the finite part of the plane $\frac{x}{a} + \frac{y}{b} + \frac{z}{c} = 1$ cut by three coordinate planes.

9. Find the quality of the smaller part surrounded by the sphere $x^2 + y^2 + z^2 = 2$ and the cone $z = \sqrt{x^2 + y^2}$. We have known that the object density is in proportion to the distance between the point and center of sphere, and is equal to 1 on the spherical surface.

10. Find the rotary inertia of the cylinder with the height h and bottom radius a to the diameter of the bottom.

11. Suppose there is a homogeneous semicircle slice with the weight n on the xOy plane, and the closed region $D=\{(x,y)\mid x^2+y^2\leqslant R^2, y\geqslant 0\}$. There is a particle P with the quality m on the straight line perpendicular to the center O, and $OP=a$. Find the gravitation of the semicircle slice against the particle P.

Answers

Chapter 8

Exercises 8-1

1. (1) $\sum_{n=1}^{\infty} \dfrac{n!}{n^2+1}$; (2) $\dfrac{1}{2} + \sum_{n=2}^{\infty} (-1)^n \cdot \dfrac{n}{2^n}$;

   (3) $\sum_{n=1}^{\infty} (-1)^{n-1} \dfrac{a^{n+1}}{2n+1}$; (4) $\sum_{n=1}^{\infty} \dfrac{a^{\frac{n}{2}}}{2 \cdot 4 \cdot 6 \cdots 2n}$.

2. (1) $u_n = \dfrac{2}{n(n+1)}$; (2) $\dfrac{2}{1 \cdot 2} + \dfrac{2}{2 \cdot 3} + \dfrac{2}{3 \cdot 4} + \dfrac{2}{4 \cdot 5} + \dfrac{2}{5 \cdot 6} + \cdots$; (3) $s = 2$.

3. (1) $s_n = \dfrac{1}{2}\left(1 - \dfrac{1}{2n+1}\right)$; (2) $s = \dfrac{1}{2}$.

4. (1) divergent; (2) convergent; (3) divergent.

5. (1) $\dfrac{3}{4}$; (2) $\dfrac{\pi}{4}$; (3) $s_n = \sqrt{n+1} - 1$.

6. (1) convergent; (2) divergent; (3) divergent; (4) convergent;
   (5) $a > 1$, convergent; $0 < a \leqslant 1$, divergent; (6) convergent.

Exercises 8-2

1. (1) divergent; (2) divergent; (3) divergent; (4) convergent; (5) convergent.
2. (1) convergent; (2) convergent; (3) when $a < b$, convergent; when $a \geqslant b$, divergent.
3. (1) convergent; (2) divergent; (3) convergent; (4) convergent.
4. (1) convergent; (2) divergent; (3) convergent; (4) convergent;
   (5) convergent (Hint: u_n and the general term of the p-series with $p = \dfrac{3}{2}$ are the same order infinitesimal);
   (6) when $0 < a < 1$, convergent, when $a > 1$, divergent, when $a = 1$, if $s > 1$, convergent, if $s \leqslant 1$, divergent.

Exercises 8-3

1. (1) conditional convergent; (2) absolute convergent;
   (3) conditional convergent; (4) absolute convergent;
   (5) absolute convergent; (6) absolute convergent.
2. (1) 0; (2) 0.

Exercises 8-4

1. (1) $e^{-2} < x < e^2$;
   (2) when $x = 1$, $|x| < 1$ the general term does not tend to 0, divergent; when $|x| > 1$, by $0 < \left|\dfrac{1}{1+x^n}\right| < \dfrac{1}{|x|^n - 1}$ using d'Alembert method for $\sum_{n=1}^{\infty} \dfrac{1}{|x|^n - 1}$, convergent domain $|x| > 1$.

2. (1) $-2 < x < 2$; (2) $-4 < x < -4$.

3. (1) $R = \dfrac{1}{2}, \left(-\dfrac{1}{2}, \dfrac{1}{2}\right)$; (2) $R = \dfrac{1}{2}, \left[-\dfrac{1}{2}, \dfrac{1}{2}\right]$; (3) $R = 1, (0, 2]$;

   (4) $R = 1, [-1, 1)$; (5) $R = \sqrt{2}, (-\sqrt{2}, \sqrt{2})$; (6) $R = 1, \left(-\infty, -\dfrac{1}{2}\right]$.

4. $\dfrac{1}{2}\ln\dfrac{1+x}{1-x}$.   5. $\dfrac{x}{(1-x)^2}, -1 < x < 1.$ 6. $0.$

Exercises 8-5

1. $x^2 e^x = \sum\limits_{n=2}^{\infty} \dfrac{1}{(n-2)!} x^n$, $x \in (-\infty, +\infty)$.

2. $a^x = e^{x\ln a} = \sum\limits_{n=0}^{\infty} \dfrac{\ln^n a}{n!} x^n$, $x \in (-\infty, +\infty)$.

3. from $e^x + e^{-x} = 2\sum\limits_{n=0}^{\infty} \dfrac{x^{2n}}{(2n)!}$, we get $\sum\limits_{n=0}^{\infty} \dfrac{x^{2n}}{(2n)!} = \dfrac{1}{2}(e^x + e^{-x})$, $x \in (-\infty, +\infty)$.

4. (1) $\ln(a+x) = \ln a + \sum\limits_{n=0}^{\infty} (-1)^n \dfrac{x^{n+1}}{(n+1)a^{n+1}}$ $(-a < x \leqslant a)$;

   (2) find the derivative of the both sides of the expansion of $\dfrac{1}{1-x}$;

   (3) $\dfrac{x}{\sqrt{1+x^2}} = x + \sum\limits_{n=1}^{\infty} (-1)^n \dfrac{2(2n)!}{(n!)^2} \left(\dfrac{x}{2}\right)^{2n+1}$ $(-1 < x < 1)$;

   (4) $\ln(x + \sqrt{x^2+1}) = x + \sum\limits_{n=1}^{\infty} (-1)^n \dfrac{(2n-1)!! x^{2n+1}}{(2n)!!(2n+1)}$, $x \in [-1, 1]$.

   Hint: by the integral $\int_0^x \dfrac{dt}{\sqrt{t^2+1}}$.

5. $\cos x = \dfrac{1}{2} \sum\limits_{n=0}^{\infty} (-1)^n \left[\dfrac{\left(x+\dfrac{\pi}{3}\right)^{2n}}{(2n)!} + \sqrt{3} \cdot \dfrac{\left(x+\dfrac{\pi}{3}\right)^{2n+1}}{(2n+1)!} \right]$, $x \in (-\infty, +\infty)$.

6. $f(x) = \sum\limits_{n=0}^{\infty} \left(\dfrac{1}{2^{n+1}} - \dfrac{1}{3^{n+1}}\right)(x+4)^n$ $(-6 < x < 2)$.

7. $s = \dfrac{1}{\sqrt{2}} \ln(\sqrt{2}+1)$.   8. $s = 1$.

Quiz

1. $s = 1$.
2. (1) conditional convergent;   (2) conditional convergent.
3. when $\alpha > 1$, divergent; when $\alpha = 1$, conditional convergent; when $0 < \alpha < 1$, divergent.
4. (1) $R = 1, [-1, 1)$;   (2) $R = 1, [1, 3]$.
5. (1) $s(x) = \begin{cases} 1 + \dfrac{1-x}{x} \ln(1-x), & x \in [-1, 0) \cup (0, 1), \\ 0, & x = 0, \\ 1, & x = 1; \end{cases}$

   (2) $s(x) = 2x\arctan x - \ln(1+x^2)$, $x \in [-1, 1]$.

6. (1) $f(x) = -\dfrac{1}{5} \sum\limits_{n=0}^{\infty} \left[\dfrac{1}{3^{n+1}} + \dfrac{(-1)^n}{2^{n+1}}\right](x+4)^n$, $x \in (-6, -2)$;

   (2) $f(x) = \sum\limits_{n=0}^{\infty} \dfrac{(-1)^n}{2n+1} x^{2n+1} + \dfrac{\pi}{4}$, $x \in [-1, 1]$.

Exercises

1. (1) when $0 \leqslant \beta < 1$, α is any real number, or $\beta = 1, \alpha < -1$, convergent; when $\beta > 1$, α is any real number, or $\beta = 1, \alpha \geqslant -1$, divergent.

   (2) convergent.   (3) divergent.

2. when $m > \dfrac{1}{2}$, convergent; when $m \leqslant \dfrac{1}{2}$, divergent.

3. by comparison test, we get $u_n < \dfrac{1}{n^2}$, so the series is convergent.

Answers

4. (1) convergent;   (2) divergent.
5. (1) absolute convergent;   (2) conditional convergent.
6. (1) 0;   (2) $\sqrt[4]{8}$. Hint: rewrite into $2^{\frac{1}{3}+\frac{2}{3^2}+\cdots+\frac{n}{3^n}+\cdots}$.
7. (1) when $0<a\leqslant 1$, the convergent domain is $(-1,1)$; when $a>1$, the convergent domian is $[-1,1]$;

   (2) $\left[-3,\frac{1}{3}\right)$;   (3) $\left(-\frac{1}{e},\frac{1}{e}\right)$;   (4) $(-\sqrt{2},\sqrt{2})$.

8. (1) $s(x)=-\frac{x}{2}\ln(1-x^2)$ $(-1<x<1)$;   (2) $s(x)=\frac{2x}{(1-x)^3}$ $(-1<x<1)$;

   (3) $s(x)=\frac{x+2x^3}{(1-x^2)^2}$ $(-1<x<1)$.

9. (1) $s=\frac{3}{2}\sqrt{e}-1$;   (2) 2e.

10. (1) $\sum_{n=1}^{\infty}\frac{nx^{n-1}}{2^{n+1}}, x\in(-2,2)$;   (2) $1-x+x^8-x^9+x^{16}-x^{17}+\cdots$ $(-1<x<1)$.

Chapter 9

Exercises 9-1

1. (1) 1;   (2) 2;   (3) 3;   (4) 1;   (5) 3;   (6) 2.
2. yes, special solution;   no;   yes, general solution;   no;   yes, general solution.
4. (1) $y=x^2+C$;   (2) $y=x^2+3$;   (3) $y=x^2+4$;   (4) $y=x^2+\frac{5}{3}$.

5. $m\frac{\mathrm{d}v}{\mathrm{d}t}=k_1v+k_2t$.   6. $\frac{\mathrm{d}y}{\mathrm{d}x}=2\frac{y}{x}$.

Exercises 9-2

1. (1) $-\frac{1}{1+y}=x^3+C$;   (2) $y=\frac{1}{1+\ln|1+x|}$;

   (3) $2e^{3x}-3e^{-y^2}=C$;   (4) $y=2(1+x^2)$.

2. (1) $\arctan\frac{y}{x}+\ln\sqrt{x^2+y^2}=0$;   (2) $y^2=2x^2\ln x$;

   (3) $y^2=2x^2(C-\ln x)$;   (4) $1+\ln\frac{y}{x}=Cy$.

3. (1) $2y=C(x+1)^2+(x+1)^4$;   (2) $y=x^2(1+Ce^{\frac{1}{x}})$;

   (3) $y=e^{x^2}+\frac{1}{2}x^2$;   (4) $y=\frac{1}{x}+\frac{x^3}{4}$;

   (5) $y^{-1}=(C+x)\cos x$;   (6) $x=Cy+\frac{1}{2}y^3$.

4. $y=(1+x)e^x$.   5. $y=2(e^x-x-1)$.
6. (1) $y=\tan(x+C)$;   (2) $y=\arctan(x+y)+C$;

   (3) $x^2+y^2-2xy+4y+10x=C$.

Exercises 9-3

1. (1) $y=C_1x-\frac{3}{4}x^2+\frac{1}{2}x^2\ln x+C_2$;   (2) $y=C_1e^x-\frac{1}{2}x^2-x+C_2$;

   (3) $C_1y^2-1=(C_1x+C_2)^2$;   (4) $\frac{1}{2}y^2=C_1x+C_2$.

2. (1) $y=\frac{1}{8}e^{2x}-\frac{1}{4}e^2x^2+\frac{1}{4}e^2x-\frac{1}{8}e^2$;   (2) $y=x^3+3x+1$;

   (3) $y=-\frac{1}{a}\ln|ax+1|$;   (4) $y=\sqrt{2x-x^2}$.

3. $y = \dfrac{x^4}{12} + \dfrac{x}{6} + \dfrac{11}{4}$.

Exercises 9-4

1. (1),(2),(5) linear independent;(3),(4) linear dependent.

2. $y = (C_1 + C_2 x) e^{x^2}$.

3. $x = \cos 2t + \dfrac{1}{2} \sin 2t$.

5. $y = \cos 3x + \dfrac{9}{32} \sin 3x + \dfrac{1}{32}(4x\cos x + \sin x)$.

Exercises 9-5

1. (1) $y = C_1 e^{-x} + C_2 e^{-5x}$;
   (2) $y = C_1 e^{2x} + C_2 e^{-2x}$;
   (3) $y = (C_1 + C_2 x) e^{-2x}$;
   (4) $y = e^{-x}(C_1 \cos 2x + C_2 \sin 2x)$;
   (5) $y = C_1 \cos 3x + C_2 \sin 3x$;
   (6) $y = C_1 e^{-x} + C_2 e^{-x} + C_3 e^{-2x} + C_4 e^{-2x}$.

2. (1) $y = -e^{2x} + 2e^{4x}$;
   (2) $y = (1-x) e^{3x}$;
   (3) $y = e^{-3x}(3\cos 2x + 4\sin 2x)$.

3. $y = \cos 3x - \dfrac{1}{3} \sin 3x$.

4. (1) $y^* = (ax+b) e^{-x}$;
   (2) $y^* = x(ax^2 + bx + c) e^{-x}$;
   (3) $y^* = x^2 (ax+b) e^{2x}$;
   (4) $y^* = e^{2x}[(ax+b)\cos x + (cx+d)\sin x]$;
   (5) $y^* = x e^x [(ax+b)\cos 2x + (cx+d)\sin 2x]$.

5. (1) $-\dfrac{9}{8}(1 - e^{2x}) - \dfrac{x}{4}(x+5)$;
   (2) $y = C_1 e^{-x} + C_2 e^{-2x} + \left(\dfrac{3}{2} x^2 - 3x\right) e^{-x}$;
   (3) $y = -5e^x + \dfrac{7}{2} e^{2x} + \dfrac{5}{2}$;
   (4) $y = -\cos x - \dfrac{1}{3} \sin x + \dfrac{1}{3} \sin 2x$;
   (5) $y = C_1 \sin x + C_2 \cos x - 2x\cos x$;
   (6) $y = C_1 \cos x + C_2 \sin x + \dfrac{e^x}{2} + \dfrac{x}{2} \sin x$.

6. $f(x) = -\dfrac{1}{2}(3\cos x + \sin x) + \dfrac{3}{2} e^x$.

7. $\varphi(x) = \dfrac{1}{2}(\cos x + \sin x + e^x)$.

8. $y = C_1 e^{2x} + C_2 e^{3x} - \dfrac{4}{3} - 2x + 3e^x$.

Quiz

1. (1) $\sqrt{1 - y^2} = \arcsin x + C$;
   (2) $x = y\left(1 - \dfrac{1}{2} \ln |y|\right)^2$;
   (3) $y = \dfrac{e^{-x}}{x}(1 + x^3)$;
   (4) $\dfrac{1}{y} = \dfrac{C}{x} + \dfrac{1}{x^2}$;
   (5) $y = \dfrac{C_1}{2} x^2 + C_1^2 x + C_2$;
   (6) $y = \tan\left(x + \dfrac{\pi}{4}\right)$.

2. $f(x) = \sqrt{x}$.

3. (1) $y = \dfrac{1}{5} e^{3x} + \dfrac{4}{5} e^{-2x}$.
   (2) $y = (C_1 + C_2 x) e^{-3x}$.
   (3) when $a \neq 0$, $y = C_1 e^x + C_2 e^{-x} - \dfrac{1}{a^2}(x+1)$; when $a = 0$, $y = C_1 + C_2 x + \dfrac{1}{6} x^3 + \dfrac{1}{2} x^2$.
   (4) $y = \dfrac{1}{6} x^3 e^x$.
   (5) $y = C_1 \cos 2x + C_2 \sin 2x + \dfrac{1}{4} x \sin 2x$.
   (6) $y = e^x (C_1 \cos x + C_2 \sin x) + 2x e^x \sin x$.

Exercises

1. (1) 3.   (2) ① $y = C e^{-\int P(x) dx}$; ② $y = e^{-\int P(x) dx}\left(\int Q(x) e^{\int P(x) dx} dx + C\right)$.   (3) 1.
   (4) $y = C_1 (x-1) + C_2 (x^2 - 1) + 1$.

2. (1) $\dfrac{y}{x} + \dfrac{1}{2} y^2 = C$;
   (2) $x - \sqrt{xy} = C$;

Answers

(3) $(e^y-1)(e^x+1)=C$;

(4) $y=C\cos x+\sin x$;

(5) $y=(x-2)(C+x^2-4x)$;

(6) $y=Ce^{-f(x)}+f(x)-1$;

(7) $y=\dfrac{1}{x}e^{Cx}$;

(8) $y^{-2}=Ce^{x^2}+x^2+1$;

(9) $y=\dfrac{C_1 e^{C_1 x+C_2}}{1-e^{C_1 x+C_2}}$;

(10) $y=C_1+C_2 e^x+C_3 e^{-2x}+\left(\dfrac{1}{6}x^2-\dfrac{4}{9}x\right)e^x-x^2-x$;

(11) $y=e^{-x}(C_1\cos 2x+C_2\sin 2x)-\dfrac{4}{17}\cos 2x+\dfrac{1}{17}\sin 2x$;

(12) $y=(C_1+C_2 x)e^{2x}+C_3 e^{-2x}+\dfrac{1}{4}+e^x+\dfrac{1}{2}x^2 e^{2x}$.

3. (1) $x(1+2\ln y)-y^2=0$;

(2) $y=\dfrac{5}{2}e^x-2e^{2x}+\dfrac{1}{2}e^{3x}$;

(3) $y=2\arctan e^x$;

(4) $y=xe^{-x}+\dfrac{1}{2}\sin x$.

4. $f(x)=\cos x+\sin x$.

5. $s=\dfrac{mg}{k}\left(t+\dfrac{m}{k}e^{\frac{k}{m}t}-\dfrac{m}{k}\right)$.

Chapter 10

Exercises 10-1

1. (1) two components are 0; (2) one components is 0; (3) $y=\pm 3$; (4) $z=\pm 5$.
2. A is on the plane xOz; B is on the plane yOz; C is on the z-axis; D is on the y-axis.
3. A: IV; B: V; C: VIII; D: III.
4. P: (1) $(2,-3,1),(-2,-3,-1),(2,3,-1)$;

   (2) $(2,3,1),(-2,-3,1),(-2,3,-1)$; (3) $(-2,3,1)$.

   M: (1) $(a,b,-c),(-a,b,c),(a,-b,c)$;

   (2) $(a,-b,-c),(-a,b,-c),(-a,-b,c)$; (3) $(-a,-b,-c)$.
5. Hint: $|CA|=|CB|=\sqrt{6}$.
6. $(0,1,-2)$.
7. $x^2+y^2+z^2-2x-6y+4z=0$.
8. the center is $(-1,2,0)$; radius is 3.

Exercises 10-2

1. $4e_1+e_3$; $-2e_1+4e_2-3e_3$; $-3e_1+10e_2-7e_3$.
2. Hint: $\overrightarrow{AB}+\overrightarrow{BC}+\overrightarrow{CD}=2a+10b=2\overrightarrow{AB}$. 3. $B(-2,4,-3)$.
4. $\overrightarrow{P_1P_2}=(-2,-2,-2)$; $5\overrightarrow{P_1P_2}=(-10,-10,-10)$.
5. $|a|=\sqrt{3},|b|=\sqrt{38},|c|=3;a°=\left(\dfrac{\sqrt{3}}{3},\dfrac{\sqrt{3}}{3},\dfrac{\sqrt{3}}{3}\right),b°=\left(\dfrac{2}{\sqrt{38}},\dfrac{-3}{\sqrt{38}},\dfrac{5}{\sqrt{38}}\right)$,

   $c°=\left(\dfrac{-2}{3},\dfrac{-1}{3},\dfrac{2}{3}\right);a=\sqrt{3}a°,b=\sqrt{38}b°,c=3c°$.
6. $A(-1,2,4)$; $B(8,-4,-2)$.
7. (1) $3,5i+j+7k$; (2) $-18,10i+2j+14k$;

   (3) $\cos(\widehat{a,b})=\dfrac{3}{2\sqrt{21}}$, $\sin(\widehat{a,b})=\dfrac{5}{2\sqrt{7}}$, $\tan(\widehat{a,b})=\dfrac{5\sqrt{3}}{3}$.
8. (1) $l=10$; (2) $l=-2$. 9. (1) $-8j-24k$; (2) $-j-k$.
10. (1) $3\sqrt{6}$; (2) $\dfrac{3\sqrt{21}}{7},\dfrac{3\sqrt{6}}{\sqrt{77}}$.

Exercises 10-3

1. $3x-7y+5z-4=0$. 2. $11x-17y-13z+3=0$.

3. the plane parallel to the x-axis: $z=1$; the plane parallel to the y-axis: $z=1$; the plane parallel to the z-axis: $x+y-1=0$.

4. (1) parallel to the z-axis; (2) passing through the origin; (3) parallel to the plane yOz; (4) passing through the y-axis.

5. (1) $\left(\dfrac{2}{7},\dfrac{3}{7},\dfrac{6}{7}\right)$, $\cos\alpha=\dfrac{2}{7}$, $\cos\beta=\dfrac{3}{7}$, $\cos r=\dfrac{6}{7}$;

   (2) $\left(\dfrac{1}{3},-\dfrac{2}{3},\dfrac{2}{3}\right)$, $\cos\alpha=\dfrac{1}{3}$, $\cos\beta=-\dfrac{2}{3}$, $\cos r=\dfrac{2}{3}$.

6. (1) $\dfrac{\pi}{4}$; (2) $\arccos\dfrac{8}{21}$.

7. (1) $l-3m-9=0$; (2) $m=3, l=-4$.

8. (1) $\dfrac{1}{3}$; (2) 0; (3) $\dfrac{16}{\sqrt{14}}$.

9. (1) $x=-y=z$; (2) $\dfrac{x-2}{3}=\dfrac{y-5}{5}=\dfrac{z-8}{5}$;

   (3) $\dfrac{x-2}{1}=\dfrac{y+8}{2}=\dfrac{z-3}{-3}$; (4) $\dfrac{x-1}{1}=\dfrac{y}{1}=\dfrac{z+2}{2}$.

10. $\dfrac{x-\frac{11}{3}}{1}=\dfrac{y+\frac{7}{3}}{-1}=\dfrac{z}{-1}$. 11. $\arccos\dfrac{72}{77}$.

12. $\dfrac{x-1}{1}=\dfrac{y}{\frac{5}{2}}=\dfrac{z+2}{1}$.

13. $16x-14y-11z-65=0$. 14. $\dfrac{\pi}{6}$.

15. $x-y-3z-7=0$. 16. $-x+y+z+2=0$.

Exercises 10-4

1. $3x^2+3y^2+3z^2-8x-14y+4z-21=0$.

2. (1) elliptic cylinder; (2) parabolic cylinder; (3) hyperbolic cylinder; (4) plane.

3. (1) $\dfrac{x^2}{4}+\dfrac{y^2+z^2}{9}=1$; (2) $x^2+y^2-z^2=1$;

   (3) $y^2+z^2=5x$; (4) $4(x^2+z^2)-9y^2=36$.

4. (1) straight line; (2) hyperbolic.

5. (1) $\begin{cases} x=3z+1, \\ y=\left(\dfrac{z}{2}+1\right)^2; \end{cases}$ (2) $\dfrac{x^2}{18}+\dfrac{y^2}{50}+\dfrac{z^2}{16}=1$.

6. (1) $\begin{cases} x=\dfrac{3}{\sqrt{2}}\cos t, \\ y=\dfrac{3}{\sqrt{2}}\cos t, \\ z=3\sin t \end{cases}$ $(0\leqslant t\leqslant 2\pi)$; (2) $\begin{cases} x=1+\sqrt{3}\cos\theta, \\ y=\sqrt{3}\sin\theta, \\ z=0 \end{cases}$ $(0\leqslant\theta\leqslant 2\pi)$.

7. the project cylinder on the plane xOy is $2x^2+4y^2-7x-8y+5xy+1=0$;

   the project cylinder on the plane xOy is $\begin{cases} 2x^2+4y^2-7x-8y+5xy+1=0, \\ z=0; \end{cases}$

   the project cylinder on the plane yOz is $y^2+2z^2+3z-yz-4=0$;

   the project cylinder on the plane yOz is $\begin{cases} y^2+2z^2+3z-yz-4=0, \\ x=0; \end{cases}$

   the project cylinder on the plane xOz is $x^2+4z^2-2x+3xz-3=0$;

Answers

the project cylinder on the plane xOz is $\begin{cases} x^2+4z^2-2x+3xz-3=0, \\ y=0. \end{cases}$

8. (1) ellipsoid; (2) hyperboloid of one sheet; (3) hyperboloid of two sheet; (4) paraboloid; (5) hyperboloid of two sheet; (6) hyperbolic paraboloid.

9. (1) ellipse; (2) circle; (3) hyperbola; (4) circle.

Quiz

1. (1) $(1,4,4)$; (2) $|\overrightarrow{AB}|=\sqrt{13}$; (3) $\overrightarrow{AB}°=\left(0,\dfrac{2}{\sqrt{13}},\dfrac{3}{\sqrt{13}}\right)$.

2. (1) $(-1,-16,3)$; (2) $\mathbf{a}\cdot\mathbf{b}=11, \mathbf{a}\times\mathbf{b}=(-7,-5,-1)$; (3) -21.

3. $\dfrac{\sqrt{2}}{2}$.   4. $x-3y+4z-13=0$.   5. $9y-z-2=0$.

6. $\dfrac{x-1}{1}=\dfrac{y-1}{2}=\dfrac{z-1}{-1}$.   7. $(-3,-3,-3)$.

8. (1) cylindrical surface; (2) hyperboloid of one sheet; (3) ellipsoidal; (4) paraboloid; (5) hyperbola; (6) straight line in space.

9. the projective curvilinear equation on the plane xOy is $\begin{cases} 3x^2+4y^2-2x-2y+2xy=0, \\ z=0; \end{cases}$

the projective curvilinear equation on the plane yOz is $\begin{cases} 5y^2+3z^2-4y-4z+4yz+1=0, \\ x=0; \end{cases}$

the projective curvilinear equation on the plane xOz is $\begin{cases} 5x^2+4z^2-6x-6z+6xz+2=0, \\ y=0. \end{cases}$

Exercises

2. (1) $(3,-2,0)$;   (2) $\left(\dfrac{2}{\sqrt{14}},\dfrac{-3}{\sqrt{14}},\dfrac{-1}{\sqrt{14}}\right)$.

3. (1) $\mathbf{a}\cdot\mathbf{b}=11, \mathbf{a}\times\mathbf{b}=8\mathbf{i}-6\mathbf{j}-2\mathbf{k}, (\mathbf{a}+\mathbf{b})\cdot(\mathbf{a}-\mathbf{b})=-16$;

   (2) $\left(\dfrac{4}{\sqrt{26}},\dfrac{-3}{\sqrt{26}},\dfrac{-1}{\sqrt{26}}\right)$.

4. $S_{\triangle ABC}=2\sqrt{10}$; the height on the side $AB=\dfrac{4\sqrt{10}}{\sqrt{11}}$.

5. $V=\dfrac{58}{3}, h=\dfrac{29}{7}$.   6. $3x+26y+5z-2=0$.

7. (1) $k=1$;   (2) $k=-\dfrac{7}{3}$.

8. (1) $\begin{cases} z=2y, \\ x=0; \end{cases}$ (2) $\dfrac{x-3}{4}=\dfrac{y+3}{1}=\dfrac{z-2}{17}$.

9. (1) $\left(-4,\dfrac{9}{2},\dfrac{3}{2}\right)$;   (2) $(1,0,1)$.

10. $\dfrac{x+1}{9}=\dfrac{y}{5}=\dfrac{z-4}{-7}$.   11. $x^2+y^2+z^2=9$.

12. (1) sphere; (2) cylinder; (3) hyperbolic paraboloid; (4) hyperboloid of two sheet.

13. (1) projected cylinder $x^2+y^2=1$, projected curve $\begin{cases} x^2+y^2=1, \\ z=0; \end{cases}$

    (2) projected cylinder $x^2+y^2=2y$, projected curve $\begin{cases} x^2+y^2=2y, \\ z=0. \end{cases}$

Chapter 11

Exercises 11-1

1. (1) bounded domain; (2) bounded closed domain; (3) bounded closed domain; (4) unbounded domain.

2. (1) $\{(x,y)|x^2+y^2>1\}$; (2) $\{(x,y)|1\leqslant x^2+y^2\leqslant 7\}$; (3) $\{(x,y)|\sqrt{y}\leqslant x, 0\leqslant y<+\infty\}$; (4) $\{(x,y)|2k\pi\leqslant x^2+y^2\leqslant(2k+1)\pi, k=0,1,2,\cdots\}$; (5) $\{(x,y)|y-x>0, x>0, x^2+y^2<1\}$; (6) $\{(x,y)|r^2<x^2+y^2\leqslant R^2\}$.

3. $f(-y,x)=\dfrac{x^2-y^2}{2xy}$, $f\left(\dfrac{1}{x},\dfrac{1}{y}\right)=\dfrac{-x^2+y^2}{2xy}$, $f[x,f(x,y)]=\dfrac{4x^4y^2-(x^2-y^2)^2}{4x^2y(x^2-y^2)}$.

4. $f(x)=x(x+2)$; $z=\sqrt{y}+x-1$.

5. (1) 0; (2) $-\dfrac{1}{4}$; (3) 0; (4) ∞; (5) e^k; (6) 0.

7. $f(x)$ is continuous in domain.   8. $x^2+y^2=\left(k+\dfrac{1}{2}\right)\pi, k=0,\pm1,\pm2,\cdots$.

9. $f(x,y)$ is continuous at point $(0,0)$.   10. don't exist.

Exercises 11-2

1. (1) $z'_x=y+\dfrac{1}{y}$, $z'_y=x-\dfrac{x}{y^2}$;   (2) $z'_x=\dfrac{y^2}{(x^2+y^2)^{\frac{3}{2}}}$, $z'_y=\dfrac{-xy}{(x^2+y^2)^{\frac{3}{2}}}$;

   (3) $z'_x=\dfrac{1}{1+(x-y^2)^2}$, $z'_y=\dfrac{-2y}{1+(x-y^2)^2}$;

   (4) $z'_x=\sin(x+y)+x\cos(x+y)$, $z'_y=x\cos(x+y)$;

   (5) $z'_x=\dfrac{2x}{y}\sec^2\dfrac{x^2}{y}$, $z'_y=-\dfrac{x^2}{y^2}\sec^2\dfrac{x^2}{y}$;

   (6) $z'_x=y^2(1+xy)^{y-1}$, $z'_y=(1+xy)^y\left[\ln(1+xy)+\dfrac{xy}{1+xy}\right]$;

   (7) $u'_x=yz(xy)^{z-1}$, $u'_y=xz(xy)^{z-1}$, $u'_z=(xy)^z\ln xy$;

   (8) $u'_x=\dfrac{z}{y}\left(\dfrac{x}{y}\right)^{z-1}$, $u'_y=-\dfrac{z}{y^2}\left(\dfrac{x}{y}\right)^{z-1}$, $u'_z=\left(\dfrac{x}{y}\right)^z\ln\left(\dfrac{x}{y}\right)$;

   (9) $u'_x=e^{x(x^2+y^2+z^2)}(3x^2+y^2+z^2)$, $u'_y=e^{x(x^2+y^2+z^2)}2xy$, $u'_z=e^{x(x^2+y^2+z^2)}2xz$;

   (10) $u'_x=\dfrac{z(x-y)^{z-1}}{1+(x-y)^{2z}}$, $u'_y=\dfrac{-z(x-y)^{z-1}}{1+(x-y)^{2z}}$, $u'_z=\dfrac{(x-y)^z\ln(x-y)}{1+(x-y)^{2z}}$.

2. $1, \dfrac{1}{2}, \dfrac{1}{2}$.   4. $\dfrac{x}{\sqrt{1+x^2}}$.   5. $\dfrac{\pi}{6}$.

6. (1) $\dfrac{\partial^2 z}{\partial x^2}=\dfrac{-y}{(\sqrt{2xy+y^2})^3}$, $\dfrac{\partial^2 z}{\partial y^2}=\dfrac{-x^2}{(y+2xy)^{\frac{3}{2}}}$, $\dfrac{\partial^2 z}{\partial x\partial y}=\dfrac{xy}{(2xy+y^2)^{\frac{3}{2}}}$;

   (2) $\dfrac{\partial^2 z}{\partial x^2}=\dfrac{2x}{(1+x^2)^2}$, $\dfrac{\partial^2 z}{\partial y^2}=\dfrac{2y}{(1+y^2)^2}$, $\dfrac{\partial^2 z}{\partial x\partial y}=0$;

   (3) $\dfrac{\partial^2 z}{\partial x^2}=y^x(\ln y)^2$, $\dfrac{\partial^2 z}{\partial y^2}=x(x-1)y^{x-2}$, $\dfrac{\partial^2 z}{\partial x\partial y}=xy^{x-1}\ln y$;

   (4) $\dfrac{\partial^2 z}{\partial x^2}=2a^2\cos 2(ax+by)$, $\dfrac{\partial^2 z}{\partial y^2}=2b^2\cos 2(ax+by)$, $\dfrac{\partial^2 z}{\partial x\partial y}=2ab\cos 2(ax+by)$.

7. $\dfrac{\partial^3 u}{\partial x\partial y\partial z}=\alpha\beta\gamma x^{\alpha-1}y^{\beta-1}z^{\gamma-1}$.

9. (1) $dz=2xy^3 dx+3x^2y^2 dy$;   (2) $dz=\dfrac{4xy}{(x^2+y^2)^2}(ydx-xdy)$;

   (3) $dz=y^2 x^{y-1}dx+x^y(1+y\ln x)dy$;   (4) $dz=\sin 2xdx-\sin 2ydy$;

   (5) $dz=0$;

Answers

(6) $dz = \left(xy+\dfrac{x}{y}\right)^{z-1}\left[\left(y+\dfrac{1}{y}\right)z\,dx+\left(1-\dfrac{1}{y^4}\right)xz\,dy+\left(xy+\dfrac{x}{y}\right)\ln\left(xy+\dfrac{x}{y}\right)dz\right].$

10. $df(3,4,5)=\dfrac{1}{25}(5dz-4dy-3dx).$ 11. $\Delta z \approx 0.028\,252;\ dz \approx 0.027\,778.$

12. $du\big|_{(1,1,1)}=dx-dy.$ 13. (1) 108.972; (2) 2.95.

14. (1) $f_x(x,y)=2x\sin\dfrac{1}{\sqrt{x^2+y^2}}-\dfrac{x}{\sqrt{x^2+y^2}}\cos\dfrac{1}{\sqrt{x^2+y^2}},$

$f_y(x,y)=2y\sin\dfrac{1}{\sqrt{x^2+y^2}}-\dfrac{y}{\sqrt{x^2+y^2}}\cos\dfrac{1}{\sqrt{x^2+y^2}};$

(2) $f_x(x,y),f_y(x,y)$ are discontinuous at $(0,0)$; (3) $f(x,y)$ is differentiable at $(0,0)$.

15. 7.6 m. 16. $34\,560$ g.

19. $\dfrac{dz}{dt}=-e^t-e^{-t}.$

20. $\dfrac{\partial z}{\partial x}=3x^2\sin y\cos y(\cos y-\sin y),$

$\dfrac{\partial z}{\partial y}=-2x^3\sin y\cos y(\sin y+\cos y)+x^3(\sin^3 y+\cos^3 y).$

21. $\dfrac{dz}{dx}=\dfrac{e^x(1+x)}{1+x^2e^{2x}}.$

23. (1) $\dfrac{\partial u}{\partial x}=2xf',\ \dfrac{\partial u}{\partial y}=-2f',\ \dfrac{\partial u}{\partial z}=2zf';$

(2) $\dfrac{\partial u}{\partial x}=\dfrac{1}{y}f'_1,\ \dfrac{\partial u}{\partial y}=-\dfrac{x}{y^2}f'_1+\dfrac{1}{z}f'_2,\ \dfrac{\partial u}{\partial z}=-\dfrac{y}{z^2}f'_2;$

(3) $\dfrac{\partial u}{\partial x}=2xf'_1+yf'_2+yzf'_3,\ \dfrac{\partial u}{\partial y}=2yf'_1+xf'_2+xzf'_3,\ \dfrac{\partial u}{\partial z}=xyf'_3.$

24. (1) $\dfrac{\partial^2 z}{\partial x^2}=f''_{11}+\dfrac{2}{y}f''_{12}+\dfrac{1}{y^2}f''_{22},\ \dfrac{\partial^2 z}{\partial x\partial y}=-\dfrac{x}{y^2}\left(f''_{11}+\dfrac{1}{y}f''_{12}\right)-\dfrac{1}{y^2}f'_2,\ \dfrac{\partial^2 z}{\partial y^2}=\dfrac{2x}{y^3}f'_2+\dfrac{x^2}{y^4}f''_{22};$

(2) $\dfrac{\partial^2 z}{\partial x^2}=2yf'_2+y^4f''_{11}+4xy^3f''_{12}+4x^2y^2f''_{22},$

$\dfrac{\partial^2 z}{\partial x\partial y}=2yf'_1+2x^4f'_2+2xy^3f''_{11}+2x^3yf''_{22}+5x^2y^2f''_{12},$

$\dfrac{\partial^2 z}{\partial y^2}=2xf'_1+4x^2y^2f''_{11}+4x^3yf''_{12}+x^4f''_{22}.$

28. $\dfrac{dy}{dx}=-\dfrac{x}{y},\ \dfrac{d^2y}{dx^2}=-\dfrac{y^2+x^2}{y^3}.$ 29. $\dfrac{\partial z}{\partial x}=-\dfrac{z}{x},\ \dfrac{\partial z}{\partial y}=\dfrac{(2xyz-1)z}{(2xz-2xyz+1)y}.$

32. $\dfrac{dy}{dx}=\dfrac{-2x-24xz}{10y+36zy^2},\ \dfrac{dz}{dx}=\dfrac{-10x-3xy}{5+18zy}.$

33. $\dfrac{\partial z}{\partial x}=(v\cos v-n\sin v)e^{-u},\ \dfrac{\partial z}{\partial y}=(u\cos v+v\sin v)e^{-u}.$

34. $\dfrac{\partial z}{\partial x}=f'_x(x,y)+3x^2g'_u(u,v)+yx^{y-1}g'_v(u,v),\ \dfrac{\partial z}{\partial y}=f'_y(x,y)+x^y\ln x\cdot g'_v(u,v).$

36. (1) $\dfrac{EQ_x}{EP_x}=-1,\ \dfrac{EQ_y}{EP_y}=-0.6;$ (2) 0.75.

Exercises 11-3

1. $1+2\sqrt{3}.$ 2. $\dfrac{98}{13}.$

3. $\dfrac{\partial f}{\partial l}=\cos\alpha-\sin\alpha;$ when $\alpha=\dfrac{\pi}{4},\ \dfrac{\partial f}{\partial l}$ reaches the maximum; when $\alpha=\dfrac{5}{4}\pi,\ \dfrac{\partial f}{\partial l}$ reaches the minimum;

when $\alpha=\dfrac{3}{4}\pi$ 或 $\dfrac{7}{4}\pi,\ \dfrac{\partial f}{\partial l}=0.$

4. when $\theta=\dfrac{\pi}{4},\ \dfrac{\partial z}{\partial l}\bigg|_{(1,2)}=\dfrac{\sqrt{2}}{3};$ when $\theta=\dfrac{5}{4}\pi,\ \dfrac{\partial z}{\partial l}\bigg|_{(1,2)}=\dfrac{\sqrt{2}}{3}.$

5. at $l^\circ = \dfrac{1}{\sqrt{14}}(1,2,3), \dfrac{\partial u}{\partial l}$ reaches the maximum value $\sqrt{14}$;

at $l^\circ = -\dfrac{1}{\sqrt{14}}(1,2,3), \dfrac{\partial u}{\partial l}$ reaches the minimum value $-\sqrt{14}$.

6. $\left.\dfrac{\partial u}{\partial n}\right|_{(x_0,y_0,z_0)} = \pm\dfrac{x_0+y_0+z_0}{\sqrt{x_0^2+y_0^2+z_0^2}}$.

7. **grad** $f(1,1,1) = 6\boldsymbol{i}+3\boldsymbol{j}$, **grad** $f(2,2,2) = 9\boldsymbol{i}+8\boldsymbol{j}+6\boldsymbol{k}$.

8. $|\textbf{grad } u| = \dfrac{1}{r_0^2}, \cos(\textbf{grad } u, x) = -\dfrac{x_0}{r_0}, \cos(\textbf{grad } u, y) = -\dfrac{y_0}{r_0}, \cos(\textbf{grad } u, z) = -\dfrac{z_0}{r_0}$.

Exercises 11-4

1. tangent: $\dfrac{x-\frac{1}{2}}{1} = \dfrac{y-2}{-4} = \dfrac{z-1}{8}$, normal plane: $2x-8y+16z-1=0$.

2. tangent: $\dfrac{x-a\cos\alpha\cos t_0}{\cos\alpha\sin t_0} = \dfrac{y-a\sin\alpha\cos t}{\sin\alpha\sin t_0} = \dfrac{z-a\sin t_0}{-\cos t_0}$,

   normal plane: $\cos\alpha\sin t_0(x-a\cos\alpha\cos t_0) + \sin\alpha\sin t_0(y-\sin\alpha\cos t_0) - \cos t_0(z-a\sin t_0) = 0$.

3. $P_1(-1,1,-1)$ or $P_2\left(-\dfrac{1}{3},\dfrac{1}{9},-\dfrac{1}{27}\right)$.

4. tangent: $\dfrac{x-1}{1} = \dfrac{y+2}{0} = \dfrac{z-1}{-1}$, normal plane: $x-z=0$.

5. point $(-3,-1,3)$, normal equation: $\dfrac{x+3}{1} = \dfrac{y+1}{3} = \dfrac{z-3}{1}$.

6. $\cos r = \dfrac{3}{\sqrt{22}}$. 7. $\left(\pm\dfrac{a^2}{\sqrt{a^2+b^2+c^2}}, \pm\dfrac{b^2}{\sqrt{a^2+b^2+c^2}}, \pm\dfrac{c^2}{\sqrt{a^2+b^2+c^2}}\right)$.

8. $x+y+z = \sqrt{a^2+b^2+c^2}$ or $x-y-z = -\sqrt{a^2+b^2+c^2}$.

Exercises 11-5

1. local minimum $f(2,-1)=-28$, local maximum $f(-2,1)=28$.
2. local minimum $f(5,2)=30$. 3. local maximum $f(-4,-2)=8e^{-2}$.
4. local minimum $z\left(\dfrac{ab^2}{a^2+b^2},\dfrac{a^2b}{a^2+b^2}\right) = \dfrac{a^2b^2}{a^2+b^2}$. 5. $d = \dfrac{7}{8}\sqrt{2}$.
6. $\left(\dfrac{a}{\sqrt{3}}, \dfrac{b}{\sqrt{3}}, \dfrac{c}{\sqrt{3}}\right)$. 7. $r = \dfrac{1}{\sqrt[3]{2\pi}}, h = \dfrac{2}{\sqrt[3]{2\pi}}$.
8. $\left(\dfrac{8}{5}, \dfrac{16}{5}\right)$. 9. $a = 6\left(\dfrac{qx}{py}\right)^y, b = 6\left(\dfrac{py}{qx}\right)^x$.
10. $x=250, y=50$, that is, when we hire 250 workers we could get the maximum output $L(250,50) = 16\ 719$.

Exercises 11-6

1. $f(x,y) = 5 + 2(x-1)^2 - (x-1)(y+2) - (y+2)^2$.

2. $f(x,y) = y + \dfrac{1}{2!}(2xy-y^2) + \dfrac{1}{3!}(3x^2y-3xy^2+2y^3) + \dfrac{1}{4!}e^{\theta x}\left[\ln(1+\theta)x^4 + \dfrac{4}{1+\theta y}x^3y - \dfrac{6}{(1+\theta y)^2}x^2y^2 + \dfrac{8}{(1+\theta y)^3}xy^3 - \dfrac{6}{(1+\theta y)^4}y^4\right], 0<\theta<1$.

3. $f(x,y) = 1 + (x+y) + \dfrac{1}{2!}(x^2+2xy+y^2) + \cdots + \dfrac{1}{n!}\left\{x^n + nx^{n-1}y + \dfrac{n(n-1)}{2!}x^{n-2}y^2 + \cdots + y^n\right.$

   $\left. + \dfrac{1}{(n+1)!}e^{\theta x+y}[x^{n+1} + (n+1)x^ny + \cdots + y^{n+1}]\right\}, 0<\theta<1$.

Answers

Quiz

1. (1) C;   (2) B;   (3) B.

2. (1) $x>0, y>1$ or $x<0, 0<y<1$; (2) $(3,-1)$; (3) $y-z=2$; (4) $\dfrac{\cos x}{1+e^z}$.

3. (1) $z'_x = (e^v \sin u + u e^v \cos u) y + u e^v + u e^v \sin u$, $z'_y = e^v(x \sin u + x u \cos u + u \sin u)$;

   (2) $-\dfrac{1}{2}$; (3) $\dfrac{x-1}{1}=\dfrac{y+1}{-2}=\dfrac{z-1}{3}$ 或 $\dfrac{x-\frac{1}{3}}{1}=\dfrac{y+\frac{1}{9}}{-\frac{2}{3}}=\dfrac{z-\frac{1}{27}}{\frac{1}{3}}$.

4. $f(x)=x^2+2x, z=\sqrt{y}+x-1$. 5. $z\left(\dfrac{ab^2}{a^2+b^2}, \dfrac{a^2 b}{a^2+b^2}\right)=\dfrac{a^2 b^2}{a^2+b^2}$. 6. $\dfrac{\partial u}{\partial n}=\pm\sqrt{5}$.

7. $a=-\dfrac{11}{4}$ or $a=\dfrac{13}{4}$.   8. $C_1 e^u + C_2 e^{-u}$ (C_1, C_2 be any constants).

Exercises

1. (1) $x^2+y^2>1$; (2) sufficient, necessary. 2. (1) B; (2) D. 3. 0. 5. $-1, -1$.

6. (1) $z_x = y + \dfrac{x}{x^2+y^2}, z_y = x + \dfrac{y}{x^2+y^2}$,

   $z_{xx} = \dfrac{y^2-x^2}{(x^2+y^2)^2}, z_{xy} = 1 - \dfrac{2xy}{(x^2+y^2)^2}, z_{yy} = \dfrac{x^2-y^2}{(x^2+y^2)^2}$;

   (2) $u_x = y x^{y-1}, u_y = x^y \ln x, u_{xx} = y(y-1) x^{y-2}, u_{yy} = x^y \ln^2 x, u_{xy} = x^{y-1}(1+y \ln x)$.

7. (1) $dz = \dfrac{x^2+y^2}{(x^2-y^2)^2}(-y\,dx + x\,dy)$; (2) $du = (1+\ln x)dx + (1+\ln y)dy + (1+\ln z)dz$.

8. $\dfrac{dz}{dx} = \dfrac{\partial F}{\partial u}\varphi'(x) + \dfrac{\partial F}{\partial v}\psi'(x) + \dfrac{\partial F}{\partial x}$. 10. $\dfrac{\partial^2 z}{\partial x^2} = \dfrac{y^2}{x^3} f'' + \dfrac{2}{y} \varphi''$; $\dfrac{\partial^2 z}{\partial x \partial y} = -\dfrac{y}{x^3} f'' - \dfrac{2x}{y^2} \varphi''$.

11. $z_{\text{local minimum}}(2a-b, 2b-a) = 3(ab - a^2 - b^2)$.

13. $\dfrac{x-8}{8} = \dfrac{y-1}{0} = \dfrac{z-2\ln 2}{1}$; $8(x-8)+(z-2\ln 2)=0$.

14. maximum $\sqrt{14}$, minimum $-\sqrt{14}$.   15. $R=\sqrt{\dfrac{S}{3\pi}}$, $h=\dfrac{2}{3}\sqrt{\dfrac{3S}{\pi}}$.

16. $x=15$, $y=10, 15$ thousand Yuan.   17. $p_1 = 80$, $p_2 = 120$.

Chapter 12

Exercises 12-1

2. (1) $\geqslant$; (2) $\leqslant$. 3. negative. 4. (1) $0 \leqslant I \leqslant \pi^2$; (2) $36\pi \leqslant I \leqslant 100\pi$.

Exercises 12-2

1. (1) $\dfrac{\pi^2}{4}$; (2) $(e-1)^2$; (3) -2; (4) $\dfrac{9}{8}\ln 3 - \ln 2 - \dfrac{1}{2}$.

2. (1) $\dfrac{1}{6} a^2 b^2 (a-b)$; (2) $2\dfrac{3}{5}$; (3) $\dfrac{1}{21} p^5$; (4) $\dfrac{64}{15}$.

3. (1) $\displaystyle\int_0^1 dx \int_{x-1}^{1-x} f(x,y) dy = \int_{-1}^0 dy \int_0^{1+y} f(x,y) dx + \int_0^1 dy \int_0^{1-y} f(x,y) dx$;

   (2) $\displaystyle\int_{-\sqrt{2}}^{\sqrt{2}} dx \int_{x^2}^{4-x^2} f(x,y) dy = \int_0^2 dy \int_{-\sqrt{y}}^{\sqrt{y}} f(x,y) dx + \int_2^4 dy \int_{-\sqrt{4-y}}^{\sqrt{4-y}} f(x,y) dx$;

   (3) $\displaystyle\int_1^3 dx \int_x^{3x} f(x,y) dy = \int_1^3 dy \int_1^y f(x,y) dx + \int_3^9 dy \int_{\frac{y}{3}}^3 f(x,y) dx$;

   (4) $\displaystyle\int_{-a}^a dx \int_{-\frac{b}{a}\sqrt{a^2-x^2}}^{\frac{b}{a}\sqrt{a^2-x^2}} f(x,y) dy = \int_{-b}^b dy \int_{-\frac{a}{b}\sqrt{b^2-y^2}}^{\frac{a}{b}\sqrt{b^2-y^2}} f(x,y) dx$.

6. (1) $\int_0^1 dx \int_{x^2}^x f(x,y)dy$; (2) $\int_0^a dy \int_{-y}^{\sqrt{y}} f(x,y)dx$;

(3) $\int_{\sqrt{2}}^{\sqrt{3}} dy \int_0^{\sqrt{y^2-2}} f(x,y)dx + \int_{\sqrt{3}}^2 dy \int_0^{\sqrt{4-y^2}} f(x,y)dx$;

(4) $\int_0^1 dy \int_{e^y}^e f(x,y)dx$; (5) $\int_{-\frac{1}{4}}^0 dy \int_{-\frac{1}{2}-\frac{1}{2}\sqrt{1+4y}}^{-\frac{1}{2}+\frac{1}{2}\sqrt{1+4y}} f(x,y)dx + \int_0^2 dy \int_{y-1}^{-\frac{1}{2}+\frac{1}{2}\sqrt{1+4y}} f(x,y)dx$;

(6) $\int_0^1 dy \int_y^{2-y} f(x,y)dx$.

7. $\dfrac{153}{20}$.   8. 6π.   9. $\dfrac{1}{2}\pi a^4$.

10. (1) $\int_0^{\frac{\pi}{2}} d\theta \int_0^R f(r)rdr$; (2) $\int_0^{\frac{\pi}{2}} d\theta \int_0^{2R\sin\theta} f(r\cos\theta, r\sin\theta)rdr$;

(3) $\int_{\frac{\pi}{4}}^{\frac{\pi}{3}} d\theta \int_0^{2\sec\theta} f(r)rdr$; (4) $\int_0^{\frac{\pi}{2}} d\theta \int_0^{\frac{1}{\cos\theta+\sin\theta}} f(r\cos\theta, r\sin\theta)rdr$;

(5) $\int_0^{\frac{\pi}{4}} d\theta \int_{\sec\theta\cdot\tan\theta}^{\sec\theta} f(r\cos\theta, r\sin\theta)rdr$; (6) $\int_0^{\frac{\pi}{2}} d\theta \int_{\frac{1}{\cos\theta+\sin\theta}}^1 f(r\cos\theta, r\sin\theta)rdr$.

11. (1) $\pi(e^{R^2}-1)$; (2) $-6\pi^2$; (3) $\pi^2 a^2$; (4) $\dfrac{\pi}{4}(2\ln 2 - 1)$; (5) $\dfrac{2}{45}(\sqrt{2}-1)$; (6) $\dfrac{3}{4}\pi a^4$.

12. (1) $\dfrac{\pi}{4}\left(\dfrac{\pi}{2}-1\right)$; (2) $\dfrac{\pi^2}{6}$; (3) $\dfrac{8}{3}$.   13. $\dfrac{\pi^5}{40}$.   14. $\dfrac{2}{3}\pi$.

*15. (1) $\dfrac{7}{3}\ln 2$; (2) $\dfrac{e-1}{2}$; (3) $\dfrac{1}{2}\pi ab$; (4) $\dfrac{1}{2}\sin 1$.

Exercises 12-3

1. (1) $\int_{-1}^1 dx \int_{x^2}^1 dy \int_0^{x^2+y^2} f(x,y,z)dz$; (2) $\int_0^1 dx \int_0^{1-x} dy \int_0^{xy} f(x,y,z)dz$;

(3) $\int_0^1 dx \int_0^{\sqrt{1-x^2}} dy \int_0^{\sqrt{1-x^2-y^2}} f(x,y,z)dz$; (4) $\int_{-1}^1 dx \int_{-\sqrt{1-x^2}}^{\sqrt{1-x^2}} dy \int_{x^2+2y^2}^{2-x^2} f(x,y,z)dz$.

2. (1) $\dfrac{1}{48}$; (2) $\dfrac{1}{2}\left(\ln 2 - \dfrac{5}{8}\right)$; (3) $\dfrac{\pi}{4}$; (4) 0; (5) $\dfrac{59}{480}\pi R^5$; (6) $\dfrac{a^9}{36}$.

3. $\dfrac{3}{2}$.   5. (1) $\dfrac{\pi}{10}h^5$; (2) $\dfrac{7}{12}\pi$.   6. (1) $\dfrac{8}{9}a^2$; (2) $\dfrac{7}{6}\pi a^4$.

7. (1) $\dfrac{3}{16}\pi R^4$; (2) $\dfrac{14}{3}\pi$; (3) $\dfrac{8}{5}\pi$; (4) $\dfrac{4}{15}\pi(A^5-a^5)$.

8. (1) $\dfrac{5}{6}\pi a^3$; (2) $\dfrac{\pi}{6}$; (3) $\dfrac{2\pi}{3}(5\sqrt{5}-4)$.   9. $\dfrac{27}{57}$.

Exercises 12-4

1. $2\pi R^2$.   2. $4a^2\left(\dfrac{\pi}{2}-1\right)$.   3. $\dfrac{16\pi}{3}a^2$.   4. $\left(\dfrac{7}{6}, 0\right)$.   5. $\bar{x}=\dfrac{35}{48}, \bar{y}=\dfrac{35}{54}$.   6. $\left(0,0,\dfrac{3}{8}c\right)$.

7. (1) $\left(0,0,\dfrac{3}{4}\right)$; (2) $\left(\dfrac{2}{5}a, \dfrac{2}{5}a, \dfrac{7}{30}a^2\right)$; (3) $\left(0,0,\dfrac{3(A^4-a^4)}{8(A^3-a^3)}\right)$.

8. $\left(\dfrac{5}{9}, \dfrac{5}{9}, \dfrac{5}{9}\right)$.   9. $\dfrac{8}{15}\pi R^2$.   10. $\dfrac{8}{5}a^4$.

11. (1) $\dfrac{8}{3}a^4$; (2) $\bar{x}=\bar{y}=0, \bar{z}=\dfrac{7}{15}a^2$; (3) $\dfrac{112}{45}a^6\rho$.

12. $F_x=F_y=0, F_z=-2\pi G\rho[\sqrt{(h-a)^2+R^2}-\sqrt{R^2+a^2}+h]$.

*Exercises 12-5

1. (1) $\dfrac{\pi}{4}$; (2) 1.

Answers

2. (1) $\dfrac{1}{3}\cos x(\cos x - \sin x)(1+2\sin 2x)$; (2) $\dfrac{2}{x}\ln(1+x^2)$;

   (3) $\ln\sqrt{\dfrac{x^2+1}{x^4+1}} + 3x^2\arctan x^2 - 2x\arctan x$;

   (4) $2x e^{-x^5} - e^{-x^3} - \displaystyle\int_x^{x^2} y^2 e^{-xy^2}\,dy$.

3. $3f(x)+2xf'(x)$. 4. $\pi\ln\dfrac{1+a}{2}$. 5. (1) $\dfrac{\pi}{2}\ln(1+\sqrt{2})$; (2) $\arctan(1+b)-\arctan(1+a)$.

Quiz

1. (1) C; (2) C; (3) C; (4) B.

2. (1) $\displaystyle\int_1^2 dx\int_{1-x}^0 f(x,y)\,dy$; (2) $\displaystyle\int_0^{\pi/4} d\theta\int_0^{\sec\theta} f(r^2)r\,dr$;

   (3) $\displaystyle\int_0^{2\pi} d\theta\int_0^\pi \sin\varphi\,d\varphi\int_0^1 f(r\cos\theta\cdot\sin\varphi)r^2\,dr$; (4) π.

3. (1) $-\dfrac{9}{4}$; (2) $\dfrac{\pi}{3}$; (3) $\dfrac{1}{2}A^2$. 4. $\dfrac{4}{5}\pi abc$.

5. $\dfrac{8}{3}$. 6. $\pi\left[\dfrac{h^3}{3}+hf(0)\right]$. 7. 4. 8. $\dfrac{256}{3}\pi$.

Exercises

1. (1) C; (2) C.

2. (1) $\dfrac{1}{2}\left(1-\dfrac{1}{e}\right)$; (2) $\dfrac{1}{8}$; (3) $e-2$; (4) $\dfrac{\pi}{4}R^4 + 9\pi R^2$.

3. (1) $\displaystyle\int_0^a dy\int_{a-\sqrt{a^2-y^2}}^y f(x,y)\,dx$; (2) $\displaystyle\int_{-1}^0 dy\int_{-\sqrt{1-y^2}}^{\sqrt{1-y^2}} f(x,y)\,dx + \int_0^1 dy\int_{-\sqrt{1-y}}^{\sqrt{1-y}} f(x,y)\,dx$;

   (3) $\displaystyle\int_0^{1/3} dx\int_0^{\sqrt{3}x} f(x,y)\,dy + \int_{1/3}^{\sqrt{2}} dx\int_0^1 f(x,y)\,dy + \int_{\sqrt{2}}^{\sqrt{3}} dx\int_0^{\sqrt{3-x^2}} f(x,y)\,dy$.

4. $\displaystyle\int_0^{\pi/4} d\theta\int_0^{a\sec\theta} r^2\,dr = \dfrac{a^3}{6}[\sqrt{2}+\ln(\sqrt{2}-1)]$. 6. $\displaystyle\int_{-1}^1 dx\int_{x^2}^1 dy\int_0^{x^2+y^2} f(x,y,z)\,dz$.

7. (1) $\dfrac{1}{8}$; (2) $\dfrac{8}{9}a^2$; (3) 336π. 8. $\dfrac{1}{2}\sqrt{a^2b^2+b^2c^2+a^2c^2}$.

9. $\dfrac{4\pi}{5}(\sqrt{2}-1)$. 10. $\dfrac{\pi h a^2}{12}(3a^2+4h^2)$.

11. $F_x=0, F_y=\dfrac{4Gm_1m_2}{\pi R^2}\left(\ln\dfrac{R+\sqrt{R^2+a^2}}{a}-\dfrac{R}{\sqrt{R^2+a^2}}\right)$,

    $F_z=\dfrac{2Gm_1m_2}{R^2}\left(1-\dfrac{a}{\sqrt{R^2+a^2}}\right)$.